An Introduction to
Statistical Computing

WILEY SERIES IN COMPUTATIONAL STATISTICS

Consulting Editors:

Paolo Giudici
University of Pavia, Italy

Geof H. Givens
Colorado State University, USA

Bani K. Mallick
Texas A & M University, USA

Wiley Series in Computational Statistics is comprised of practical guides and cutting edge research books on new developments in computational statistics. It features quality authors with a strong applications focus. The texts in the series provide detailed coverage of statistical concepts, methods and case studies in areas at the interface of statistics, computing, and numerics.

With sound motivation and a wealth of practical examples, the books show in concrete terms how to select and to use appropriate ranges of statistical computing techniques in particular fields of study. Readers are assumed to have a basic understanding of introductory terminology.

The series concentrates on applications of computational methods in statistics to fields of bioinformatics, genomics, epidemiology, business, engineering, finance and applied statistics.

Titles in the Series

Biegler, Biros, Ghattas, Heinkenschloss, Keyes, Mallick, Marzouk, Tenorio, Waanders, Willcox – Large-Scale Inverse Problems and Quantification of Uncertainty

Billard and Diday – Symbolic Data Analysis: Conceptual Statistics and Data Mining

Bolstad – Understanding Computational Bayesian Statistics

Borgelt, Steinbrecher and Kruse – Graphical Models, 2e

Dunne – A Statistical Approach to Neutral Networks for Pattern Recognition

Liang, Liu and Carroll – Advanced Markov Chain Monte Carlo Methods

Ntzoufras – Bayesian Modeling Using WinBUGS

Tufféry – Data Mining and Statistics for Decision Making

An Introduction to Statistical Computing

A Simulation-based Approach

Jochen Voss

School of Mathematics, University of Leeds, UK

Library of Congress Cataloging-in-Publication Data

Voss, Jochen.
  An introduction to statistical computing : a simulation-based approach / Jochen Voss. – First edition.
    pages cm. – (Wiley series in computational statistics)
  Includes bibliographical references and index.
  ISBN 978-1-118-35772-9 (hardback)
  1. Mathematical statistics–Data processing.   I. Title.
  QA276.4.V66 2013
  519.501′13–dc23

                                                                        2013019321

A catalogue record for this book is available from the British Library.

ISBN: 978-1-118-35772-9

Typeset in 10/12pt Times by Aptara Inc., New Delhi, India
Printed and bound in Malaysia by Vivar Printing Sdn Bhd

1   2014

Contents

List of algorithms   ix

Preface   xi

Nomenclature   xiii

1 Random number generation   1
  1.1 Pseudo random number generators   2
    1.1.1 The linear congruential generator   2
    1.1.2 Quality of pseudo random number generators   4
    1.1.3 Pseudo random number generators in practice   8
  1.2 Discrete distributions   8
  1.3 The inverse transform method   11
  1.4 Rejection sampling   15
    1.4.1 Basic rejection sampling   15
    1.4.2 Envelope rejection sampling   18
    1.4.3 Conditional distributions   22
    1.4.4 Geometric interpretation   26
  1.5 Transformation of random variables   30
  1.6 Special-purpose methods   36
  1.7 Summary and further reading   36
  Exercises   37

2 Simulating statistical models   41
  2.1 Multivariate normal distributions   41
  2.2 Hierarchical models   45
  2.3 Markov chains   50
    2.3.1 Discrete state space   51
    2.3.2 Continuous state space   56
  2.4 Poisson processes   58
  2.5 Summary and further reading   67
  Exercises   67

3   Monte Carlo methods   **69**

3.1   Studying models via simulation   69

3.2   Monte Carlo estimates   74

    3.2.1   Computing Monte Carlo estimates   75

    3.2.2   Monte Carlo error   76

    3.2.3   Choice of sample size   80

    3.2.4   Refined error bounds   82

3.3   Variance reduction methods   84

    3.3.1   Importance sampling   84

    3.3.2   Antithetic variables   88

    3.3.3   Control variates   93

3.4   Applications to statistical inference   96

    3.4.1   Point estimators   97

    3.4.2   Confidence intervals   100

    3.4.3   Hypothesis tests   103

3.5   Summary and further reading   106

Exercises   106

4   Markov Chain Monte Carlo methods   **109**

4.1   The Metropolis–Hastings method   110

    4.1.1   Continuous state space   110

    4.1.2   Discrete state space   113

    4.1.3   Random walk Metropolis sampling   116

    4.1.4   The independence sampler   119

    4.1.5   Metropolis–Hastings with different move types   120

4.2   Convergence of Markov Chain Monte Carlo methods   125

    4.2.1   Theoretical results   125

    4.2.2   Practical considerations   129

4.3   Applications to Bayesian inference   137

4.4   The Gibbs sampler   141

    4.4.1   Description of the method   141

    4.4.2   Application to parameter estimation   146

    4.4.3   Applications to image processing   151

4.5   Reversible Jump Markov Chain Monte Carlo   158

    4.5.1   Description of the method   160

    4.5.2   Bayesian inference for mixture distributions   171

4.6   Summary and further reading   178

4.6   Exercises   178

5   Beyond Monte Carlo   **181**

5.1   Approximate Bayesian Computation   181

    5.1.1   Basic Approximate Bayesian Computation   182

    5.1.2   Approximate Bayesian Computation with regression   188

5.2   Resampling methods   192

	5.2.1	Bootstrap estimates	192
	5.2.2	Applications to statistical inference	197
5.3		Summary and further reading	209
	Exercises		209

6 Continuous-time models — **213**

6.1		Time discretisation	213
6.2		Brownian motion	214
	6.2.1	Properties	216
	6.2.2	Direct simulation	217
	6.2.3	Interpolation and Brownian bridges	218
6.3		Geometric Brownian motion	221
6.4		Stochastic differential equations	224
	6.4.1	Introduction	224
	6.4.2	Stochastic analysis	226
	6.4.3	Discretisation schemes	231
	6.4.4	Discretisation error	236
6.5		Monte Carlo estimates	243
	6.5.1	Basic Monte Carlo	243
	6.5.2	Variance reduction methods	247
	6.5.3	Multilevel Monte Carlo estimates	250
6.6		Application to option pricing	255
6.7		Summary and further reading	259
	Exercises		260

Appendix A Probability reminders — **263**

A.1	Events and probability	263
A.2	Conditional probability	266
A.3	Expectation	268
A.4	Limit theorems	269
A.5	Further reading	270

Appendix B Programming in R — **271**

B.1		General advice	271
B.2		R as a Calculator	272
	B.2.1	Mathematical operations	273
	B.2.2	Variables	273
	B.2.3	Data types	275
B.3		Programming principles	282
	B.3.1	Don't repeat yourself!	283
	B.3.2	Divide and conquer!	286
	B.3.3	Test your code!	290
B.4		Random number generation	292
B.5		Summary and further reading	294
	Exercises		294

Appendix C   Answers to the exercises   **299**

   C.1   Answers for Chapter 1   299
   C.2   Answers for Chapter 2   315
   C.3   Answers for Chapter 3   319
   C.4   Answers for Chapter 4   328
   C.5   Answers for Chapter 5   342
   C.6   Answers for Chapter 6   350
   C.7   Answers for Appendix B   366

References   **375**

Index   **379**

List of algorithms

Random number generation
alg. 1.2   linear congruential generator   2
alg. 1.13   inverse transform method   12
alg. 1.19   basic rejection sampling   15
alg. 1.22   envelope rejection sampling   19
alg. 1.25   rejection sampling for conditional distributions   22

Simulating statistical models
alg. 2.9   mixture distributions   47
alg. 2.11   componentwise simulation   49
alg. 2.22   Markov chains with discrete state space   53
alg. 2.31   Markov chains with continuous state space   58
alg. 2.36   Poisson process   61
alg. 2.41   thinning method for Poisson processes   65

Monte Carlo methods
alg. 3.8   Monte Carlo estimate   75
alg. 3.22   importance sampling   85
alg. 3.26   antithetic variables   89
alg. 3.31   control variates   93

Markov Chain Monte Carlo methods
alg. 4.2   Metropolis–Hastings method for continuous state space   110
alg. 4.4   Metropolis–Hastings method for discrete state space   113
alg. 4.9   random walk Metropolis   117
alg. 4.11   independence sampler   119
alg. 4.12   Metropolis–Hastings method with different move types   121
alg. 4.27   Gibbs sampler   142
alg. 4.31   Gibbs sampler for the Ising model   155
alg. 4.32   Gibbs sampler in image processing   158
alg. 4.36   reversible jump Markov Chain Monte Carlo   165

Beyond Monte Carlo

alg. 5.1   basic Approximate Bayesian Computation   182
alg. 5.6   Approximate Bayesian Computation with regression   191
alg. 5.11   general bootstrap estimate   196
alg. 5.15   bootstrap estimate of the bias   200
alg. 5.18   bootstrap estimate of the standard error   202
alg. 5.20   simple bootstrap confidence interval   205
alg. 5.21 BC_a bootstrap confidence interval   207

Continuous-time models

alg. 6.6   Brownian motion   217
alg. 6.12   Euler–Maruyama scheme   232
alg. 6.15   Milstein scheme   235
alg. 6.26   multilevel Monte Carlo estimates   251
alg. 6.29   Euler–Maruyama scheme for the Heston model   256

Preface

This is a book about exploring random systems using computer simulation and thus, this book combines two different topic areas which have always fascinated me: the mathematical theory of probability and the art of programming computers. The method of using computer simulations to study a system is very different from the more traditional, purely mathematical approach. On the one hand, computer experiments normally can only provide approximate answers to quantitative questions, but on the other hand, results can be obtained for a much wider class of systems, including large and complex systems where a purely theoretical approach becomes difficult.

In this text we will focus on three different types of questions. The first, easiest question is about the normal behaviour of the system: what is a typical state of the system? Such questions can be easily answered using computer experiments: simulating a few random samples of the system gives examples of typical behaviour. The second kind of question is about variability: how large are the random fluctuations? This type of question can be answered statistically by analysing large samples, generated using repeated computer simulations. A final, more complicated class of questions is about exceptional behaviour: how small is the probability of the system behaving in a specified untypical way? Often, advanced methods are required to answer this third type of question. The purpose of this book is to explain how such questions can be answered. My hope is that, after reading this book, the reader will not only be able to confidently use methods from statistical computing for answering such questions, but also to adjust existing methods to the requirements of a given problem and, for use in more complex situations, to develop new specialised variants of the existing methods.

This text originated as a set of handwritten notes which I used for teaching the 'Statistical Computing' module at the University of Leeds, but now is greatly extended by the addition of many examples and more advanced topics. The material we managed to cover in the 'Statistical Computing' course during one semester is less than half of what is now the contents of the book! This book is aimed at postgraduate students and their lecturers; it can be used both for self-study and as the basis of taught courses. With the inclusion of many examples and exercises, the text should also be accessible to interested undergraduate students and to mathematically inclined researchers from areas outside mathematics.

Only very few prerequisites are required for this book. On the mathematical side, the text assumes that the reader is familiar with basic probability, up to and including the law of large numbers; Appendix A summarises the required results. As a consequence of the decision to require so little mathematical background, some of the finer mathematical subtleties are not discussed in this book. Results are presented in a way which makes them easily accessible to readers with limited mathematical background, but the statements are given in a form which allows the mathematically more knowledgeable reader to easily add the required detail on his/her own. (For example, I often use phrases such as 'every set $A \subseteq \mathbb{R}^d$' where full mathematical rigour would require us to write 'every *measurable* set $A \subseteq \mathbb{R}^d$'.) On the computational side, basic programming skills are required to make use of the numerical methods introduced in this book. While the text is written independent of any specific programming language, the reader will need to choose a language when implementing methods from this book on a computer. Possible choices of programming language include Python, Matlab and C/C++. For my own implementations, provided as part of the solutions to the exercises in Appendix C, I used the R programming language; a short introduction to programming with R is provided in Appendix B.

Writing this book has been a big adventure for me. When I started this project, more than a year ago, my aim was to cover enough material so that I could discuss the topics of multilevel Monte Carlo and reversible jump Markov Chain Monte Carlo methods. I estimated that 350 pages would be enough to cover this material but it quickly transpired that I had been much too optimistic: my estimates for the final page count kept rising and even after several rounds of throwing out side-topics and generally tightening the text, the book is still stretching this limit! Nevertheless, the text now covers most of the originally planned topics, including multilevel Monte Carlo methods near the very end of the book. Due to my travel during the last year, parts of this book have been written on a laptop in exciting places. For example, the initial draft of section 1.5 was written on a coach travelling through the beautiful island of Kyushu, halfway around the world from where I live! All in all, I greatly enjoyed writing this book and I hope that the result is useful to the reader.

This book contains an accompanying website. Please visit www.wiley.com/go/statistical_computing

Jochen Voss

Leeds, March 2013

Nomenclature

For reference, the following list summarises some of the notation used throughout this book.

$\emptyset$	the empty set
$\mathbb{N}$	the natural numbers: $\mathbb{N} = \{1, 2, 3, \ldots\}$
$\mathbb{N}_0$	the non-negative integers: $\mathbb{N} = \{0, 1, 2, \ldots\}$
$\mathbb{Z}$	the integers: $\mathbb{Z} = \{\ldots, -2, -1, 0, 1, 2, \ldots\}$
$n \bmod m$	the remainder of the division of n by m, in the range $0, 1, \ldots, m-1$
δ_{kl}	the Kronecker delta: $\delta_{kl} = 1$ if $k = l$ and $\delta_{kl} = 0$ otherwise
$\mathbb{R}$	the real numbers
$\lceil x \rceil$	the number $x \in \mathbb{R}$ 'rounded up', that is the smallest integer greater than or equal to x
$(a_n)_{n\in\mathbb{N}}$	a sequence of (possibly random) numbers: $(a_n)_{n\in\mathbb{N}} = (a_1, a_2, \ldots)$
$\mathcal{O}(\cdot)$	the big $\mathcal{O}$ notation, introduced in definition 3.16
$[a, b]$	an interval of real numbers: $[a, b] = \{x \in \mathbb{R} \mid a \le x \le b\}$
$\{a, b\}$	the set containing a and b
$A^{\complement}$	the complement of a set: $A^{\complement} = \{x \mid x \notin A\}$.
$A \times B$	the Cartesian product of the sets A and B: $A \times B = \{(a, b) \mid a \in A, b \in B\}$
$\mathbb{1}_A(x)$	the indicator function of the set A: $\mathbb{1}_A(x) = 1$ if $x \in A$ and 0 otherwise (see section A.3)
$\mathcal{U}[0, 1]$	the uniform distribution on the interval $[0, 1]$
$\mathcal{U}\{-1, 1\}$	the uniform distribution on the two-element set $\{-1, 1\}$
$\mathrm{Pois}(\lambda)$	the Poisson distribution with parameter λ
$X \sim \mu$	indicates that a random variable X is distributed according to a probability distribution μ
$\lvert S \rvert$	the number of elements in a finite set S; in section 1.4 also the volume of a subsets $S \subseteq \mathbb{R}^d$
$\mathbb{R}^S$	space of vectors where the components are indexed by elements of S (see section 2.3.2)
$\mathbb{R}^{S\times S}$	space of matrices where rows and columns are indexed by elements of S (see section 2.3.2)

1

Random number generation

The topic of this book is the study of statistical models using computer simulations. Here we use the term 'statistical models' to mean any mathematical models which include a random component. Our interest in this chapter and the next is in simulation of the random component of these models. The basic building block of such simulations is the ability to generate random numbers on a computer, and this is the topic of the present chapter. Later, in Chapter 2, we will see how the methods from Chapter 1 can be combined to simulate more complicated models.

Generation of random numbers, or more general random objects, on a computer is complicated by the fact that computer programs are inherently deterministic: while the output of computer program may look random, it is obtained by executing the steps of some algorithm and thus is totally predictable. For example the output of a program computing the decimal digits of the number

$$\pi = 3.14159265358979323846264338327950288419716939937510\cdots$$

(the ratio between the perimeter and diameter of a circle) looks random at first sight, but of course π is not random at all! The output can only start with the string of digits given above and running the program twice will give the same output twice.

We will split the problem of generating random numbers into two distinct subproblems: first we will study the problem of generating any randomness at all, concentrating on the simple case of generating independent random numbers, uniformly distributed on the interval $[0, 1]$. This problem and related concerns will be discussed in Section 1.1. In the following sections, starting with Section 1.2, we will study the generation of random numbers from different distributions, using the independent, uniformly distributed random numbers obtained in the previous step as a basis.

An Introduction to Statistical Computing: A Simulation-based Approach, First Edition. Jochen Voss.
© 2014 John Wiley & Sons, Ltd. Published 2014 by John Wiley & Sons, Ltd.

1.1 Pseudo random number generators

There are two fundamentally different classes of methods to generate random numbers:

(a) True random numbers are generated using some physical phenomenon which is random. Generating such numbers requires specialised hardware and can be expensive and slow. Classical examples of this include tossing a coin or throwing dice. Modern methods utilise quantum effects, thermal noise in electric circuits, the timing of radioactive decay, etc.

(b) Pseudo random numbers are generated by computer programs. While these methods are normally fast and resource effective, a challenge with this approach is that computer programs are inherently deterministic and therefore cannot produce 'truly random' output.

In this text we will only consider pseudo random number generators.

Definition 1.1   A pseudo random number generator (PRNG) is an algorithm which outputs a sequence of numbers that can be used as a replacement for an independent and identically distributed (i.i.d.) sequence of 'true random numbers'.

1.1.1 The linear congruential generator

This section introduces the linear congruential generator (LCG), a simple example of a PRNG. While this random number generator is no longer of practical importance, it shares important characteristics with the more complicated generators used in practice today and we study it here as an accessible example. The LCG is given by the following algorithm.

Algorithm 1.2   (linear congruential generator)
    input:
        $m > 1$ (the *modulus*)
        $a \in \{1, 2, \ldots, m - 1\}$ (the *multiplier*)
        $c \in \{0, 1, \ldots, m - 1\}$ (the *increment*)
        $X_0 \in \{0, 1, \ldots, m - 1\}$ (the *seed*)
    output:
        a sequence $X_1, X_2, X_3, \ldots$ of pseud random numbers
    1: **for** $n = 1, 2, 3, \ldots$ **do**
    2:    $X_n \leftarrow (a X_{n-1} + c) \bmod m$
    3:    output X_n
    4: **end for**

In the algorithm, 'mod' denotes the modulus for integer division, that is the value $n \bmod m$ is the remainder of the division of n by m, in the range $0, 1, \ldots, m - 1$. Thus the sequence generated by algorithm 1.2 consists of integers X_n from the range $\{0, 1, 2, \ldots, m - 1\}$. The output depends on the parameters m, a, c and on the seed X_0. We will see that, if m, a and c are carefully chosen, the resulting sequence behaves 'similar' to a sequence of independent, uniformly distributed random variables. By choosing different values for the seed X_0, different sequences of pseudo random numbers can be obtained.

Example 1.3   For parameters $m = 8, a = 5, c = 1$ and seed $X_0 = 0$, algorithm 1.2 gives the following output:

n	$5X_{n-1} + 1$	X_n
1	1	1
2	6	6
3	31	7
4	36	4
5	21	5
6	26	2
7	11	3
8	16	0
9	1	1
10	6	6

The output $1, 6, 7, 4, 5, 2, 3, 0, 1, 6, \ldots$ shows no obvious pattern and could be considered to be a sample of a random sequence.

While the output of the LCG looks random, from the way it is generated it is clear that the output has several properties which make it different from truly random sequences. For example, since each new value of X_n is computed from X_{n-1}, once the generated series reaches a value X_n which has been generated before, the output starts to repeat. In example 1.3 this happens for $X_8 = X_0$ and we get $X_9 = X_1, X_{10} = X_2$ and so on. Since X_n can take only m different values, the output of a LCG starts repeating itself after at most m steps; the generated sequence is eventually periodic.

Sometimes the periodicity of a sequence of pseudo random numbers can cause problems, but on the other hand, if the period length is longer than the amount of random numbers we use, periodicity cannot affect our result. For this reason, one needs to carefully choose the parameters m, a and c in order to achieve a long enough period. In particular m, since it is an upper bound for the period length, needs to be chosen large. In practice, typical values of m are on the order of $m = 2^{32} \approx 4 \cdot 10^9$ and a and c are then chosen such that the generator actually achieves the maximally possible period length of m. A criterion for the choice of m, a and c is given in the following theorem (Knuth, 1981, Section 3.2.1.2).

Theorem 1.4   The LCG has period m if and only if the following three conditions are satisfied:

(a) m and c are relatively prime;

(b) $a - 1$ is divisible by every prime factor of m;

(c) if m is a multiple of 4, then $a - 1$ is a multiple of 4.

In the situation of the theorem, the period length does not depend on the seed X_0 and usually this parameter is left to be chosen by the user of the PRNG.

Example 1.5   Let $m = 2^{32}$, $a = 1\,103\,515\,245$ and $c = 12\,345$. Since the only prime factor of m is 2 and c is odd, the values m and c are relatively prime and condition (a) of the theorem is satisfied. Similarly, condition (b) is satisfied, since $a - 1$ is even and thus divisible by 2. Finally, since m is a multiple of 4, we have to check condition (c) but, since $a - 1 = 1\,103\,515\,244 = 275\,878\,811 \cdot 4$, this condition also holds. Therefore the LCG with these parameters m, a and c has period 2^{32} for every seed X_0.

1.1.2   Quality of pseudo random number generators

PRNGs used in modern software packages such as R or Matlab are more sophisticated (and more complicated) than the LCG presented in Section 1.1.1, but they still share many characteristics of the LCG. We will see that no PRNG can produce a perfect result, but the random number generators used in practice, for example the Mersenne Twister algorithm (Matsumoto and Nishimura, 1998), are good enough for most purposes. In this section we will discuss criteria for the quality of the output of general PRNGs, and will illustrate these criteria using the LCG as an example.

1.1.2.1   Period length of the output

We have seen that the output of the LCG is eventually periodic, with a period length of at most m. This property that the output is eventually periodic is shared by all PRNGs implemented in software. Most PRNGs used in practice have a period length which is much larger than the amount of random numbers a computer program could ever use in a reasonable time. For this reason, periodicity of the output is not a big problem in practical applications of PRNGs. The period length is a measure for the quality of a PRNG.

1.1.2.2   Distribution of samples

The output of almost all PRNGs is constructed so that it can be used as a replacement for an i.i.d. sample of *uniformly distributed* random numbers. Since the output takes values in a finite set $S = \{0, 1, \ldots, m - 1\}$, in the long run, for every set $A \subseteq S$ we should have

$$\frac{\#\{i \mid 1 \leq i \leq N, X_i \in A\}}{N} \approx \frac{\#A}{\#S}, \tag{1.1}$$

where $\#A$ stands for the number of elements in a finite set A.

Uniformity of the output can be tested using statistical tests like the chi-squared test or the Kolmogorov–Smirnov test (see e.g. Lehmann and Romano, 2005, Chapter 14).

One peculiarity when applying statistical tests for the distribution of samples to the output of a PRNG is that the test may fail in two different ways: The output could either have the wrong distribution (i.e. not every value appears with the same probability), or the output could be too regular. For example, the sequence $X_n = n \bmod m$ hits every value equally often in the long run, but it shows none of the fluctuations which are typical for a sequence of real random numbers. For this reason, statistical tests should be performed as two-sided tests when the distribution of the output of a PRNG is being tested.

Example 1.6   Assume that we have a PRNG with $m = 1024$ possible output values and that we perform a chi-squared test for the hypothesis

$$P\left(X_i \in \{64j, 64j + 1, \ldots, 64j + 63\}\right) = 1/16$$

for $j = 0, 1, \ldots, 15$.

If we consider a sample $X_1, X_2, \ldots, X_N$, the test statistic of the chi-squared test is computed from the observed numbers of samples in each block, given by

$$O_j = \#\left\{i \mid 64j \leq X_i < 64(j + 1)\right\}.$$

The expected count for block j, assuming that (1.1) holds, is

$$E_j = N \cdot 64/1024 = N/16$$

for $j = 0, 1, \ldots, 15$ and the test statistic of the corresponding chi-squared test is

$$Q = \sum_{j=0}^{15} \frac{(O_j - E_j)^2}{E_j}.$$

For large sample size N, and under the hypothesis (1.1), the value Q follows a χ^2-distribution with 15 degrees of freedom. Some quantiles of this distribution are:

q	6.262	7.261	$\cdots$	24.996	27.488
$P(Q \leq q)$	0.025	0.05	$\cdots$	0.95	0.975

Thus, for a one-sided test with significance level $1 - \alpha = 95\%$ we would reject the hypothesis if $Q > 24.996$. In contrast, for a two-sided test with significance level $1 - \alpha = 95\%$, we would reject the hypothesis if either $Q < 6.262$ or $Q > 27.488$.

We consider two different test cases: first, if $X_n = n \bmod 1024$ for $n = 1, 2, \ldots, N = 10^6$, we find $Q = 0.244368$. Since the series is very regular, the value of Q is very low. The one-sided test would accept this sequence as being uniformly distributed, whereas the two-sided test would reject the sequence.

Secondly, we consider $X_n = n \bmod 1020$ for $n = 1, 2, \ldots, N = 10^6$. Since this series never takes the values 1021 to 1023, the distribution is wrong and we expect a large value of Q. Indeed, for this case we get $Q = 232.5864$ and thus both versions of the test reject this sequence.

Random number generators used in practice, and even the LCG for large enough values of m, pass statistical tests for the distribution of the output samples without problems.

1.1.2.3 Independence of samples

Another aspect of the quality of PRNGs is the possibility of statistical dependence between consecutive samples. For example, in the LCG each output sample is a deterministic function of the previous sample and thus consecutive samples are clearly dependent. To some extent this problem is shared by all PRNGs.

An easy way to visualise the dependence between pairs of consecutive samples is a scatter plot of the points (X_i, X_{i+1}) for $i = 1, 2, \ldots, N - 1$. A selection of such plots is shown in Figure 1.1. Figure 1.1(a) illustrates what kind of plot one would expect if $X_i \sim \mathcal{U}[0, 1]$ was a true i.i.d. sequence. The remaining panels correspond to different variants of the LCG. Figure 1.1(b) (using $m = 81$) clearly illustrates that each X_i can only be followed by exactly one value X_{i+1}. While the same is true for Figure 1.1(c) and (d) (using $m = 1024$ and $m = 2^{32}$, respectively), the dependence is much convoluted there and in particular the structure of Figure 1.1(d) is visually indistinguishable from the structure of Figure 1.1(a).

One method for constructing PRNGs where X_{i+1} is not a function of X_i is to use a function $f(X_i)$ of the state, instead of the state X_i itself, as the output of the PRNG. Here, $f: \{0, 1, \ldots, m - 1\} \to \{0, 1, \ldots, \tilde{m} - 1\}$ is a map where $\tilde{m} < m$ and where the same number of pre-images is mapped to each output value. Then a uniform distribution of X_i will be mapped to a uniform distribution for $f(X_i)$ but the output $f(X_{i+1})$ is not a function of the previous output $f(X_i)$. This allows to construct random number generators with some degree of independence between consecutive values.

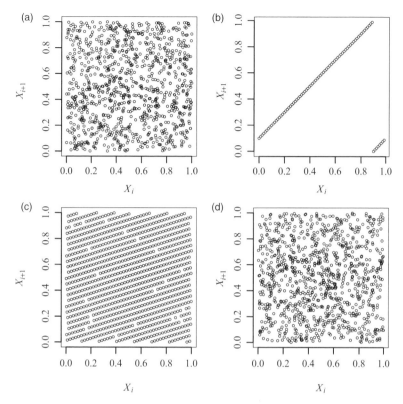

Figure 1.1   Scatter plots to illustrate the correlation between consecutive outputs X_i and X_{i+1} of different pseudo random number generators. The random number generators used are the `runif` *function in R (a), the LCG with $m = 81$, $a = 1$ and $c = 8$ (b), the LCG with $m = 1024$, $a = 401$, $c = 101$ (c) and finally the LCG with parameters $m = 2^{32}$, $a = 1\,664\,525$, $c = 1\,013\,904\,223$ (d). Clearly the output in the second and third example does not behave like a sequence of independent random variables.*

One way to quantify the independence of the output samples of a PRNG is the following criterion.

Definition 1.7   A periodic sequence $(X_n)_{n \in \mathbb{N}}$ with values in a finite set S and period length P is k-dimensionally equidistributed, if every possible subsequence $x = (x_1, \ldots, x_k) \in S^k$ of length k occurs equally often in the sequence X, that is if

$$N_x = \# \left\{ i \mid 0 \leq i < P, X_{i+1} = x_i, \ldots, X_{i+k} = x_k \right\}$$

does not depend on x.

A random number generator is good, if the output is k-dimensionally equidistributed for large values of k.

1.1.3   Pseudo random number generators in practice

This section contains advice on using PRNGs in practice.

First, it is normally a bad idea to implement your own PRNG: finding a good algorithm for pseudo random number generation is a difficult problem, and even when an algorithm is available, given the nature of the generated output, it can be a challenge to spot and remove all mistakes in the implementation. Therefore, it is advisable to use a well-established method for random number generation, typically the random number generator built into a well-known software package or provided by a well-established library.

A second consideration concerns the rôle of the seed. While different PRNGs differ greatly in implementation details, they all use a seed (like the value X_0 in algorithm 1.2) to initialise the state of the random number generator. Often, when non-predictability is required, it is useful to set the seed to some volatile quantity (like the current time) to get a different sequence of random numbers for different runs of the program. At other times it can be more useful to get reproducible results, for example to aid debugging or to ensure repeatability of published results. In these cases, the seed should be set to a known, fixed value.

Finally, PRNGs like the LCG described above often generate a sequence which behaves like a sequence of independent random numbers, uniformly distributed on a finite set $\{0, 1, \ldots, m - 1\}$ for a big value of m. In contrast, most applications require a sequence of independent, $\mathcal{U}[0, 1]$-distributed random variables, that is a sequence of i.i.d. values which are uniformly distributed on the real interval $[0, 1]$. We can obtain a sequence $(U_n)_{n \in \mathbb{N}}$ of pseudo random numbers to replace an i.i.d. sequence of $\mathcal{U}[0, 1]$ random variables by setting

$$U_n = \frac{X_n + 1}{m + 1},$$

where $(X_n)_{n \in \mathbb{N}}$ is the output of the PRNG. The output U_n can only take the m different values

$$\frac{1}{m + 1}, \frac{2}{m + 1}, \ldots, \frac{m}{m + 1}$$

and thus U_n is not exactly uniformly distributed on the continuous interval $[0, 1]$. But, since the possible values are evenly spaced inside the interval $[0, 1]$ and since each of these values has the same probability, the distribution of U_n is a reasonable approximation to a uniform distribution on $[0, 1]$. This is particularly true since computers can only represent finitely many real numbers exactly.

This concludes our discussion of how a replacement for an i.i.d. sequence of $\mathcal{U}[0, 1]$-distributed random numbers can be generated on a computer.

1.2   Discrete distributions

Building on the methods from Section 1.1, in this and the following sections we will study methods to transform an i.i.d. sequence of $\mathcal{U}[0, 1]$-distributed random variables into an i.i.d. sequence from a prescribed target distribution. The methods from the previous section were inexact, since the output of a PRNG is not 'truly random'. In contrast, the transformations described in this and the following sections can be carried out with complete mathematical rigour. We will discuss different methods for generating samples from a given distribution, applicable to different classes of target distributions. In this section we concentrate on the simplest case where the target distribution only takes finitely or countably infinitely many values.

As a first example, we consider the uniform distribution on the set $A = \{0, 1, \ldots, n - 1\}$, denoted by $\mathcal{U}\{0, 1, \ldots, n - 1\}$. Since the set A has n elements, a random variable X with $X \sim \mathcal{U}\{0, 1, \ldots, n - 1\}$ satisfies

$$P(X = k) = \frac{1}{n}$$

for all $k \in A$. To generate samples from such a random variable X, at first it may seem like a good idea to just use a PRNG with state space A, for example the LCG with modulus $m = n$. But considering the fact that the maximal period length of a PRNG is restricted to the size of the state space, it becomes clear that this is not a good idea. Instead we will follow the approach to first generate a continuous sample $U \sim \mathcal{U}[0, 1]$ and then to transform this sample into the required discrete uniform distribution. A method to implement this idea is described in the following lemma.

Lemma 1.8 Let $U \sim \mathcal{U}[0, 1]$ and $n \in \mathbb{N}$. Define a random variable X by $X = \lfloor nU \rfloor$, where $\lfloor \cdot \rfloor$ denotes rounding down. Then $X \sim \mathcal{U}\{0, 1, \ldots, n - 1\}$.

Proof By the definition of X we have

$$P(X = k) = P(\lfloor nU \rfloor = k) = P(nU \in [k, k + 1)) = P\left(U \in \left[\frac{k}{n}, \frac{k + 1}{n}\right)\right)$$

for all $k = 0, 1, \ldots, n - 1$.

The uniform distribution $\mathcal{U}[0, 1]$ is characterised by the fact that $U \sim \mathcal{U}[0, 1]$ satisfies

$$P(U \in [a, b]) = b - a$$

for all $0 \leq a \leq b \leq 1$. Also, since U is a continuous distribution, we have $P(U = x) = 0$ for all $x \in [0, 1]$ and thus the boundary points of the interval $[a, b]$ can be included or excluded without changing the probability. Using these results, we find

$$P(X = k) = P\left(U \in \left[\frac{k}{n}, \frac{k + 1}{n}\right)\right) = \frac{k + 1}{n} - \frac{k}{n} = \frac{1}{n}$$

for all $k = 0, 1, \ldots, n - 1$. This completes the proof.   □

Another common problem related to discrete distributions is the problem of constructing random events which occur with a given probability p. Such events will, for example, be needed in the rejection algorithms considered in Section 1.4. There are many fascinating aspects to this problem, but here we will restrict ourselves to the simplest case where the probability p is known explicitly and where we have access to $\mathcal{U}[0, 1]$-distributed random variables. This case is considered in the following lemma.

Lemma 1.9 Let $p \in [0, 1]$ and $U \sim \mathcal{U}[0, 1]$ and define the event E as $E = \{U \le p\}$. Then $P(E) = p$.

Proof We have

$$P(E) = P(U \le p) = P\left(U \in [0, p]\right) = p - 0 = p.$$

This completes the proof.   □

The idea underlying lemmas 1.8 and 1.9 can be generalised to sample from arbitrary distributions on a finite set A. Let $A = \{a_1, \ldots, a_n\}$ where $a_i \ne a_j$ for $i \ne j$ and let $p_1, \ldots, p_n \ge 0$ be given with $\sum_{i=1}^{n} p_i = 1$. Assume that we want to generate random values $X \in A$ with $P(X = a_i) = p_i$ for $i = 1, 2, \ldots, n$. Since the p_i sum up to 1, we can split the unit interval $[0, 1]$ into disjoint sub-intervals lengths $p_1, \ldots, p_n$.

With this arrangement, if we choose $U \in [0, 1]$ uniformly, the value of U lies in the ith subinterval with probability p_i. Thus, we can choose X to be the a_i corresponding to the subinterval which contains U. This idea is formalised in the following lemma.

Lemma 1.10   Assume $A = \{a_i \mid i \in I\}$ where either $I = \{1, 2, \ldots, n\}$ for some $n \in \mathbb{N}$ or $I = \mathbb{N}$, and where $a_i \ne a_j$ whenever $i \ne j$. Let $p_i \ge 0$ be given for $i \in I$ with $\sum_{i \in I} p_i = 1$. Finally let $U \sim \mathcal{U}[0, 1]$ and define

$$K = \min\left\{k \in I \,\middle|\, \sum_{i=1}^{k} p_i \ge U\right\}. \tag{1.2}$$

Then $X = a_K \in A$ satisfies $P(X = a_k) = p_k$ for all $k \in I$.

Proof We have

$$P(X = a_k) = P(K = k) = P\left(\sum_{i=1}^{k-1} p_i < U, \sum_{i=1}^{k} p_i \ge U\right)$$

$$= P\left(U \in \left(\sum_{i=1}^{k-1} p_i, \sum_{i=1}^{k} p_i\right]\right) = \sum_{i=1}^{k} p_i - \sum_{i=1}^{k-1} p_i = p_k$$

for all $k \in I$, where we interpret the sum $\sum_{i=1}^{0} p_i$ for $k = 1$ as 0. This completes the proof.   □

The numerical method described by lemma 1.10 requires that we find the index K of the subinterval which contains U. The most efficient way to do this is to find a function φ which maps the boundaries of the subintervals to consecutive integers and then to consider the rounded value $\lfloor \varphi(I) \rfloor$. This approach is taken in lemma 1.8 and also in the following example.

Example 1.11   The geometric distribution, describing the number X of individual trials with probability p until the first success, has probability weights $P(X = i) = p^{i-1}(1 - p) = p_i$ for $i \in \mathbb{N}$. We can use lemma 1.10 with $a_i = i$ for all $i \in \mathbb{N}$ to generate samples from this distribution.

For the weights p_i, the value sum in equation (1.2) can be determined explicitly: using the formula for geometric sums we find

$$\sum_{i=1}^{k} p_i = (1 - p) \sum_{i=1}^{k} p^{i-1} = (1 - p)\frac{1 - p^k}{1 - p} = 1 - p^k.$$

Thus, we can rewrite the event $\sum_{i=1}^{k} p_i \geq U$ as follows:

$$\left\{ U \leq \sum_{i=1}^{k} p_i \right\} = \left\{ U \leq 1 - p^k \right\}$$
$$= \left\{ p^k \leq 1 - U \right\}$$
$$= \left\{ k \log(p) \leq \log(1 - U) \right\}$$
$$= \left\{ k \geq \frac{\log(1 - U)}{\log(p)} \right\}.$$

In the last expression, we had to change the $\leq$ sign into a $\geq$ sign, since we divided by the negative number $\log(p)$. By definition, the K from equation (1.2) is the smallest integer such that $\sum_{i=1}^{k} p_i \geq U$ is satisfied and thus the smallest integer greater than or equal to $\log(1 - U)/\log(p)$. Thus, the value

$$X = a_K = K = \left\lceil \frac{\log(1 - U)}{\log(p)} \right\rceil,$$

where $\lceil \cdot \rceil$ denotes the operation of rounding up a number to the nearest integer, is geometrically distributed with parameter p.

1.3   The inverse transform method

The inverse transform method is a method which can be applied when the target distribution is one-dimensional, that is to generate samples from a prescribed target

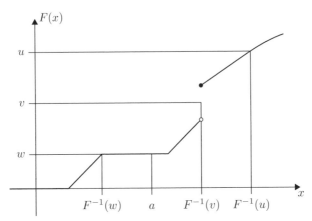

Figure 1.2   Illustration of the inverse F^{-1} of a CDF F. At level u the function F is continuous and injective; here F^{-1} coincides with the usual inverse of a function. The value v falls in the middle of a jump of F and thus has no preimage; $F^{-1}(v)$ is the preimage of the right-hand limit of F and $F(F^{-1}(v)) \neq v$. At level w the function F is not injective, several points map to w; the preimage $F^{-1}(w)$ is the left-most of these points and we have, for example, $F^{-1}(F(a)) \neq a$.

distribution on the real numbers $\mathbb{R}$. The method uses the cumulative distribution function (CDF) (see Section A.1) to specify the target distribution and can be applied for distributions which have no density.

Definition 1.12   Let F be a distribution function. Then the *inverse* of F is defined by

$$F^{-1}(u) = \inf \left\{ x \in \mathbb{R} \mid F(x) \geq u \right\}$$

for all $u \in (0, 1)$.

The definition of the inverse of a distribution function is illustrated in Figure 1.2. In the case where F is bijective, that is when F is strictly monotonically increasing and has no jumps, F^{-1} is just the usual inverse of a function. In this case we can find $F^{-1}(u)$ by solving the equation $F(x) = u$ for x. The following algorithm can be used to generate samples from a given distribution, whenever the inverse F^{-1} of the distribution function can be determined.

Algorithm 1.13   (inverse transform method)
    input:
        the inverse F^{-1} of a CDF F
    randomness used:
        $U \sim \mathcal{U}[0, 1]$

output:

  $X \sim F$

1: generate $U \sim \mathcal{U}[0, 1]$

2: return $X = F^{-1}(U)$

This algorithm is very simple and it directly transforms $\mathcal{U}[0, 1]$-distributed samples into samples with distribution function F. The following proposition shows that the samples X generated by algorithm 1.13 have the correct distribution.

Proposition 1.14   Let $F: \mathbb{R} \to [0, 1]$ be a distribution function and $U \sim \mathcal{U}[0, 1]$. Define $X = F^{-1}(U)$. Then X has distribution function F.

Proof   Using the definitions of X and F^{-1} we find

$$P(X \leq a) = P\left(F^{-1}(U) \leq a\right) = P\left(\inf\{ x \mid F(x) \geq U \} \leq a\right).$$

Since $\inf\{ x \mid F(x) \geq U \} \leq a$ holds if and only if $F(a) \geq U$, we can conclude

$$P(X \leq a) = P\left(F(a) \geq U\right) = F(a)$$

where the final equality comes from the definition of the uniform distribution on the interval $[0, 1]$. $\qquad\square$

Example 1.15   The exponential distribution $\mathrm{Exp}(\lambda)$ has density

$$f(x) = \begin{cases} \lambda e^{-\lambda x} & \text{if } x \geq 0 \text{ and} \\ 0 & \text{otherwise.} \end{cases}$$

Using integration, we find the corresponding CDF as

$$F(a) = \int_{-\infty}^{a} f(x)\,dx = \int_{0}^{a} \lambda e^{-\lambda x}\,dx = \left. -e^{-\lambda x} \right|_{x=0}^{a} = 1 - e^{-\lambda a}$$

for all $a \geq 0$. Since this function is strictly monotonically increasing and continuous, F^{-1} is the usual inverse of F. We have

$$1 - e^{-\lambda x} = u \quad \Longleftrightarrow \quad -\lambda x = \log(1 - u) \quad \Longleftrightarrow \quad x = -\frac{\log(1 - u)}{\lambda}$$

and thus $F^{-1}(u) = -\log(1 - u)/\lambda$ for all $u \in (0, 1)$. Now assume $U \sim \mathcal{U}[0, 1]$. Then proposition 1.14 gives that $X = -\log(1 - U)/\lambda$ is $\mathrm{Exp}(\lambda)$-distributed. Thus we have found a method to transform $\mathcal{U}[0, 1]$ random variables into $\mathrm{Exp}(\lambda)$-distributed random variables. The method can be further simplified by using the observation that U and $1 - U$ have the same distribution: if $U \sim \mathcal{U}[0, 1]$, then $-\log(U)/\lambda \sim \mathrm{Exp}(\lambda)$.

Example 1.16   The *Rayleigh distribution* with parameter $\sigma > 0$ has density

$$f(x) = \begin{cases} \dfrac{x}{\sigma^2} e^{-x^2/2\sigma^2} & \text{if } x \geq 0 \text{ and} \\ 0 & \text{otherwise.} \end{cases}$$

For this distribution we find

$$F(a) = \int_0^a \frac{x}{\sigma^2} e^{-x^2/2\sigma^2}\, dx = -e^{-x^2/2\sigma^2}\Big|_{x=0}^a = 1 - e^{-a^2/2\sigma^2}$$

for all $a \geq 0$. Solving the equation $u = F(x) = 1 - e^{-a^2/2\sigma^2}$ for x we find the inverse $F^{-1}(u) = x = \sqrt{-2\sigma^2 \log(1-u)}$. By proposition 1.14 we know that $X = \sqrt{-2\sigma^2 \log(1-U)}$ has density f if we choose $U \sim \mathcal{U}[0,1]$. As in the previous example, we can also write U instead of $1 - U$.

Example 1.17   Let X have density

$$f(x) = \begin{cases} 3x^2 & \text{for } x \in [0,1] \text{ and} \\ 0 & \text{otherwise.} \end{cases}$$

Then

$$F(a) = \int_{-\infty}^a f(x)\, dx = \begin{cases} 0 & \text{if } a < 0 \\ a^3 & \text{if } 0 \leq a < 1 \text{ and} \\ 1 & \text{for } 1 \leq a. \end{cases}$$

Since F maps $(0,1)$ into $(0,1)$ bijectively, F^{-1} is given by the usual inverse function and consequently $F^{-1}(u) = u^{1/3}$ for all $u \in (0,1)$. Thus, by proposition 1.14, if $U \sim \mathcal{U}[0,1]$, the cubic root $U^{1/3}$ has the same distribution as X.

Example 1.18   Let X be discrete with $P(X=0) = 0.6$ and $P(X=1) = 0.4$. Then

$$F(a) = \begin{cases} 0 & \text{if } a < 0 \\ 0.6 & \text{if } 0 \leq a < 1 \text{ and} \\ 1 & \text{if } 1 \leq a. \end{cases}$$

Using the definition of F^{-1} we find

$$F^{-1}(u) = \begin{cases} 0 & \text{if } 0 < u \leq 0.6 \text{ and} \\ 1 & \text{if } 0.6 < u < 1. \end{cases}$$

By proposition 1.14 we can construct a random variable X with the correct distribution from $U \sim \mathcal{U}[0, 1]$, by setting

$$
X = \begin{cases} 0 & \text{if } U \leq 0.6 \text{ and} \\ 1 & \text{if } U > 0.6. \end{cases}
$$

The inverse transform method can always be applied when the inverse F^{-1} is easy to evaluate. For some distributions like the normal distribution this is not the case, and the inverse transform method cannot be applied directly. The method can be applied (but may not be very useful) for discrete distributions such as in example 1.18. The main restriction of the inverse transform method is that distribution functions only exist in the one-dimensional case. For distributions on $\mathbb{R}^d$ where $d > 1$, more sophisticated methods are required.

1.4   Rejection sampling

The rejection sampling method is a more advanced, and very popular, method for random number generation. Several aspects make this method different from basic methods such as inverse transform method discussed in the previous section. First, rejection sampling is not restricted to $\mathcal{U}[0, 1]$-distributed input samples. The method is often used in multi-stage approaches where different methods are used to generate samples of approximately the correct distribution and then rejection sampling is used to convert these samples to follow the target distribution exactly. Secondly, while we state the method here only for distributions on the Euclidean space $\mathbb{R}^d$, the rejection sampling method can be generalised to work on very general spaces. Finally, a random and potentially large number of input samples is required to generate one output sample in the rejection method. As a consequence, the efficiency of the method becomes a concern.

1.4.1   Basic rejection sampling

In this section we introduce the fundamental idea that all rejection algorithms are based on. We start by presenting the basic algorithm which forms the prototype of the methods presented later.

Algorithm 1.19   (basic rejection sampling)
  input:
    a probability density g (the *proposal density*),
    a function p with values in [0,1] (the *acceptance probability*)
  randomness used:
    X_n i.i.d. with density g (the *proposals*),
    $U_n \sim \mathcal{U}[0, 1]$ i.i.d.

output:
    a sequence of i.i.d. random variables with density

$$f(x) = \frac{1}{Z} p(x)g(x) \quad \text{where} \quad Z = \int p(x)g(x)\,dx. \tag{1.3}$$

1: **for** $n = 1, 2, 3, \ldots$ **do**
2:    generate X_n with density g
3:    generate $U_n \sim \mathcal{U}[0, 1]$
4:    **if** $U_n \leq p(X_n)$ **then**
5:        output X_n
6:    **end if**
7: **end for**

The effect of the random variables U_n in the algorithm is to randomly decide whether to output or to ignore the value X_n: the value X_n is output with probability $p(X_n)$, and using the trick from lemma 1.9 we use the event $\{U \leq p(X_n)\}$ to decide whether or not to output the value. In the context of rejection sampling, the random variables X_n are called *proposals*. If the proposal X_n is chosen for output, that is if $U_n \leq p(X_n)$, we say that X_n is *accepted*, otherwise we say that X_n is *rejected*.

Proposition 1.20   For $k \in \mathbb{N}$, let X_{N_k} denote the kth output of algorithm 1.19. Then the following statements hold:

(a) The elements of the sequence $(X_{N_k})_{k\in\mathbb{N}}$ are i.i.d. with density f given by (1.3).

(b) Each proposal is accepted with probability Z; the number of proposals required to generate each X_{N_k} is geometrically distributed with mean $1/Z$.

Proof   For fixed n, the probability of accepting X_n is

$$P\left(U_n \leq p(X_n)\right) = \int p(x)g(x)\,dx = Z, \tag{1.4}$$

where Z is the constant defined in equation (1.3). Since the decisions whether to accept X_n for different n are independent, the time until the first success is geometrically distributed with mean $1/Z$ as required. This completes the proof of the second statement.

For the proof of the first statement, first note that the indices $N_1, N_2, N_3, \ldots$ of the accepted X_n are random. If we let $N_0 = 0$, we can write

$$N_k = \min\left\{n \in \mathbb{N} \mid n > N_{k-1}, U_n \leq p(X_n)\right\}$$

for all $k \in \mathbb{N}$. If we consider the distribution of X_{N_k} conditional on the value of N_{k-1}, we find

$$P\left(X_{N_k} \in A \,\middle|\, N_{k-1} = n\right)$$

$$= \sum_{m=1}^{\infty} P\left(N_k = n + m, X_{n+m} \in A \,\middle|\, N_{k-1} = n\right)$$

$$= \sum_{m=1}^{\infty} P\left(U_{n+1} > p(X_{n+1}), \ldots, U_{n+m-1} > p(X_{n+m-1}),\right.$$

$$\left. U_{n+m} \le p(X_{n+m}), X_{n+m} \in A \,\middle|\, N_{k-1} = n\right)$$

$$= \sum_{m=1}^{\infty} P\left(U_{n+1} > p(X_{n+1})\right) \cdots P\left(U_{n+m-1} > p(X_{n+m-1})\right) \cdot$$

$$P\left(U_{n+m} \le p(X_{n+m}), X_{n+m} \in A\right).$$

Here we used the fact that all the probabilities considered in the last expression are independent of the value of N_{k-1}. Similar to (1.4) we find

$$P\left(U_n \le p(X_n), X_n \in A\right) = \int_A p(x) g(x) \, dx$$

and consequently we have

$$P\left(X_{N_k} \in A \,\middle|\, N_{k-1} = n\right)$$

$$= \sum_{m=1}^{\infty} (1 - Z)^{m-1} \int_A p(x) g(x) \, dx$$

$$= \frac{1}{Z} \int_A p(x) g(x) \, dx$$

by the geometric series formula. Since the right-hand side does not depend on n, we can conclude

$$P\left(X_{N_k} \in A\right) = \frac{1}{Z} \int_A p(x) g(x) \, dx$$

and thus we find that X_{N_k} has density pg/Z.

To see that the X_{N_k} are independent we need to show that

$$P\left(X_{N_1} \in A_1, \ldots, X_{N_k} \in A_k\right) = \prod_{i=1}^{k} P\left(X_{N_i} \in A_i\right)$$

for all sets $A_1, \ldots, A_k$ and for all $k \in \mathbb{N}$. This can be done by summing up the probabilities for the cases $N_1 = n_1, \ldots, N_k = n_k$, similar to the first part of the proof, but we omit this tedious calculation here.   □

Example 1.21 Let $X \sim \mathcal{U}[-1, +1]$ and accept X with probability

$$p(X) = \sqrt{1 - X^2}.$$

Then, by proposition 1.20, the accepted samples have density

$$f(x) = \frac{1}{Z} p(x) g(x) = \frac{1}{Z} \cdot \sqrt{1 - x^2} \cdot \frac{1}{2} \mathbb{1}_{[-1, +1]}(x),$$

where we use the indicator function notation from equation (A.7) to get the abbreviation

$$\mathbb{1}_{[-1, +1]}(x) = \begin{cases} 1 & \text{if } x \in [-1, +1] \text{ and} \\ 0 & \text{otherwise} \end{cases}$$

and

$$Z = \int_{\mathbb{R}} \sqrt{1 - x^2} \cdot \frac{1}{2} \mathbb{1}_{[-1, +1]}(x) \, dx = \frac{1}{2} \int_{-1}^{1} \sqrt{1 - x^2} \, dx = \frac{1}{2} \cdot \frac{\pi}{2} = \frac{\pi}{4}.$$

Combining these results, we find that the density f of accepted samples is given by

$$f(x) = \frac{2}{\pi} \sqrt{1 - x^2} \, \mathbb{1}_{[-1, +1]}(x).$$

The graph of the density f forms a semicircle and the resulting distribution is known as Wigner's semicircle distribution.

One important property of the rejection algorithm 1.19 is that none of the steps in the algorithm makes any reference to the normalisation constant Z. Thus, we do not need to compute the value of Z in order to apply this algorithm. We will see that this fact, while looking like a small detail at first glance, is extremely useful in practical applications.

1.4.2   Envelope rejection sampling

The basic rejection sampling algorithm 1.19 from the previous section is usually applied by choosing the acceptance probabilities p so that the density f of the output values, given by (1.3), coincides with a given target distribution. The resulting algorithm can be written as in the following.

Algorithm 1.22   (envelope rejection sampling)
  input:
    a function f with values in $[0, \infty)$ (the non-normalised *target density*),
    a probability density g (the *proposal density*),
    a constant $c > 0$ such that $f(x) \leq c\,g(x)$ for all x
  randomness used:
    X_n i.i.d. with density g (the *proposals*),
    $U_n \sim \mathcal{U}[0, 1]$ i.i.d.
  output:
    a sequence of i.i.d. random variables with density

$$\tilde{f}(x) = \frac{1}{Z_f} f(x) \quad \text{where} \quad Z_f = \int f(x)\,dx$$

1: **for** $n = 1, 2, 3, \ldots$ **do**
2:     generate X_n with density g
3:     generate $U_n \sim \mathcal{U}[0, 1]$
4:     **if** $cg(X_n)U_n \leq f(X_n)$ **then**
5:         output X_n
6:     **end if**
7: **end for**

The assumption in the algorithm is that we can already sample from the distribution with probability density g, but we would like to generate samples from the distribution with density $\tilde{f}$ instead. Normally, f will be chosen to be a probability density and in this case we have $\tilde{f} = f$, but in some situations the normalising constant Z_f is difficult to obtain and due to the distinction between f and $\tilde{f}$, in these situations the algorithm can still be applied. The rejection mechanism employed in algorithm 1.22 is illustrated in Figure 1.3. The function cg is sometimes called an 'envelope' for f.

Proposition 1.23   Let X_{N_k} for $k \in \mathbb{N}$ denote the kth output value of algorithm 1.22 with (non-normalised) target density f. Then the following statements hold:

(a) The elements of the sequence $(X_{N_k})_{k \in \mathbb{N}}$ are i.i.d. with density $\tilde{f}$.

(b) Each proposal is accepted with probability Z_f/c; the number $M_k = N_k - N_{k-1}$ of proposals required to generate each X_{N_k} is geometrically distributed with mean $E(M_k) = c/Z_f$.

Proof   Algorithm 1.22 coincides with algorithm 1.19 where the acceptance probability p is chosen as

$$p(x) = \begin{cases} \frac{f(x)}{cg(x)} & \text{if } g(x) > 0 \text{ and} \\ 1 & \text{otherwise.} \end{cases}$$

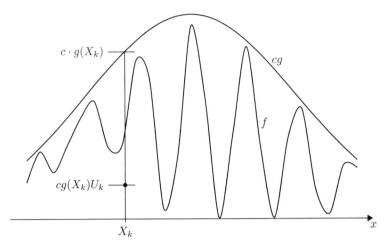

Figure 1.3   Illustration of the envelope rejection sampling method from algorithm 1.22. The proposal $(X_k, cg(X_k) U_k)$ is accepted, if it falls into the area underneath the graph of f. In Section 1.4.4 we will see that the proposal is distributed uniformly on the area under the graph of cg.

In this situation, the normalisation constant Z from (1.3) is given by:

$$Z = \int p(x)g(x)\,dx = \int \frac{f(x)}{cg(x)} g(x)\,dx = \frac{1}{c} \int f(x)\,dx = Z_f/c.$$

From proposition 1.20 we then know that the output of algorithm 1.19 is an i.i.d. sequence with density

$$\frac{1}{Z}pg = \frac{c}{Z_f} \frac{f}{cg} g = \frac{1}{Z_f} f$$

and that the required number of proposals to generate one output sample is geometrically distributed with mean $1/Z = c/Z_f$.   □

Example 1.24   We can use rejection sampling to generate samples from the half-normal distribution with density

$$f(x) = \begin{cases} \frac{2}{\sqrt{2\pi}} \exp\left(-\frac{x^2}{2}\right) & \text{if } x \geq 0 \text{ and} \\ 0 & \text{otherwise.} \end{cases} \tag{1.5}$$

If we assume that the proposals are $\text{Exp}(\lambda)$-distributed, then the density of the proposals is

$$g(x) = \begin{cases} \lambda \exp(-\lambda x) & \text{if } x \geq 0 \text{ and} \\ 0 & \text{otherwise.} \end{cases}$$

In order to apply algorithm 1.22 we need to determine a constant $c > 0$ such that $f(x) \leq cg(x)$ for all $x \in \mathbb{R}$. For $x < 0$ we have $f(x) = g(x) = 0$. For $x \geq 0$ we have

$$\frac{f(x)}{g(x)} = \frac{2}{\sqrt{2\pi}\lambda} \exp\left(-\frac{x^2}{2} + \lambda x\right).$$

It is easy to check that the quadratic function $-x^2/2 + \lambda x$ attains its maximum at $x = \lambda$. Thus we have

$$\frac{f(x)}{g(x)} \leq c^*$$

for all $x \geq 0$, where

$$c^* = \frac{2}{\sqrt{2\pi}\lambda} \exp\left(-\frac{\lambda^2}{2} + \lambda \cdot \lambda\right) = \sqrt{\frac{2}{\pi\lambda^2}} \exp\left(\lambda^2/2\right).$$

Consequently, any $c \geq c^*$ satisfies the condition $f \leq cg$. From proposition 1.23 we know that the average number of proposals required for generating one sample, and thus the computational cost, is proportional to c. Thus we should choose c as small as possible and $c = c^*$ is the optimal choice.

Given our choice of g and c, the acceptance criterion from algorithm 1.22 can be simplified as follows:

$$cg(x)U \leq f(x)$$
$$\iff \quad \sqrt{\frac{2}{\pi\lambda^2}} \exp\left(\frac{\lambda^2}{2}\right) \lambda \exp(-\lambda x)U \leq \frac{2}{\sqrt{2\pi}} \exp\left(-\frac{x^2}{2}\right)$$
$$\iff \quad U \leq \exp\left(-\frac{x^2}{2} + \lambda x - \frac{\lambda^2}{2}\right)$$
$$\iff \quad U \leq \exp\left(-\frac{1}{2}(x - \lambda)^2\right).$$

This leads to the following algorithm for generating samples from the half-normal distribution:

1: **for** $n = 1, 2, 3, \ldots$ **do**
2:    generate $X_n \sim \text{Exp}(\lambda)$
3:    generate $U_n \sim \mathcal{U}[0, 1]$

4:    **if** $U_n \leq \exp(-\frac{1}{2}(X_n - \lambda)^2)$ **then**
5:        output X_n
6:    **end if**
7: **end for**

Finally, since the density f is the density of a standard-normal distribution conditioned on being positive, and since the normal distribution is symmetric, we can generate standard normal distributed values by randomly choosing X_n or $-X_n$, both with probability $1/2$, for each accepted sample.

In algorithm 1.22, we can choose the density g of the proposal distribution in order to maximise efficiency of the method. The only constraint is that we need to be able to find the constant c. This condition implies, for example, that the support of g cannot be smaller than the support of f, that is we need $g(x) > 0$ whenever $f(x) > 0$. The average cost of generating one sample is given by the average number of proposals required times the cost for generating each proposal. Therefore the algorithm is efficient, if the following two conditions are satisfied:

(a) There is an efficient method to generate the proposals X_i. This affects the choice of the proposal density g.

(b) The average number c/Z_f of proposals required to generate one sample is small. This number is influenced by the value of c and, since the possible choices of c depend on g, also by the proposal density g.

1.4.3 Conditional distributions

The conditional distribution $P_{X|X \in A}$ corresponds to the remaining randomness in X when we already know that $X \in A$ occurred (see equation (A.4) for details). Sampling from a conditional distribution can be easily done by rejection sampling. The basic result is the following.

Algorithm 1.25   (rejection sampling for conditional distributions)
    input:
        a set A with $P(X \in A) > 0$
    randomness used:
        a sequence X_n of i.i.d. copies of X (the *proposals*)
    output:
        a sequence of i.i.d. random variables with distribution $P_{X|X \in A}$
    1: **for** $n = 1, 2, 3, \ldots$ **do**
    2:    generate X_n
    3:    **if** $X_n \in A$ **then**
    4:        output X_n
    5:    **end if**
    6: **end for**

Proposition 1.26   Let X be a random variable and let A be a set. Furthermore, let X_{N_k} for $k \in \mathbb{N}$ denote the kth output value of algorithm 1.25. Then the following statements hold:

(a) The elements of the sequence $(X_{N_k})_{k \in \mathbb{N}}$ are i.i.d. and satisfy

$$P(X_{N_k} \in B) = P(X \in B \mid X \in A)$$

for all $k \in \mathbb{N}$ and all sets B.

(b) The number $M_k = N_k - N_{k-1}$ of proposals required to generate each X_{N_k} is geometrically distributed with mean $E(M_k) = 1/P(X \in A)$.

Proof   Algorithm 1.25 is a special case of algorithm 1.19 where the acceptance probability is chosen as $p(x) = \mathbb{1}_A(x)$. For this choice of p, the decision whether or not to accept the proposal given the value of X_n is deterministic and thus we can omit generation of the auxiliary random variables U_n in the algorithm.

Now assume that the distribution of X has a density g. Using equation (1.3) we then find

$$Z = \int \mathbb{1}_A(x)g(x)\,dx = P(X \in A)$$

and by proposition 1.20 we have

$$P(X_{N_k} \in B) = \frac{\int_B \mathbb{1}_A(x)g(x)\,dx}{Z} = \frac{P(X \in B \cap A)}{P(X \in A)} = P(X \in B \mid X \in A).$$

A similar proof gives the result in the case where X does not have a density. This completes the proof of the first statement of the proposition. The second statement is a direct consequence of proposition 1.20.   □

The method presented in algorithm 1.25 works well if $p = P(X \in A)$ is not too small; the time required for producing a single output sample is proportional to $1/p$.

Example 1.27   We can use algorithm 1.25 to generate samples $X \sim \mathcal{N}(0, 1)$, conditioned on $X \geq a$. We simply have to repeat the following two steps until enough samples are output:

(a) generate $X \sim \mathcal{N}(0, 1)$;

(b) if $X \geq a$, output X.

The efficiency of this method depends on the value of a. The following table shows the average number $\mathbb{E}(N_a)$ of samples required to generate one output sample for different values of a, rounded to the nearest integer:

a	1	2	3	4	5	6
$\mathbb{E}(N_a)$	6	44	741	31 574	3 488 556	1 013 594 692

The table shows that the method will be slow even for moderate values of a. For $a \geq 5$ the required number of samples is so large that the method will likely be no longer practical.

For conditions with very small probabilities, rejection sampling can still be used to generate samples from the conditional distribution, but we have to use the full rejection sampling algorithm 1.22 instead of the simplified version from algorithm 1.25. This is illustrated in the following example.

Example 1.28   We can use algorithm 1.22 to generate samples from the conditional distribution of $X \sim \mathcal{N}(0, 1)$, conditioned on $X \geq a > 0$. The density of the conditional distribution is

$$\tilde{f}(x) = \frac{1}{Z} \exp(-x^2/2) \mathbb{1}_{[a,\infty]}(x) = \frac{1}{Z} f(x),$$

where Z is the normalising constant (we have included the pre-factor $1/\sqrt{2\pi}$ into Z to simplify notation).

We can sample from this distribution using proposals of the form $X = \tilde{X} + a$ where $\tilde{X} \sim \mathrm{Exp}(\lambda)$. This proposal distribution has density

$$g(x) = \lambda \exp\left(-\lambda(x - a)\right) \mathbb{1}_{[a,\infty]}(x)$$

and we need to find a constant $c > 0$ such that $f(x) \leq cg(x)$ for all $x \geq a$. Also, we can still choose the parameter λ and, in order to maximise efficiency of the method, we should choose a value of λ such that the shape of g is as similar to the shape of f as possible. In order to achieve this, we choose c and λ so that at $x = a$ both the values and the derivatives of f and cg coincide (see Figure 1.4 for illustration). This leads to the conditions $e^{-a^2/2} = f(a) = cg(a) = c\lambda$ and $-ae^{-a^2/2} = f'(a) = cg'(a) = -c\lambda^2$ and solving these two equations for the two unknowns c and λ gives $\lambda = a$ and $c = e^{-a^2/2}/a$.

Figure 1.4 indicates that for this choice of λ and c, condition $f \leq cg$ will be satisfied. Indeed we find

$$\frac{f(x)}{cg(x)} = \frac{\exp(-x^2/2)}{1/a \exp(-a^2/2) \cdot a \exp\left(-a(x - a)\right)}$$

$$= \exp\left(-x^2/2 + ax - a^2/2\right) = \exp\left(-\frac{(x - a)^2}{2}\right)$$

$$\leq 1$$

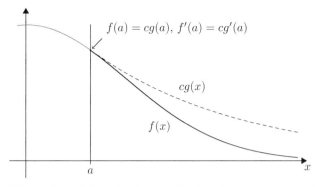

Figure 1.4   Illustration of the rejection mechanism from example 1.28. The graph shows the (scaled) proposal density cg, enveloping the (non-normalised) target density f.

and thus $f(x) \leq cg(x)$ for all $x \geq a$. Thus, we can apply algorithm 1.22 with proposal density g to generate samples from the distribution with density $\tilde{f}$. The resulting method consists of the following steps:

(a) generate $\tilde{X} \sim \text{Exp}(a)$ and $U \sim \mathcal{U}[0, 1]$;

(b) let $X = \tilde{X} + a$;

(c) if $U \leq \exp(-(X - a)^2/2)$, output X.

From proposition 1.23 we know that the average number M_a of proposals required to generate one sample is

$$\mathbb{E}(M_a) = \frac{c}{\int_{\mathbb{R}} f(x)\,dx} = \frac{\exp(-a^2/2)/a}{\int_a^\infty \exp(-x^2/2)\,dx} = \frac{\exp(-a^2/2)}{a\sqrt{2\pi}\,(1 - \Phi(a))},$$

where

$$\Phi(a) = \frac{1}{\sqrt{2\pi}} \int_{-\infty}^{a} \exp(-x^2/2)\,dx$$

is the CDF of the standard normal distribution. The following table lists the value of $\mathbb{E}(M_a)$, rounded to three significant digits, for different values of a:

a	1	2	3	4	5	6
$\mathbb{E}(M_a)$	1.53	1.19	1.09	1.06	1.04	1.03

The table clearly shows that the resulting algorithm works well for large values of a: the steps required to generate one proposal are more complicated than for the method from example 1.27, but significantly fewer proposals are required.

1.4.4   Geometric interpretation

The rejection sampling method can be applied not only to the generation of random numbers, but also to the generation of random objects in arbitrary spaces. To illustrate this, in this section we consider the problem of sampling from the uniform distribution on subsets of the Euclidean space $\mathbb{R}^d$. We then use the resulting techniques to give an alternative proof of proposition 1.23, based on geometric arguments.

We write $|A|$ for the d-dimensional volume of a set $A \subseteq \mathbb{R}^d$. Then the cube $Q = [a, b]^3 \subseteq \mathbb{R}^3$ has volume $|Q| = (b - a)^3$, the unit circle $C = \{x \in \mathbb{R}^2 \mid x_1^2 + x_2^2 \leq 1\}$ has two-dimensional 'volume' π (area) and the line segment $[a, b] \subseteq \mathbb{R}$ has one-dimensional 'volume' $b - a$ (length). For more general sets A, the volume can be found by integration: we have

$$|A| = \int_{\mathbb{R}^d} \mathbb{1}_A(x)\,dx = \int \cdots \int \mathbb{1}_A(x_1, \ldots, x_d)\,dx_d \cdots dx_1.$$

Definition 1.29   A random variable X with values in $\mathbb{R}^d$ is uniformly distributed on a set $A \subseteq \mathbb{R}^d$ with $0 < |A| < \infty$, if

$$P(X \in B) = \frac{|A \cap B|}{|A|}$$

for all $B \subseteq \mathbb{R}^d$. As for real intervals, we use the notation $X \sim \mathcal{U}(A)$ to indicate that X is uniformly distributed on A.

The intuitive meaning of X being uniformly distributed on a set A is that X is a random element of A, and that all regions of A are hit by X equally likely. The probability of X falling into a subset of A only depends on the volume of this subset, but not on the location inside A.

Let $X \sim \mathcal{U}(A)$. From the definition we can derive simple properties of the uniform distribution: first we have

$$P(X \in A) = \frac{|A \cap A|}{|A|} = 1$$

and if A and B are disjoint we find

$$P(X \in B) = \frac{|A \cap B|}{|A|} = \frac{|\emptyset|}{|A|} = 0.$$

For general $B \subseteq \mathbb{R}^d$ we get

$$P(X \notin B) = P(X \in \mathbb{R}^d \setminus B)$$
$$= \frac{|A \cap (\mathbb{R}^d \setminus B)|}{|A|} = \frac{|A \setminus B|}{|A|} = \frac{|A| - |A \cap B|}{|A|} = 1 - \frac{|A \cap B|}{|A|}.$$

Lemma 1.30   Let $A \subseteq \mathbb{R}^d$ be a set with volume $0 < |A| < \infty$. Then the uniform distribution $\mathcal{U}(A)$ has probability density $f = \mathbb{1}_A/|A|$ on $\mathbb{R}^d$.

Proof   Let $X \sim \mathcal{U}(A)$. For $B \subseteq \mathbb{R}^d$ we have

$$P(X \in B) = \frac{|A \cap B|}{|A|} = \frac{1}{|A|} \int_{\mathbb{R}^d} \mathbb{1}_{A \cap B}(x)\, dx = \int_{\mathbb{R}^d} \mathbb{1}_B(x) \frac{\mathbb{1}_A(X)}{|A|}\, dx$$

and thus X has the given density f. $\square$

Lemma 1.31   Let X be uniformly distributed on a set A, and let B be a set with $|A \cap B| > 0$. Then the conditional distribution $P_{X|X \in B}$ of X conditioned on the event $X \in B$ coincides with the uniform distribution on $A \cap B$.

Proof   From the definition of the uniform distribution we get

$$\begin{aligned} P(X \in C \,|\, X \in B) &= \frac{P(X \in B \cap C)}{P(X \in B)} \\ &= \frac{|A \cap B \cap C|/|A|}{|A \cap B|/|A|} = \frac{|(A \cap B) \cap C|}{|A \cap B|}. \end{aligned}$$

Since this is the probability of a $\mathcal{U}(A \cap B)$-distributed random variable to hit the set C, the statement is proved. $\square$

By combining the result of lemma 1.31 with the method given in algorithm 1.25, we can sample from the uniform distribution of every set which can be covered by a (union of) rectangles. This is illustrated in the following example.

Example 1.32   (uniform distribution on the circle) Let $X_n, Y_n \sim \mathcal{U}[-1, +1]$ be i.i.d. By exercise E1.10 the pairs (X_n, Y_n) are then uniformly distributed on the square $A = [0, 1] \times [0, 1]$. Now let $(Z_k)_{k \in \mathbb{N}}$ be the subsequence of all pairs (X_{n_k}, Y_{n_k}) which satisfy the condition

$$X_n^2 + Y_n^2 \leq 1.$$

Then $(Z_k)_{k \in \mathbb{N}}$ is an i.i.d. sequence, uniformly distributed on the unit circle $B = \{x \in \mathbb{R}^2 \mid |x| \leq 1\}$. The probability p to accept each sample is given by

$$p = P((X_n, V_n) \in B) = \frac{|B|}{|A|} = \frac{\pi\, 1^2}{2^2} = \frac{\pi}{4} \approx 78.5\%$$

and the number of proposals required to generate one sample is, on average, $1/p \approx 1.27$.

To conclude this section, we give an alternative proof of proposition 1.23. This proof is based on the geometric approach taken in this section and uses a connection between distributions with general densities on $\mathbb{R}^d$ and uniform distributions on $\mathbb{R}^{d+1}$, given in the following result.

Lemma 1.33   Let $f: \mathbb{R}^d \to [0, \infty)$ be a probability density and let

$$A = \{(x, y) \in \mathbb{R}^d \times [0, \infty) \mid 0 \le y < f(x)\} \subseteq \mathbb{R}^{d+1}.$$

Then $|A| = 1$ and the following two statements are equivalent:

(a) (X, Y) is uniformly distributed on A.

(b) X is distributed with density f on $\mathbb{R}^d$ and $Y = f(X)U$ where $U \sim \mathcal{U}[0, 1]$, independently of X.

Proof   The volume of the set A can be found by integrating the 'height' $f(x)$ over all of $\mathbb{R}^d$. Since f is a probability density, we get

$$|A| = \int_{\mathbb{R}^d} f(x)\, dx = 1.$$

Assume first that (X, Y) is uniformly distributed on A and define $U = Y/f(X)$. Since $(X, Y) \in A$, we have $f(X) > 0$ with probability 1 and thus there is no problem in dividing by $f(X)$. Given sets $C \subseteq \mathbb{R}^d$ and $D \subseteq \mathbb{R}$ we find

$$\begin{aligned}
P\left(X \in C, U \in D\right) &= P\left((X, Y) \in \{(x, y) \mid x \in C, y/f(x) \in D\}\right) \\
&= \left|A \cap \{(x, y) \mid x \in C, y/f(x) \in D\}\right| \\
&= \int_{\mathbb{R}^d} \int_0^{f(x)} \mathbb{1}_C(x)\mathbb{1}_D\left(y/f(x)\right)\, dy\, dx.
\end{aligned}$$

Using the substitution $u = y/f(x)$ in the inner integral we get

$$\begin{aligned}
P\left(X \in C, U \in D\right) &= \int_{\mathbb{R}^d} \int_0^1 \mathbb{1}_C(x)\mathbb{1}_D(u)f(x)\, du\, dx \\
&= \int_C f(x)\, dx \cdot \int_D \mathbb{1}_{[0,1]}(u)\, du.
\end{aligned}$$

Therefore X and U are independent with densities f and $\mathbb{1}_{[0,1]}$, respectively.

For the converse statement assume now that the random variables X with density f and $U \sim \mathcal{U}[0, 1]$ are independent, and let $Y = f(X)U$. Furthermore let $C \subseteq \mathbb{R}^d$, $D \subseteq [0, \infty)$ and $B = C \times D$. Then we get

$$
\begin{aligned}
P\left((X, Y) \in B\right) &= P\left(X \in C, Y \in D\right) \\
&= \int_C P(Y \in D \mid X = x) f(x) \, dx \\
&= \int_C P\left(f(x)U \in D\right) f(x) \, dx \\
&= \int_C \frac{|D \cap [0, f(x)]|}{f(x)} f(x) \, dx \\
&= \int_C |D \cap [0, f(x)]| \, dx.
\end{aligned}
$$

On the other hand we have

$$
\begin{aligned}
|A \cap B| &= \int_{\mathbb{R}^d} \int_0^{f(x)} \mathbb{1}_B(x, y) \, dy \, dx \\
&= \int_{\mathbb{R}^d} \mathbb{1}_C(x) \int_0^{f(x)} \mathbb{1}_D(y) \, dy \, dx \\
&= \int_C |D \cap [0, f(x)]| \, dx
\end{aligned}
$$

and thus $P((X, Y) \in B) = |A \cap B|$. This shows that (X, Y) is uniformly distributed on A. $\qquad\square$

An easy application of lemma 1.33 is to convert a uniform distribution of a subset of $\mathbb{R}^2$ to a distribution on $\mathbb{R}$ with a given density $f : [a, b] \to \mathbb{R}$. For simplicity, we assume first that f lives on a bounded interval $[a, b]$ and satisfies $f(x) \le M$ for all $x \in [a, b]$. We can generate samples from the distribution with density f as follows:

(a) Let (X_k, Y_k) are be i.i.d., uniformly distributed on the rectangle $R = [a, b] \times [0, M]$.

(b) Consider the set $A = \{(x, y) \in R \mid y \le f(x)\}$ and let $N = \min\{k \in \mathbb{N} \mid X_k \in B\}$. By lemma 1.31, (X_N, Y_N) is uniformly distributed on A.

(c) By lemma 1.33, the value X_N is distributed with density f.

This procedure is illustrated in Figure 1.5.

In the situation of algorithm 1.22, that is when f is defined on an unbounded set, we cannot use proposals which are uniformly distributed on a rectangle anymore. A solution to this problem is to use lemma 1.33 a second time to obtain a suitable proposal distribution. This approach provides an alternative way of understanding algorithm 1.22: in the situation of algorithm 1.22, X_k has density g and U_k is uniformly distributed on $[0, 1]$. Then we know from lemma 1.33 that the

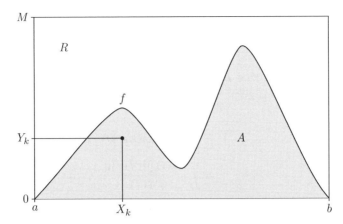

Figure 1.5   Illustration of the rejection sampling method where the graph of the target density is contained in a rectangle $R = [a, b] \times [0, M]$. In this case the proposals are uniformly distributed on the rectangle R and a proposal is accepted if it falls into the shaded region.

pair $(X_k, g(X_k)U_k)$ is uniformly distributed on the set $\{(x, v) \mid 0 \leq v < g(x)\}$. Consequently, $Z_k = (X_k, cg(X_k)U_k)$ is uniformly distributed on $A = \{(x, y) \mid 0 \leq y < cg(x)\}$. By lemma 1.31 and proposition 1.26, the accepted values are uniformly distributed on the set $B = \{(x, y) \mid 0 \leq y < f(x)\} \subseteq A$ and, applying lemma 1.33 again, we find that the X_k, conditional on being accepted, have density f. This argument can be made into an alternative proof of proposition 1.23.

1.5   Transformation of random variables

Samples from a wide variety of distributions can be generated by considering deterministic transformations of random variables. The inverse transform method, introduced in Section 1.3, is a special case of this technique where we transform a uniformly distributed random variable using the inverse of a CDF. In this section, we consider more general transformations.

The fundamental question we have to answer in order to generate samples by transforming a random variable is the following: if X is a random variable with values in $\mathbb{R}^d$ and a given distribution, and if $\varphi : \mathbb{R}^d \to \mathbb{R}^d$ is a function, what is the distribution of $\varphi(X)$? This question is answered in the following theorem.

Theorem 1.34   (transformation of random variables) Let $A, B \subseteq \mathbb{R}^d$ be open sets, $\varphi : A \to B$ be bijective and differentiable with continuous partial derivatives, and let X be a random variable with values in A. Furthermore let $g : B \to [0, \infty)$ be a probability density and define $f : \mathbb{R}^d \to \mathbb{R}$ by

$$f(x) = \begin{cases} g(\varphi(x)) \cdot |\det D\varphi(x)| & \text{if } x \in A \text{ and} \\ 0 & \text{otherwise.} \end{cases} \quad (1.6)$$

Then f is a probability density and the random variable X has density f if and only if $\varphi(X)$ has density g.

The matrix $D\varphi$ used in the theorem is the Jacobian of φ, as given in the following definition.

Definition 1.35   Let $\varphi : \mathbb{R}^d \to \mathbb{R}^d$ be differentiable. Then the *Jacobian matrix $D\varphi$* is the $d \times d$ matrix consisting of the partial derivatives of φ: for $i, j = 1, 2, \ldots, d$ we have $D\varphi(x)_{ij} = \frac{\partial \varphi_i}{\partial x_j}(x)$.

Theorem 1.34 is a consequence of the substitution rule for integrals. Before we give the proof of theorem 1.34, we first state the substitution rule in the required form.

Lemma 1.36   (substitution rule for integrals) Let $A, B \subseteq \mathbb{R}^d$ be open sets, $f : B \to \mathbb{R}$ integrable, and $\varphi: A \to B$ be bijective and differentiable with continuous partial derivatives. Then

$$\int_B f(y)\, dy = \int_A f(\varphi(x))\, |\det D\varphi(x)|\, dx$$

where $D\varphi$ denotes the Jacobian matrix of φ.

A proof of lemma 1.36 can, for example, be found in the book by Rudin (1987, theorem 7.26). Using lemma 1.36 we can now give the proof of the transformation rule for random variables.

Proof   (of theorem 1.34). By definition, the function f is positive and using lemma 1.36, we get

$$\int_{\mathbb{R}^d} f(x)\, dx = \int_A g(\varphi(x)) \cdot |\det D\varphi(x)|\, dx = \int_B g(y)\, dy = 1.$$

Thus f is a probability density.

Now assume that X is distributed with density f and let $C \subseteq B$. Then, by equation (A.8):

$$P(\varphi(X) \in C) = \int_A \mathbb{1}_C(\varphi(x)) f(x)\, dx$$
$$= \int_A \mathbb{1}_C(\varphi(x)) g(\varphi(x)) \cdot |\det D\varphi(x)|\, dx.$$

Now we can apply lemma 1.36, again, to transform the integral over A into an integral over the set B: we find

$$P(\varphi(X) \in C) = \int_B \mathbb{1}_C(y)g(y)\,dy.$$

Since this equality holds for all sets C, the random variable $\varphi(X)$ has density g. The converse statement follows by reversing the steps in this argument.   □

While theorem 1.34 is most powerful in the multidimensional case, it can be applied in the one-dimensional case, too. In this case the Jacobian matrix is a 1×1 matrix, that is a number, and we have $|\det D\varphi(x)| = |\varphi'(x)|$.

Example 1.37   (two-dimensional normal distribution) Assume that we want to sample from the two-dimensional standard normal distribution, that is from the distribution with density

$$g(x, y) = \frac{1}{2\pi} \exp\left(-\frac{x^2 + y^2}{2}\right).$$

Since g depends on (x, y) only via the squared length $x^2 + y^2$ of this vector, we try to simplify g using polar coordinates. The corresponding transformation φ is given by

$$\varphi(r, \theta) = (r \cos(\theta), r \sin(\theta))$$

for all $r > 0$, $\varphi \in (0, 2\pi)$. Note that we define φ only on the open set $A = (0, \infty) \times (0, 2\pi)$ in order to satisfy the requirement from theorem 1.34 that φ must be bijective. The resulting image set is $B = \varphi(A) = \mathbb{R}^2 \setminus \{(x, y) \mid x \geq 0, y = 0\}$, that is B is strictly smaller than $\mathbb{R}^2$ since it does not include the positive x-axis. This is not a problem, since the two-dimensional standard normal distribution hits the positive x-axis only with probability 0 and thus takes values in the set B with probability 1.

The Jacobian matrix of φ is given by

$$D\varphi(r, \theta) = \begin{pmatrix} \frac{\partial}{\partial r}\varphi_1 & \frac{\partial}{\partial \theta}\varphi_1 \\ \frac{\partial}{\partial r}\varphi_2 & \frac{\partial}{\partial \theta}\varphi_2 \end{pmatrix} = \begin{pmatrix} \cos(\theta) & -r \sin(\theta) \\ \sin(\theta) & r \cos(\theta) \end{pmatrix}$$

and thus we get $|\det D\varphi(r, \theta)| = \left| r \cos(\theta)^2 + r \sin(\theta)^2 \right| = r$. Using theorem 1.34 we have reduced the problem of sampling from a two-dimensional normal distribution to the problem of sampling from the density

$$f(r, \theta) = g\left(\varphi(r, \theta)\right) \cdot |\det D\varphi(r, \theta)| = \frac{1}{2\pi} \exp(-r^2/2) \cdot r$$

on $(0, \infty) \times (0, 2\pi)$.

The density $f(r, \theta)$ does not depend on θ and we can rewrite it as the product $f(r, \theta) = f_1(\theta) f_2(r)$ where $f_1(\theta) = 1/2\pi$ is the density of $\mathcal{U}[0, 2\pi]$ and $f_2(r) = r \exp(-r^2/2)$. From example 1.16 we know how to sample from the density f_2: if $U \sim \mathcal{U}[0, 1]$, then $R = \sqrt{-2\log(U)}$ has density f_2. Consequently, we can use the following steps to sample from the density g:

(a) Generate $\Theta \sim \mathcal{U}[0, 2\pi]$ and $U \sim \mathcal{U}[0, 1]$ independently.

(b) Let $R = \sqrt{-2\log(U)}$.

(c) Let $(X, Y) = \varphi(R, \Theta) = (R\cos(\Theta), R\sin(\Theta))$.

Then (R, Θ) has density f and, by theorem 1.34, the vector (X, Y) is standard normally distributed in $\mathbb{R}^2$. This method for converting pairs of uniformly distributed samples into pairs of normally distributed samples is called the Box–Muller transform (Box and Muller, 1958).

When the lemma is used to find sampling methods, usually g will be the given density of the distribution we want to sample from. Our task is then to find a transformation φ so that the density f described by (1.6) corresponds to a distribution we can already sample from. In this situation, φ should be chosen so that it 'simplifies' the given density g. In practice, finding a useful transformation φ often needs some experimentation.

Example 1.38   Assume we want to sample from the distribution with density $g(y) = \frac{3}{2}\sqrt{y} \cdot \mathbb{1}_{[0,1]}(y)$. We can cancel the square root from the definition of g by choosing $\varphi(x) = x^2$. Then we can apply theorem 1.34 with $A = B = [0, 1]$ and, since $|\det D\varphi(x)| = |\varphi'(x)| = 2x$, we get

$$f(x) = g\left(\varphi(x)\right) \cdot |\det D\varphi(x)| = \frac{3}{2}x \cdot 2x = 3x^2$$

for all $x \in [0, 1]$. From example 1.17 we already know how to generate samples from this density: If $U \sim \mathcal{U}[0, 1]$, then $X = U^{1/3}$ has density f and, by theorem 1.34, $Y = \varphi(X) = X^2 = U^{2/3}$ has density g.

An important application of the transformation rule from theorem 1.34 is the case where X and $\varphi(X)$ are both uniformly distributed. From the relation (1.6) we see that if X is uniformly distributed and if $|\det D\varphi|$ is constant, then $\varphi(X)$ is also uniformly distributed. For example, using this idea we can sometimes transform the problem of sampling from the uniform distribution on an unbounded set to the easier problem of sampling from the uniform distribution on a bounded set. This idea is illustrated in Figure 1.6. Combining this approach with lemma 1.33 results in the following general sampling method.

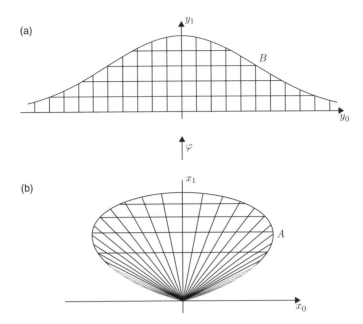

Figure 1.6   Illustration of the transformation used in the ratio-of-uniforms method. The map φ from equation (1.7) maps the bounded set shown in (b) into the unbounded set in (a). The areas shown in grey in (b) map into the tails in (a) (not displayed). Since φ preserves area (up to a constant), the uniform distribution on the set in (b) is mapped into the uniform distribution on the set in (a).

Theorem 1.39   (ratio-of-uniforms method) Let $f : \mathbb{R}^d \to \mathbb{R}_+$ be such that $Z = \int_{\mathbb{R}^d} f(x)\,dx < \infty$ and let X be uniformly distributed on the set

$$A = \left\{ (x_0, x_1, \ldots, x_d) \,\middle|\, x_0 > 0, \frac{x_0^{d+1}}{d+1} < f\left(\frac{x_1}{x_0}, \ldots, \frac{x_d}{x_0}\right) \right\} \subseteq \mathbb{R}_+ \times \mathbb{R}^d.$$

Then the vector

$$Y = \left(\frac{X_1}{X_0}, \ldots, \frac{X_d}{X_0} \right)$$

has density $\frac{1}{Z} f$ on $\mathbb{R}^d$.

Proof   The proof is an application of the transformation rule for random variables. To see this, consider the set

$$B = \left\{ (y_0, y_1, \ldots, y_d) \,\middle|\, 0 < y_0 < f(y_1, \ldots, y_d)/Z \right\} \subseteq \mathbb{R}_+ \times \mathbb{R}^d$$

and define a transformation $\varphi : \mathbb{R}_+ \times \mathbb{R}^d \to \mathbb{R}_+ \times \mathbb{R}^d$ by

$$\varphi(x_0, x_1, \ldots, x_d) = \left(\frac{Z x_0^{d+1}}{d+1}, \frac{x_1}{x_0}, \ldots, \frac{x_d}{x_0} \right). \tag{1.7}$$

We have $x \in A$ if and only if $\varphi(x) \in B$ and thus φ maps A onto B bijectively. Since the determinant of a triagonal matrix is the product of the diagonal elements, the Jacobian determinant of φ is given by

$$\det D\varphi(x) = \det \begin{pmatrix} Z x_0^d & & \\ -\frac{x_1}{x_0^2} & \frac{1}{x_0} & \\ \vdots & & \ddots \\ -\frac{x_d}{x_0^2} & & \frac{1}{x_0} \end{pmatrix} = Z x_0^d \frac{1}{x_0} \cdots \frac{1}{x_0} = Z$$

for all $x \in A$.

Since X is uniformly distributed on A, the density h of X satisfies

$$h(x) = \frac{1}{|A|} \mathbb{1}_A(x) = \frac{1}{Z|A|} \mathbb{1}_B (\varphi(x)) \cdot |\det D\varphi(x)|$$

and by theorem 1.34 the random variable $\varphi(X)$ then has density

$$g(y) = \frac{1}{Z|A|} \mathbb{1}_B(y)$$

for all $y \in \mathbb{R}_+ \times \mathbb{R}^d$. This density is constant on B and thus the random variable $\varphi(X)$ is uniformly distributed on B.

To complete the proof we note that the vector Y given in the statement of the theorem consists of the last d components of $\varphi(X)$. Using this observation, the claim now follows from lemma 1.33. $\qquad \square$

Example 1.40   The Cauchy distribution has density

$$f(x) = \frac{1}{\pi(1 + x^2)}.$$

For this case, the set A from theorem 1.39 is

$$A = \left\{ (x_0, x_1) \mid x_0 > 0, \frac{x_0^2}{2} \leq \frac{1}{\pi \left(1 + (\frac{x_1}{x_0})^2 \right)} \right\}$$

$$= \left\{ (x_0, x_1) \mid x_0 > 0, \frac{\pi}{2} x_0^2 \leq \frac{x_0^2}{x_0^2 + x_1^2} \right\}$$

$$= \left\{ (x_0, x_1) \mid x_0 > 0, x_0^2 + x_1^2 \leq \frac{2}{\pi} \right\},$$

that is A is a semicircle in the x_0/x_1-plane. Since we can sample from the uniform distribution on the semicircle (see exercises E1.13 and E1.14), we can use the ratio-of-uniforms method from theorem 1.39 to sample from the Cauchy distribution. The following steps are required:

(a) Generate (X_0, X_1) uniformly on the semicircle A.

(b) Return $Y = X_1/X_0$.

Note that, since only the ratio between X_1 and X_0 is returned, A can be replaced by a semicircle with arbitrary radius instead of the radius $\sqrt{2/\pi}$ found above.

1.6   Special-purpose methods

There are many specialised methods to generate samples from specific distributions. These are often faster than the generic methods described in the previous sections, but can typically only be used for a single distribution. These specialised methods (optimised for speed and often quite complex) form the basis of the random number generators built into software packages. In contrast, the methods discussed in the previous sections are general purpose methods which can be used for a wide range of distributions when no pre-existing method is available.

1.7   Summary and further reading

In this chapter we have learned about various aspects of random number generation on a computer. The chapter started by considering the differences between 'pseudo random number generators' (the ones considered in this book) and 'real random number generators' (which we will not consider further). Using the LCG as an example, we have learned about properties of pseudo number generators. In particular we considered the rôle of the 'seed' to control reproducability of the generated numbers. Going beyond the scope of this book, a lot of information about LCGs and about testing of random number generators can be found in Knuth (1981). The Mersenne Twister, a popular modern PRNG, is described in Matsumoto and Nishimura (1998).

Building on the output of pseudo number generators, the following sections considered various general purpose methods for generating samples from different distributions. The methods we discussed here are the inverse transform method, the rejection sampling method, and the ratio-of-uniforms method (a special case of the transformation method). More information about rejection sampling and its extensions can be found in Robert and Casella (2004, Section 2.3). A specialised method for generating normally distributed random variables can, for example, be found in Marsaglia and Tsang (2000). Specialised methods for generating random numbers from various distributions are, for example, covered in Dagpunar (2007, Chapter 4) and Kennedy and Gentle (1980, Section 6.5).

An expository presentation of random number generation and many more references can be found in Gentle et al. (2004, Chapter II.2).

Exercises

E1.1 Write a function to implement the LCG. The function should take a length n, the parameters m, a and c as well as the seed X_0 as input and should return a vector $X = (X_1, X_2, \ldots, X_n)$. Test your function by calling it with the parameters $m = 8$, $a = 5$ and $c = 1$ and by comparing the output with the result from example 1.3.

E1.2 Given a sequence $X_1, X_2, \ldots$ of $\mathcal{U}[0, 1]$-distributed pseudo random numbers, we can use a scatter plot of (X_i, X_{i+1}) for $i = 1, \ldots, n - 1$ in order to try to assess whether the X_i are independent.

(a) Create such a plot using the built-in random number generator of R:

```
X <- runif(1000)
plot(X[1:999], X[2:1000], asp=1)
```

Can you explain the resulting plot?

(b) Create a similar plot, using your function LCG from exercise E1.1:

```
m <- 81
a <- 1
c <- 8
seed <- 0
X <- LCG(1000, m, a, c, seed)/m
plot(X[1:999], X[2:1000], asp=1)
```

Discuss the resulting plot.

(c) Repeat the experiment from (b) using the parameters $m = 1024$, $a = 401$, $c = 101$ and $m = 2^{32}$, $a = 1\,664\,525$, $c = 1\,013\,904\,223$. Discuss the results.

E1.3 One (very early) method for pseudo random number generation is von Neumann's middle square method (von Neumann, 1951). The method works as follows: starting with $X_0 \in \{0, 1, \ldots, 99\}$, define X_n for $n \in \mathbb{N}$ to be the middle two digits of the four-digit number X_{n-1}^2. If X_{n-1}^2 does not have four digits, it is padded with leading zeros. For example, if $X_0 = 64$, we have $X_0^2 = 4096$ and thus $X_1 = 09 = 9$. In the next step, we find $X_1^2 = 81 = 0081$ and thus $X_2 = 08 = 8$.

(a) Write a function which computes X_n from X_{n-1}.

(b) The output of the middle square method has loops. For example, once we have $X_N = 0$, we will have $X_n = 0$ for all $n \geq N$. Write a program to find all cycles of the middle square method.

(c) Comment on the quality of the middle square method as a PRNG.

E1.4 Write a program which uses the inverse transform method to generate random numbers with the following density:

$$f(x) = \begin{cases} 1/x^2 & \text{if } x \geq 1 \text{ and} \\ 0 & \text{otherwise.} \end{cases}$$

To test your program, plot a histogram of 10 000 random numbers together with the density f.

E1.5 For $n \in \mathbb{N}$, let K_n denote the (random) number of accepted proposals among the first n generated proposals in algorithm 1.19. Show that, with probability 1, we have

$$\lim_{n \to \infty} \frac{1}{n} K_n = Z.$$

E1.6 Implement the rejection method from example 1.24 to generate samples from a half-normal distribution from Exp(1)-distributed proposals. Test your program by generating a histogram of the output and by comparing the histogram with the theoretical density of the half-normal distribution.

E1.7 In example 1.24 we have learned how rejection sampling can be used to convert Exp(λ)-distributed proposals into standard normally distributed samples.

(a) Extend the method to convert Exp(λ)-distributed proposals into $\mathcal{N}(0, \sigma^2)$-distributed samples.

(b) For given σ^2, determine the optimal value of the parameter λ.

E1.8 Consider algorithm 1.22 where the target distribution has density f/Z_f with

$$f(x) = \frac{1}{\sqrt{x}} \exp\left(-y^2/2x - x\right)$$

and $Z_f = \int_0^\infty f(\tilde{x}) \, d\tilde{x}$, and where the proposals are Exp(1)-distributed. Find the optimal value for the constant c from algorithm 1.22 for this example.

E1.9 Let f and g be two probability densities and $c \in \mathbb{R}$ with $f(x) \leq cg(x)$ for all x. Show that $c \geq 1$ and that $c = 1$ is only possible for $f = g$ (except possibly on sets with volume 0).

E1.10 Let $X \sim \mathcal{U}[a, b]$ and $Y \sim \mathcal{U}[c, d]$ be independent. Using the definition of the uniform distribution on a set, show that (X, Y) is uniformly distributed on the rectangle $R = [a, b] \times [c, d]$.

E1.11 Without using rejection sampling, propose a method to sample from the uniform distribution on the set

$$A = ([0, 1] \times [0, 1]) \cup ([2, 4] \times [0, 1]).$$

Write a program implementing your method.

E1.12 Without using rejection sampling, propose a method to sample from the uniform distribution on the set

$$B = ([0, 2] \times [0, 2]) \cup ([1, 3] \times [1, 3]).$$

Write a program implementing your method.

E1.13 Consider the uniform distribution on a semicircle.

(a) Explain how rejection sampling can be used to convert i.i.d. proposals $U_n \sim \mathcal{U}([-1, 1] \times [0, 1])$ into an i.i.d. sequence $(V_k)_{k \in \mathbb{N}}$ which is uniformly distributed on the semicircle $\{(x, y) \in \mathbb{R}^2 \mid x^2 + y^2 \leq 1, y \geq 0\}$. Compute the acceptance probability of the method.

(b) Write a computer program which generates 1000 samples from the uniform distribution on the semicircle, using the method from (a). Create a scatter plot showing the random points. How many proposals were needed to generate 1000 samples?

E1.14 Propose a rejection method to sample from the uniform distribution on the semicircle

$$\{(x, y) \in \mathbb{R}^2 \mid x^2 + y^2 \leq 1, y \geq 0\}$$

which has an acceptance probability of greater than 80%. Implement your method.

E1.15 Let (X, Y) be uniformly distributed on the semicircle

$$\{(x, y) \in \mathbb{R}^2 \mid x^2 + y^2 \leq 1, y \geq 0\}.$$

Find the densities of X and Y, respectively.

E1.16 Let X be a random variable on $\mathbb{R}^d$ with density $f : \mathbb{R}^d \to [0, \infty)$ and let $c \neq 0$ be a constant. Determine the density of cX.

E1.17 Let $X \sim \mathcal{N}(0, 1)$. Determine the density of $Y = (X^2 - 1)/2$.

E1.18 Write a program to implement the ratio-of-uniforms method to sample from the Cauchy distribution with density

$$f(x) = \frac{1}{\pi(1 + x^2)}.$$

2

Simulating statistical models

The output of the methods for random number generation considered in Chapter 1 is a series of independent random samples from a given distribution. In contrast, most real-world statistical models of interest will involve random samples with a non-trivial dependence structure and often samples will consist not just of a sequence of numbers, but will feature a more complicated structure. In this chapter, we will discuss some examples to show how the methods from Chapter 1 can be used as a building block to generate samples from more complex statistical models.

2.1 Multivariate normal distributions

One of the most important multivariate distributions is the multivariate normal distribution. In this section, we will derive the basic properties of the multivariate normal distribution and will discuss how to generate samples from this distribution.

Definition 2.1 Let $\mu \in \mathbb{R}^d$ be a vector and $\Sigma \in \mathbb{R}^{d \times d}$ be a symmetric, positive definite matrix. Then a random vector $X \in \mathbb{R}^d$ is normally distributed with mean μ and covariance matrix Σ, if the distribution of X has density $f \colon \mathbb{R}^d \to \mathbb{R}$ given by

$$f(x) = \frac{1}{(2\pi)^{d/2} \, |\det \Sigma|^{1/2}} \exp\left(-\frac{1}{2}(x - \mu)^\top \Sigma^{-1}(x - \mu) \right) \qquad (2.1)$$

for all $x \in \mathbb{R}^d$.

An Introduction to Statistical Computing: A Simulation-based Approach, First Edition. Jochen Voss.
© 2014 John Wiley & Sons, Ltd. Published 2014 by John Wiley & Sons, Ltd.

In this definition we consider the vector $x - \mu \in \mathbb{R}^d$ to be a $d \times 1$ matrix, and the expression $(x - \mu)^\top$ denotes the *transpose* of this vector, that is the vector $x - \mu$ interpreted as an $1 \times d$ matrix. Using this interpretation we have

$$(x - \mu)^\top \Sigma^{-1}(x - \mu) = \sum_{i,j=1}^{d} (x_i - \mu_i)(\Sigma^{-1})_{ij}(x_j - \mu_j).$$

The multivariate normal distribution from definition 2.1 is a generalisation of the one-dimensional normal distribution: If Σ is a diagonal matrix, say

$$\Sigma = \begin{pmatrix} \sigma_1^2 & 0 & \cdots & 0 \\ 0 & \sigma_2^2 & \cdots & 0 \\ \vdots & \vdots & \ddots & \vdots \\ 0 & 0 & \cdots & \sigma_d^2 \end{pmatrix},$$

then $|\det \Sigma| = \prod_{i=1}^{d} \sigma_i^2$ and

$$\Sigma^{-1} = \begin{pmatrix} 1/\sigma_1^2 & 0 & \cdots & 0 \\ 0 & 1/\sigma_2^2 & \cdots & 0 \\ \vdots & \vdots & \ddots & \vdots \\ 0 & 0 & \cdots & 1/\sigma_d^2 \end{pmatrix}$$

and thus the density f from (2.1) can be written as

$$\begin{aligned}
f(x) &= \frac{1}{(2\pi)^{d/2} \left| \prod_{i=1}^{d} \sigma_i^2 \right|^{1/2}} \exp\left(-\frac{1}{2} \sum_{i=1}^{d} (x_i - \mu_i) \frac{1}{\sigma_i^2} (x_i - \mu_i) \right) \\
&= \prod_{i=1}^{d} \frac{1}{\left(2\pi \sigma_i^2 \right)^{1/2}} \exp\left(-\frac{(x_i - \mu_i)^2}{2\sigma_i^2} \right) \\
&= \prod_{i=1}^{d} f_i(x_i),
\end{aligned}$$

where the function f_i, given by

$$f_i(x) = \frac{1}{\left(2\pi \sigma_i^2 \right)^{1/2}} \exp\left(-\frac{(x - \mu_i)^2}{2\sigma_i^2} \right)$$

for all $x \in \mathbb{R}$, is the density of the one-dimensional normal distribution with mean μ_i and variance σ_i^2. This shows that X is normally distributed on $\mathbb{R}^d$ with diagonal covariance matrix, if and only if the components X_i for $i = 1, 2, \ldots, d$ are independent and normally distributed on $\mathbb{R}$.

A sample X from a d-dimensional normal distribution with diagonal covariance matrix can be generated by generating the individual components $X_i \sim \mathcal{N}(\mu_i, \sigma_i^2)$ independently. The following lemma gives a method to transform such samples into samples from arbitrary d-dimensional normal distributions.

Lemma 2.2   Let $\mu \in \mathbb{R}^d$ and $A \in \mathbb{R}^{d \times d}$ be invertible. Define $\Sigma = AA^\top \in \mathbb{R}^d$. Furthermore, let $X = (X_1, X_2, \ldots, X_d) \in \mathbb{R}^d$ be a random vector such that $X_1, X_2, \ldots, X_d \sim \mathcal{N}(0, 1)$ are independent. Then

$$AX + \mu \sim \mathcal{N}(\mu, \Sigma)$$

on $\mathbb{R}^d$.

Proof   The result is a direct consequence of theorem 1.34: let f be the density of X and g be the density of $\mathcal{N}(\mu, \Sigma)$, that is

$$f(x) = \frac{1}{(2\pi)^{d/2}} \exp\left(-\frac{1}{2} x^\top x\right)$$

and

$$g(x) = \frac{1}{(2\pi)^{d/2} |\det \Sigma|^{1/2}} \exp\left(-\frac{1}{2}(x - \mu)^\top \Sigma^{-1}(x - \mu)\right)$$

for all $x \in \mathbb{R}^d$. Consider the transformation $\varphi(x) = Ax + \mu$. Then

$$\begin{aligned}
g(\varphi(x)) &= \frac{1}{(2\pi)^{d/2} |\det(AA^\top)|^{1/2}} \exp\left(-\frac{1}{2}(Ax)^\top (AA^\top)^{-1}(Ax)\right) \\
&= \frac{1}{(2\pi)^{d/2} |\det A \det A^\top|^{1/2}} \exp\left(-\frac{1}{2} x^\top A^\top (A^\top)^{-1} A^{-1} Ax\right) \\
&= \frac{1}{(2\pi)^{d/2} |\det A|} \exp\left(-\frac{1}{2} x^\top x\right)
\end{aligned}$$

for all $x \in \mathbb{R}^d$, since $\det A = \det A^\top$. Assume $A = (a_{ij})_{i,j=1,\ldots,d}$. Then the Jacobian $D\varphi(x)$ has components

$$(D\varphi(x))_{ij} = \frac{\partial}{\partial x_j}\left(\sum_{k=1}^{d} a_{ik} x_k\right) = a_{ij}$$

and thus $D\varphi(x) = A$ and $|\det D\varphi(x)| = |\det A|$ for all $x \in \mathbb{R}^d$. Consequently, $f(x) = g(\varphi(x))|\det D\varphi(x)|$ and, by theorem 1.34, the distribution of $\varphi(X)$ has density g. $\square$

Lemma 2.2 can be used to generate samples from a multivariate normal distribution $\mathcal{N}(\mu, \Sigma)$. This can be done by performing the following steps:

(a) Find a matrix A such that $\Sigma = AA^T$, for example using the Cholesky decomposition of A.

(b) Generate independent random values $X_1, X_2, \ldots, X_d \sim \mathcal{N}(0, 1)$.

(c) Return $AX + \mu$.

In cases where many $\mathcal{N}(\mu, \sigma)$-distributed random values need to be generated for the same covariance matrix Σ, the algorithm can be sped up by computing the matrix A only once and then storing it for later use.

For completeness, we now verify that the distribution $\mathcal{N}(\mu, \Sigma)$ given by definition 2.1 indeed has mean μ and covariance matrix Σ.

Lemma 2.3   Let $X \sim \mathcal{N}(\mu, \Sigma)$ where $\mu \in \mathbb{R}^d$ and $\Sigma = (\sigma_{ij})_{i,j=1,\ldots,n} \in \mathbb{R}^{d \times d}$. Then

$$\mathbb{E}(X_i) = \mu_i$$

and

$$\mathrm{Cov}(X_i, X_j) = \sigma_{ij}$$

for all $i, j = 1, 2, \ldots, d$.

Proof   By lemma 2.2 we can write X as $X = AZ + \mu$ where $A = (a_{ij})_{i,j=1,\ldots,d}$ is a matrix such that $AA^\top = \Sigma$ and $Z_1, Z_2, \ldots, Z_d$ are independent and standard-normally distributed. For the components of X we find

$$X_i = \sum_{j=1}^{d} a_{ij} Z_j + \mu_i$$

and thus

$$\mathbb{E}(X_i) = \sum_{j=1}^{d} a_{ij} \mathbb{E}(Z_j) + \mu_i = \mu_i$$

and

$$\mathrm{Cov}(X_i, X_j) = \mathrm{Cov}\left(\sum_{k=1}^{d} a_{ik} Z_k, \sum_{l=1}^{d} a_{jl} Z_l \right) = \sum_{k,l=1}^{d} a_{ik} a_{jl} \mathrm{Cov}(Z_k, Z_l)$$

$$= \sum_{k=1}^{d} a_{ik} a_{jk} = (AA^\top)_{ij} = \sigma_{ij}.$$

This completes the proof.   □

Example 2.4   Let $\varepsilon_i \sim \mathcal{N}(0, \sigma^2)$ be i.i.d. and define

$$X_i = \sum_{k=1}^{i} \varepsilon_k$$

for all $i \in \mathbb{N}$. Then $\mathbb{E}(X_i) = 0$ and

$$\text{Cov}(X_i, X_j) = \text{Cov}\left(\sum_{k=1}^{i} \varepsilon_k, \sum_{l=1}^{j} \varepsilon_l\right) = \sum_{k=1}^{i}\sum_{l=1}^{j} \text{Cov}(\varepsilon_k, \varepsilon_l).$$

Since $\text{Cov}(\varepsilon_k, \varepsilon_l) = \sigma^2$ if $k = l$ and $\text{Cov}(\varepsilon_k, \varepsilon_l) = 0$ otherwise, we find

$$\text{Cov}(X_i, X_j) = \min(i, j)\sigma^2$$

for all $i, j \in \mathbb{N}$. Consequently, for $d \in \mathbb{N}$, we can use lemma 2.3 to conclude that the vector $X = (X_1, X_2, \ldots, X_d)$ satisfies $X \sim \mathcal{N}(0, \Sigma)$ where

$$\Sigma = \sigma^2 \begin{pmatrix} 1 & 1 & 1 & \cdots & 1 \\ 1 & 2 & 2 & \cdots & 2 \\ 1 & 2 & 3 & \cdots & 3 \\ \vdots & \vdots & \vdots & \ddots & \vdots \\ 1 & 2 & 3 & \cdots & d \end{pmatrix}.$$

2.2   Hierarchical models

Many important statistical models have a hierarchical structure: not all random variables in the model are defined simultaneously, but instead there are several 'levels' of randomness and the distribution of the random variables in the later levels depends on the values of random variables in earlier levels. This structure is, for example, present in the following situations.

- In Bayesian models (discussed in Section 4.3) the distribution of the data depends on the value of one or more random parameters.

- In mixture models the distribution of samples depends on the random choice of mixture component.

- In Markov chains (discussed in Section 2.3) the distribution of the value at time t depends on the value of the Markov chain at time $t - 1$.

Simulating hierarchical models is often easy: the simulation procedure will be performed in steps, closely following the structure of the model. We illustrate this approach here using examples.

Example 2.5   Consider the Bayesian model where the data are described as i.i.d. samples $X_1, \ldots, X_n \sim \mathcal{N}(\mu, \sigma^2)$, and where the mean μ and the variance σ^2 are themselves assumed to be random with distributions $\sigma^2 \sim \mathrm{Exp}(\lambda)$ and $\mu \sim \mathcal{N}(\mu_0, \alpha\sigma^2)$. Since the variance σ^2 occurs in the distribution of μ, the model has the following dependence structure:

$$\sigma^2 \quad \longrightarrow \quad \mu \quad \longrightarrow \quad X_1, \ldots, X_n.$$

To generate samples from this model, we use steps corresponding to the levels in the model:

1: generate $\sigma^2 \sim \mathrm{Exp}(\lambda)$
2: generate $\mu \sim \mathcal{N}(\mu_0, \alpha\sigma^2)$
3: **for** $i = 1, \ldots, n$ **do**
4:    generate $X_i \sim \mathcal{N}(\mu, \sigma^2)$
5: **end for**

Sometimes, the hierarchical structure of a model is not immediately clear. This is for example the case for mixture distributions as given in the following definition, but we will see that for generating samples from a mixture distribution it is beneficial to introduce a hierarchical structure.

Definition 2.6   Let $P_1, \ldots, P_k$ be probability distributions on $\mathbb{R}^d$ and let $\theta_1, \ldots, \theta_k > 0$ such that $\sum_{a=1}^{k} \theta_a = 1$. Then the *mixture* P_θ of the distributions $P_1, \ldots, P_k$ with weights $\theta_1, \ldots, \theta_k$ is given by

$$P_\theta(A) = \sum_{a=1}^{k} \theta_a P_a(A)$$

for all $A \subseteq \mathbb{R}^d$.

It is important to note that a mixture is a weighted sum of distributions, not the distribution of a weighted sum of random variables. The difference is illustrated in the following example.

Example 2.7   Consider the three normal distributions $P_1 = \mathcal{N}(1, 0.01^2)$, $P_2 = \mathcal{N}(2, 0.5^2)$ and $P_3 = \mathcal{N}(5, 0.02^2)$, together with the weights $\theta_1 = 0.1$, $\theta_2 = 0.7$ and $\theta_3 = 0.02$. If $X_i \sim P_i$ for $i = 1, 2, 3$ are independent, then $X = \sum_{i=1}^{3} \theta_i X_i$ is normally distributed. In contrast, the mixture distribution $P_\theta = \sum_{i=1}^{3} \theta_i P_i$ has the density shown in Figure 2.1. The figure shows clearly that P_θ is not a normal distribution. The density shown in Figure 2.1 was determined using the following lemma.

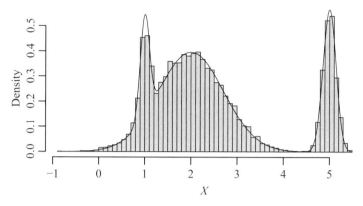

Figure 2.1   A histogram generated from 10 000 samples of the mixture model from example 2.7.

Lemma 2.8   Assume that $P_1, \ldots, P_k$ have densities $f_1, \ldots, f_k$. Then the mixture distribution P_θ also has a density which is given by

$$f_\theta = \sum_{a=1}^{k} \theta_a f_a.$$

Proof   Let $A \subseteq \mathbb{R}^d$. Then

$$P_\theta(A) = \sum_{a=1}^{k} \theta_a P_a(A) = \sum_{a=1}^{k} \theta_a \int_A f_a(x)\, dx = \int_A \sum_{a=1}^{k} \theta_a f_a(x)\, dx = \int_A f_\theta(x)\, dx.$$

Thus, f_θ is the density of P_θ.   □

At first glance, the problem of generating samples from a mixture distribution is straightforward: since we know the distribution, we could just use the methods from Chapter 1 to generate such samples. But, as Figure 2.1 shows, mixture distributions can have a complicated density and it transpires that often the easiest way to generate samples from a mixture distribution is to artificially introduce a hierarchical structure, used only for the generation of samples. This method is used in the following algorithm.

Algorithm 2.9   (mixture distributions)
    input:
        probability distributions $P_1, \ldots, P_k$
        weights $\theta_1, \ldots, \theta_k > 0$ with $\sum_{a=1}^{k} \theta_a = 1$

randomness used:

$Y \in \{1, 2, \ldots, k\}$ with $P(Y = a) = \theta_a$ for all a

samples $X \sim P_a$ for different $a \in \{1, \ldots, k\}$

output:

$X \sim P_\theta$

1: generate $Y \in \{1, 2, \ldots, k\}$ with $P(Y = a) = \theta_a$ for all a

2: generate $X \sim P_Y$

3: return X

Lemma 2.10   The sample X constructed by algorithm 2.9 is distributed according to the mixture distribution from definition 2.6, that is $X \sim P_\theta$.

Proof   Let $A \subseteq \mathbb{R}^d$. By splitting the event $\{X \in A\}$ according to the possible values of Y and using Bayes' rule (see Section A.2), we find

$$P(X \in A) = \sum_{a=1}^{k} P(X \in A, Y = a) = \sum_{a=1}^{k} P(Y = a) P(X \in A | Y = a).$$

From step 1 of algorithm 2.9 we know $P(Y = a) = \theta_a$ and from step 2 we see that, conditional on $Y = a$, we have $X \sim P_a$. Thus we get

$$P(X \in A) = \sum_{a=1}^{k} \theta_a P_a(A) = P_\theta(A).$$

Thus, $X \sim P_\theta$ as required.   □

Algorithm 2.9 showed the idea of artificially introducing a hierarchical structure to simplify sampling. We will reuse the idea, again in the context of mixture distributions, in Section 4.4.2. To conclude the present section, we will rephrase this idea in a more general (and more abstract) form: one method for generating samples from a multivariate distribution P is to simulate the components one by one. Instead of directly generating a sample $X = (X_1, X_2, \ldots, X_n) \in \mathbb{R}^n$ we can first sample X_1 from the corresponding marginal distribution and then, for $i = 2, 3, \ldots, n$, sampling X_i from the conditional distribution given the (already sampled) values $X_1, \ldots, X_{i-1}$.

In the following algorithm we assume (for simplicity) that the distribution P has a density $p : \mathbb{R}^d \to [0, \infty)$ and we use the marginal density p_{X_1} of X_1, given by

$$p_{X_1}(x_1) = \int_{\mathbb{R}} \cdots \int_{\mathbb{R}} p(x_1, x_2, \ldots, x_n) \, dx_n \cdots dx_2,$$

as well as the conditional densities $p_{X_i | X_1, \ldots, X_{i-1}}$ as defined in Section A.2.

Algorithm 2.11   (componentwise simulation)
  input:
    marginal density p_{X_1}
    conditional densities $p_{X_i|X_1,\ldots,X_{i-1}}$ for $i = 2, 3, \ldots, n$
  randomness used:
    samples from p_{X_1} and $p_{X_i|X_1,\ldots,X_{i-1}}$
  output:
    a sample $(X_1, \ldots, X_n) \sim p$
  1: generate $X_1 \sim p_{X_1}$
  2: **for** $i = 2, 3, \ldots, n$ **do**
  3:    generate $X_i \sim p_{X_i|X_1,\ldots,X_{i-1}}(\,\cdot\,|X_1, \ldots, X_{i-1})$
  4: **end for**
  5: return $(X_1, \ldots, X_n)$

Lemma 2.12   The random vector $(X_1, \ldots, X_n)$ constructed by algorithm 2.11 has density p.

Proof   For $i = 1, 2, \ldots, n$, denote the marginal density of P for the first i coordinates by $p_{X_1,\ldots,X_i}$, that is let

$$p_{X_1,\ldots,X_i}(x_1, \ldots, x_i) = \int_{\mathbb{R}} \cdots \int_{\mathbb{R}} p(x_1, \ldots, x_i, x_{i+1}, \ldots, x_n)\, dx_n \cdots dx_{i+1}$$

for all $x_1, \ldots, x_i \in \mathbb{R}$. We will use induction to prove that $(X_1, \ldots, X_i)$, as constructed by the first i iterations of the loop in algorithm 2.11, has density $p_{X_1,\ldots,X_i}$ for all $i = 1, 2, \ldots, n$.

For $i = 1$ we have $P(X_1 \in A_1) = p_{X_1}(A_1)$ by construction of X_1 in line 1 of the algorithm. Thus the statement holds for $i = 1$. Let $i > 1$ and assume we have already shown that $(X_1, \ldots, X_{i-1})$ has density $p_{X_1,\ldots,X_{i-1}}$. The construction of X_i in line 3 of the algorithm depends on the values of $X_1, \ldots, X_{i-1}$ constructed in previous steps of the algorithm. We can integrate over all possible values of $X_1, \ldots, X_{i-1}$ to find

$$\begin{aligned}
&P(X_1 \in A_1, \ldots, X_i \in A_i) \\
&= \int \int \mathbb{1}_{A_1 \times \cdots \times A_i}(x_1, \ldots, x_i) \\
&\qquad\qquad \cdot p_{X_i|X_1,\ldots,X_{i-1}}(\,\cdot\,|x_1, \ldots, x_{i-1})\, dx_i \\
&\qquad\qquad \cdot p_{X_1,\ldots,X_{i-1}}(x_1, \ldots, x_{i-1})\, dx_{i-1} \cdots dx_1 \\
&= \int \int \mathbb{1}_{A_1 \times \cdots \times A_i}(x_1, \ldots, x_i)\, p_{X_1,\ldots,X_i}(x_1, \ldots, x_i)\, dx_i \cdots dx_1.
\end{aligned}$$

This shows that $(X_1, \ldots, X_i)$ has density $p_{X_1,\ldots,X_i}$. Using induction, we get this statement for all $i = 1, \ldots, n$. Since $p_{X_1,\ldots,X_n} = p$, this completes the proof. $\qquad\square$

2.3   Markov chains

Markov chains are stochastic processes, which play an important rôle in many areas of probability, both as objects of independent mathematical interest and as a tool in other areas of mathematics. In this section we will give a very brief introduction to Markov chains, concentrating on basic properties of Markov chains and how to simulate Markov chains on a computer. Later in this book, in Chapter 4, we will see how Markov chains can be used as a tool in random number generation.

Throughout this section, we restrict ourselves to the case of discrete-time Markov chains and we omit some of the technical details.

Definition 2.13   A stochastic process $X = (X_j)_{j \in \mathbb{N}_0}$ with values in a set S is a *Markov chain*, if

$$P\left(X_j \in A_j \,\middle|\, X_{j-1} \in A_{j-1}, X_{j-2} \in A_{j-2}, \ldots, X_0 \in A_0\right)$$
$$= P\left(X_j \in A_j \,\middle|\, X_{j-1} \in A_{j-1}\right) \tag{2.2}$$

for all $A_0, A_1, \ldots, A_j \subseteq S$ and all $j \in \mathbb{N}$, that is if the distribution of X_j depends on $X_0, \ldots, X_{j-2}$ only through X_{j-1}. The set S is called the *state space* of X. The distribution of X_0 is called the *initial distribution* of X.

Often X_0 is deterministic, that is $P(X_0 = x) = 1$ for some $x \in S$; in this case x is called the *initial value* or *starting point* of X. The index j is typically interpreted as *time*.

Example 2.14   Let $(\varepsilon_j)_{j \in \mathbb{N}}$ be an i.i.d. sequence of random variables. The process X given by $X_0 = 0$ and $X_j = X_{j-1} + \varepsilon_j$ for all $j \in \mathbb{N}$ is a Markov chain. We can write X_j as

$$X_j = \sum_{i=1}^{j} \varepsilon_i.$$

A Markov chain of this type is called a *random walk*. Important special cases are $\varepsilon_j \sim \mathcal{U}(\{-1, 1\})$ (the *symmetric, simple random walk* on $\mathbb{Z}$) and $\varepsilon_j \sim \mathcal{N}(0, 1)$.

Example 2.15   Let $(\varepsilon_j)_{j \in \mathbb{N}}$ be an i.i.d. sequence of random variables with variance $\mathrm{Var}(\varepsilon_j) = 1$. Then the process X given by $X_0 = X_1 = 0$ and

$$X_j = \frac{X_{j-1} + X_{j-2}}{2} + \varepsilon_j$$

for all $j = 2, 3, \ldots$ is *not* a Markov chain.

Definition 2.16   If the transition probabilities given by the right-hand side of (2.2) do not depend on the time j, the Markov chain X is called *time-homogeneous*.

For the rest of this chapter (and nearly all of the book) we restrict ourselves to the case of time-homogeneous Markov chains.

2.3.1   Discrete state space

If the state space S is finite, for example $S = \{1, 2, \ldots, N\}$, then the transition probabilities $P(X_n \in A_n \,|\, X_{n-1} \in A_{n-1})$ in (2.2) can be described by giving the probabilities

$$p_{xy} = P\left(X_j = y \,|\, X_{j-1} = x\right)$$

of the transitions between all pairs of elements $x, y \in S$. The resulting matrix $P = (p_{xy})_{x,y \in S}$ is called the transition matrix of the Markov chain X.

When considering transition matrices, it is often convenient to label the rows and columns of the matrix P using elements of S instead of using the usual indices $\{1, 2, \ldots, n\}$. Thus, if S is the alphabet $S = \{A, B, \ldots, Z\}$ we write p_{AZ} instead of $p_{1,26}$ to denote the probability of transitions from A to Z. We write $\mathbb{R}^{S \times S}$ for the set of all matrices where the columns and rows are indexed by elements of S. Similarly, for vectors consisting of probability weights for the elements of S, for example the initial distribution of a Markov chain, it is convenient to label the components of the vector by elements of S. We write $\mathbb{R}^S$ for the set of all such vectors.

At this stage, collecting the transition probabilities into a matrix could be seen to be only a method for bookkeeping, but in the rest of this section we will see the surprising fact that the connection between Markov chains and matrices is much deeper. Many of the concepts from linear algebra, including matrix multiplication and eigenvectors, have a probabilistic interpretation in the context of transition matrices!

So far we have considered Markov chains with finite state space S, but we can also consider transition matrices for Markov chains with countably infinite state space. In these cases, the transition matrix P is an 'infinite matrix'

$$P = \begin{pmatrix} p_{11} & p_{12} & p_{13} & \cdots \\ p_{21} & p_{22} & p_{23} & \cdots \\ \vdots & \vdots & \vdots & \ddots \end{pmatrix}.$$

Markov chains with finite or countably infinite state space are called Markov chains with *discrete state space*.

Example 2.17   Consider the state space $S = \{1, 2, 3\}$ and let X be the Markov chain with $X_0 = 1$ where transitions between states have the probabilities:

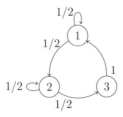

This Markov chain has transition matrix

$$P = \begin{pmatrix} \frac{1}{2} & \frac{1}{2} & 0 \\ 0 & \frac{1}{2} & \frac{1}{2} \\ 1 & 0 & 0 \end{pmatrix}.$$

Row x of this matrix, for $x = 1, 2, 3$, consists of the probabilities $p_{x,1}, p_{x,2}, p_{x,3}$ for going from state x to states 1, 2 and 3, respectively.

Example 2.18   The symmetric simple random walk X, given by

$$X_j = \sum_{i=1}^{j} \varepsilon_i$$

for all $j \in \mathbb{N}_0$ with $\varepsilon_i \sim \mathcal{U}\{-1, +1\}$ i.i.d., is a Markov chain with state space $S = \mathbb{Z}$.

Lemma 2.19   Let $P^{S \times S}$ be the transition matrix of a Markov chain with state space S. Then $P = (p_{xy})_{x,y \in S}$ has the following properties:

(a) $p_{xy} \geq 0$ for all $x, y \in S$.

(b) $\sum_{y \in S} p_{xy} = 1$ for all $x \in S$.

Proof   The claim follows directly from the definition of P.   □

Definition 2.20   A vector $\pi \in \mathbb{R}^S$ is called a *probability vector*, if $\pi_x \geq 0$ for all $x \in S$ and $\sum_{x \in S} \pi_x = 1$.

Definition 2.21   A matrix which satisfies the two conditions from lemma 2.19 is called a *stochastic matrix*.

In order to simulate paths of a Markov chain with discrete state space on a computer, we have to provide a probability vector π to specify the initial distribution and a transition matrix P to specify the transition probabilities. As in Section 2.2, the simulation of the random variables X_j is done one at a time, starting with X_0.

Algorithm 2.22   (Markov chains with discrete state space)
  input:
      a finite or countable state space S
      a probability vector $\pi \in \mathbb{R}^S$
      a stochastic matrix $P = (p_{xy})_{x,y \in S} \in \mathbb{R}^{S \times S}$
  randomness used:
      samples from discrete distributions on S
  output:
      a path of a Markov chain with initial distribution π and transition matrix P
  1: generate $X_0 \in S$ with $P(X_0 = x) = \pi_x$ for all $x \in S$
  2: output X_0
  3: **for** $j = 1, 2, 3, \ldots$ **do**
  4:    generate $X_j \in S$ with $P(X_j = x) = p_{X_{j-1},x}$ for all $x \in S$
  5:    output X_j
  6: **end for**

Lemma 2.23   Let X be a time-homogeneous Markov chain with finite state space and transition matrix P. Then

$$P\left(X_{j+k} = y \mid X_j = x\right) = (P^k)_{xy}$$

for all $j, k \in \mathbb{N}_0$ and $x, y \in S$, where $P^k = P \cdot P \cdots P$ is the kth power of the transition matrix P.

Proof   For $k = 0$, the matrix P^0 is by definition the identity matrix and the statement holds. Also, for $k = 1$ we have

$$P\left(X_{j+1} = y \mid X_j = x\right) = p_{xy} = (P^1)_{xy}$$

by the definition of the transition matrix.
  Now let $k > 1$ and assume that the statement holds for $k - 1$. Then we have

$$\begin{aligned}
& P\left(X_{j+k} = y \mid X_j = x\right) \\
&= \frac{P\left(X_{j+k} = y, X_j = x\right)}{P\left(X_j = x\right)} \\
&= \frac{1}{P\left(X_j = x\right)} \sum_{z \in S} P\left(X_{j+k} = y, X_{j+k-1} = z, X_j = x\right) \\
&= \frac{1}{P\left(X_j = x\right)} \sum_{z \in S} P\left(X_{j+k-1} = z, X_j = x\right) \\
& \qquad \cdot P\left(X_{j+k} = y \mid X_{j+k-1} = z, X_j = x\right).
\end{aligned}$$

Using the Markov property (2.2) we get

$$P\left(X_{j+k} = y \mid X_j = x\right)$$
$$= \sum_{z \in S} \frac{P\left(X_{j+k-1} = z, X_j = x\right)}{P\left(X_j = x\right)} P\left(X_{j+k} = y \mid X_{j+k-1} = z\right)$$
$$= \sum_{z \in S} P\left(X_{j+k-1} = z \mid X_j = x\right) p_{zy}$$
$$= \sum_{z \in S} (P^{k-1})_{xz} p_{zy}.$$

The last expression can be read as a matrix-matrix multiplication of P^{k-1} with P and thus we get

$$P\left(X_{j+k} = y \mid X_j = x\right) = (P^{k-1} \cdot P)_{xy} = (P^k)_{xy}$$

as required.   □

Lemma 2.24   Let X be a time-homogeneous Markov chain with finite state space and transition matrix P and initial distribution π. Then we have

$$P(X_j = y) = (\pi^\top P^j)_y \tag{2.3}$$

for all $y \in S$.

Proof   At time 0 we have

$$P(X_0 = x) = \pi_x$$

for all $x \in S$. For time $k > 0$ we can use Bayes' formula to find

$$P(X_j = y) = \sum_{x \in S} P\left(X_j = y, X_0 = x\right)$$
$$= \sum_{x \in S} P\left(X_0 = x\right) P\left(X_j = y \mid X_0 = x\right) = \sum_{x \in S} \pi_x (P^j)_{xy}$$

for all $y \in S$. If we consider the transposed vector $\pi^\top$ as a matrix with one row and $|S|$ columns, we can write the last expression as a matrix-matrix multiplication. This gives the required expression (2.3).   □

Definition 2.25   Let X be a time-homogeneous Markov chain with transition matrix P. A probability vector π is called a *stationary distribution* of X, if $\pi^\top P = \pi^\top$, that is if

$$\sum_{x \in S} \pi_x p_{xy} = \pi_y \tag{2.4}$$

for all $y \in S$.

To understand the significance of this definition, we have to recall relation (2.3): we know that

$$P(X_j = y) = (\pi^\top P^j)_y$$

for all $y \in S$. If π is a stationary distribution, we have $\pi^\top P = \pi^\top$ and then $\pi^\top P^j = \pi^\top$ for all $j \in \mathbb{N}$. Consequently, if we start the Markov chain with initial distribution π, the distribution of X_j does not change in time: we have $X_j \sim \pi$ for all $j \in \mathbb{N}$.

In general, a Markov chain may have more than one stationary distribution, but for the cases which will be of interest in this chapter, there will only be one stationary distribution.

Example 2.26   On the state space $S = \{1, 2, 3\}$, consider the Markov chain with transition matrix

$$P = \begin{pmatrix} 1/2 & 1/2 & 0 \\ 0 & 1/2 & 1/2 \\ 1/5 & 0 & 4/5 \end{pmatrix}$$

and initial distribution $\alpha = (1, 0, 0)$.

We can use equation (2.3) to get the distribution after one step: $P(X_1 = x) = (\alpha^\top P)_x$ where

$$\alpha^\top P = \begin{pmatrix} 1 & 0 & 0 \end{pmatrix} \cdot \begin{pmatrix} 1/2 & 1/2 & 0 \\ 0 & 1/2 & 1/2 \\ 1/5 & 0 & 4/5 \end{pmatrix} = \begin{pmatrix} 1/2 & 1/2 & 0 \end{pmatrix}.$$

Similarly, for X_2 we find $P(X_2 = x) = (\alpha^\top P^2)_x$ where

$$\alpha^\top P^2 = \begin{pmatrix} 1/2 & 1/2 & 0 \end{pmatrix} \cdot \begin{pmatrix} 1/2 & 1/2 & 0 \\ 0 & 1/2 & 1/2 \\ 1/5 & 0 & 4/5 \end{pmatrix} = \begin{pmatrix} 1/4 & 1/2 & 1/4 \end{pmatrix}.$$

Continuing to $X_2, X_3, \ldots$ we get

$$\alpha^\top P^3 = (0.175 \quad 0.375 \quad 0.450)$$
$$\alpha^\top P^4 = (0.178 \quad 0.275 \quad 0.548)$$
$$\vdots$$
$$\alpha^\top P^{10} = (0.223 \quad 0.222 \quad 0.555).$$

Experimenting shows that the value of $\alpha^\top P^j$ does not change significantly when j is increased further, so we can guess that this value is close to the stationary distribution of the Markov chain. Indeed, we can use equation (2.4) to verify that $\pi = (2/9, 2/9, 5/9)$ is a stationary distribution of X:

$$\pi^\top P = \begin{pmatrix} 2/9 & 2/9 & 5/9 \end{pmatrix} \cdot \begin{pmatrix} 1/2 & 1/2 & 0 \\ 0 & 1/2 & 1/2 \\ 1/5 & 0 & 4/5 \end{pmatrix} = \begin{pmatrix} 2/9 & 2/9 & 5/9 \end{pmatrix} = \pi^\top.$$

Thus, we have seen that for the Markov chain considered in this example, as $n \to \infty$, the probabilities $P(X_j = x)$ converge to the stationary probabilities π_x. Convergence results like this often, but not always, hold.

The condition for π being a stationary distribution can be rewritten by taking the transpose of the equation $\pi^\top P = \pi^\top$: a probability vector π is a stationary distribution for P if and only if

$$P^\top \pi = \pi,$$

that is if π is an eigenvector of $P^\top$ with eigenvalue 1. Since computing eigenvectors is a well-studied problem, and many software packages provide built-in functions for this purpose, this property can be used to find a stationary distribution π for a given transition matrix P.

2.3.2   Continuous state space

In this section we will briefly discuss Markov chains with *continuous state space*. Markov chains can be considered on very general state spaces, but here we restrict ourselves to the case $S = \mathbb{R}^d$. Most of the results from the previous section formally carry over to the case of continuous state space, with only changes in notation.

The concept of transition matrices in the case of continuous state space is replaced by transition kernels, as given in the following definition.

Definition 2.27   A *transition kernel* is a map $P(\cdot, \cdot)$ such that:

(a) $P(x, A) \geq 0$ for all $x \in \mathbb{R}^d$ and all $A \subseteq \mathbb{R}^d$; and

(b) $P(x, \cdot)$ is a probability distribution on $\mathbb{R}^d$ for all $x \in \mathbb{R}^d$.

This definition hides some of the technical complications associated with the study of Markov chains on a continuous state space. If full mathematical rigour is required, an additional condition, relating to 'measurability', must be included.

The idea in this definition is that x takes the rôle of the current state of the Markov chain and, for given x, the map $A \mapsto P(x, A)$ is the distribution of the next value of the Markov chain. Thus, the transition kernel of a time-homogeneous Markov chain is defined by

$$P(x, A) = P(X_j \in A | X_{j-1} = x).$$

Often, the conditional distribution of X_j, given $X_{j-1} = x$, has a density. In this case, instead of giving a transition kernel, we can describe the transitions of a Markov chain by giving a transition density. In analogy to lemma 2.19, a transition density is defined in the following.

Definition 2.28   A *transition density* is a map $p \colon \mathbb{R}^d \times \mathbb{R}^d \to \mathbb{R}$ such that:

(a) $p(x, y) \geq 0$ for all $x, y \in \mathbb{R}^d$; and

(b) $\int_{\mathbb{R}^d} p(x, y)\, dy = 1$ for all $x \in \mathbb{R}^d$.

If the Markov chain X can be described by a transition density, then we have

$$P(X_j \in A | X_{j-1} = x) = \int_A p(x, y)\, dy$$

for all $x \in \mathbb{R}^d$.

Example 2.29   On $S = \mathbb{R}$, let $X_0 = 0$ and

$$X_j = \frac{1}{2} X_{j-1} + \varepsilon_j$$

for all $j \in \mathbb{N}$, where $\varepsilon_j \sim \mathcal{N}(0, 1)$ i.i.d. is a Markov chain with state space $S = \mathbb{R}$. Given $X_{j-1} = x$, we have $X_j = \frac{1}{2}x + \varepsilon_j \sim \mathcal{N}(x/2, 1)$. Thus, the transition kernel for this Markov chain satisfies $P(x, \cdot) = \mathcal{N}(x/2, 1)$ for all $x \in \mathbb{R}^d$. Since the normal distribution has a density, the Markov chain X has a transition density p, given by

$$p(x, y) = \frac{1}{\sqrt{2\pi}} \exp\left(-\frac{1}{2}(y - x/2)^2 \right)$$

for all $x, y \in \mathbb{R}$.

As already remarked, most of the results from the previous section carry over with only changes in notation required, but the resulting notation is often cumbersome. Here we restrict ourselves to state the definition of a stationary density in analogy to definition 2.25 we have the following definition.

Definition 2.30   A probability density $\pi\colon \mathbb{R}^d \to [0, \infty)$ is a *stationary density* for a Markov chain on the state space $\mathbb{R}^d$ with transition density p, if it satisfies

$$\int_S \pi(x)p(x, y)\,dx = \pi(y)$$

for all $y \in \mathbb{R}^d$.

To conclude this section, we state the algorithm for generating paths from a time-homogeneous Markov chain with a transition density on $\mathbb{R}^d$.

Algorithm 2.31   (Markov chains with continuous state space)
   input:
      a probability density $\pi\colon \mathbb{R}^d \to [0, \infty)$
      a transition density $p\colon \mathbb{R}^d \times \mathbb{R}^d \to [0, \infty)$
   randomness used:
      one sample $X_0 \sim \pi$
      samples from the densities $p(x, \cdot)$ for $x \in \mathbb{R}^d$
   output:
      a path of a Markov chain with initial distribution π and transition matrix P
  1: generate $X_0 \sim \pi$
  2: output X_0
  3: **for** $j = 1, 2, 3, \ldots$ **do**
  4:   generate $X_j \sim p(X_{j-1}, \cdot)$
  5:   output X_j
  6: **end for**

2.4   Poisson processes

A sample of a Poisson process consists not of a single number, but of a random number of random points in a given set. Poisson processes are used to model the occurrence of events in space and/or time, when the individual events are random and independent of each other. For example, a Poisson process on a time interval $[t_1, t_2]$ could be used to model the times where individual telephone calls arrive at a call centre.

The definition of the Poisson *process* builds on the Poisson *distribution*, as given in the following definition.

Definition 2.32   A random variable X follows a *Poisson distribution* with parameter λ, if $X \in \mathbb{N}_0$ and

$$P(X = k) = \mathrm{e}^{-\lambda}\frac{\lambda^k}{k!}$$

for all $k \in \mathbb{N}_0$. The Poisson distribution with parameter λ is denoted by $\mathrm{Pois}(\lambda)$.

For reference, we state the following basic results about Poisson distributions.

Lemma 2.33   Let $X \sim \text{Pois}(\lambda)$. Then the following statements hold:

(a) $\mathbb{E}(X) = \lambda$.

(b) $\text{Var}(X) = \lambda$.

(c) If $Y \sim \text{Pois}(\mu)$, independent of X, then $X + Y \sim \text{Pois}(\lambda + \mu)$.

In the mathematical formalism, the points of a Poisson process are usually collected into a set, so that a sample of a Poisson process can be seen as a random set. This idea leads to the following definition of a Poisson process.

Definition 2.34   A *Poisson process* on a set $D \subseteq \mathbb{R}^d$ with *intensity function* $\lambda \colon \mathbb{R}^d \to [0, \infty)$ is a random set $\Pi \subseteq D$ such that the following two conditions hold:

(a) If $A \subseteq D$, then $|\Pi \cap A| \sim \text{Pois}(\Lambda(A))$ where $|\Pi \cap A|$ is the number of points of Π in A and

$$\Lambda(A) = \int_A \lambda(x)\,dx. \tag{2.5}$$

(b) If $A, B \subseteq D$ are disjoint, then $|\Pi \cap A|$ and $|\Pi \cap B|$ are independent.

The intensity function λ in the definition specifies how many points of the Poisson process, on average, are located in a given region. The process will have many points where λ is large and will have only few points where λ is small. This is illustrated in Figure 2.2. More specifically, Since the expectation of the $\text{Pois}(\lambda)$-distribution equals λ, the expected number of points in a set A is

$$\mathbb{E}\left(|\Pi \cap A|\right) = \Lambda(A).$$

A consequence of the independence statement in definition 2.34 is the following lemma, which allows to build up a Poisson process from smaller building blocks.

Lemma 2.35

(a) Let Π be a Poisson process on D with intensity λ and let $A \subseteq D$. Then the restriction $\Pi \cap A$ of the process Π to the set A is a Poisson process on A with intensity λ.

(b) Let Π_1 and Π_2 be independent Poisson processes on D_1 and D_2, respectively, and let $D_1 \cap D_2 = \emptyset$. Assume that both processes have the same intensity $\lambda \colon \mathbb{R}^d \to [0, \infty)$. Then $\Pi = \Pi_1 \cup \Pi_2$ is a Poisson process on $D = D_1 \cup D_2$ with intensity λ.

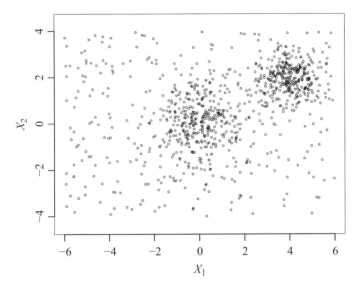

Figure 2.2   One sample of a two-dimensional Poisson process with a non-homogeneous intensity. The intensity λ is high at the location of the two visible clusters and low elsewhere. The exact setup for this figure is described in example 2.40.

Proof   The first statement is a direct consequence of definition 2.34. For the second statement, let $A \subseteq D$ and define $A_i = A \cap D_i$ for $i = 1, 2$. Since A_1 and A_2 are disjoint, the random variables $|\Pi \cap A_1| \sim \text{Pois}(\Lambda(A_1))$ and $|\Pi \cap A_2| \sim \text{Pois}(\Lambda(A_2))$ are independent. Thus, by lemma 2.33:

$$|\Pi \cap A| = |\Pi \cap A_1| + |\Pi \cap A_2| \sim \text{Pois}\left(\Lambda(A_1) + \Lambda(A_2)\right) = \text{Pois}\left(\Lambda(A)\right).$$

This shows that Π satisfies the first condition from definition 2.34. For the second condition, let A and B be disjoint. Then, the four sets $A_i = A \cap D_i$ and $B_i = B \cap D_i$ for $i = 1, 2$ are disjoint. Since Π_1 and Π_2 are Poisson processes, $|\Pi_1 \cap A_1|$ is independent of $|\Pi_1 \cap B_1|$ and $|\Pi_2 \cap A_2|$ is independent of $|\Pi_2 \cap B_2|$. By assumption, $|\Pi_1 \cap A_1|$ is independent of $|\Pi_2 \cap B_2|$ and $|\Pi_2 \cap A_2|$ is independent of $|\Pi_1 \cap B_1|$. Thus, the two values

$$|\Pi \cap A| = |\Pi_1 \cap A_1| + |\Pi_2 \cap A_2|$$

and

$$|\Pi \cap B| = |\Pi_1 \cap B_1| + |\Pi_2 \cap B_2|$$

are independent as required. $\square$

Since the definition of a Poisson process only specifies the behaviour of the number of points in a given set, but does not mention the points of Π individually, the definition cannot be directly used to simulate samples from a Poisson process. Instead, most methods for simulating a Poisson process are based on the construction described in the following algorithm and in proposition 2.37.

Algorithm 2.36   (Poisson process)
  input:
      an intensity function $\lambda \colon \mathbb{R}^d \to \mathbb{R}$
      a set $D \subseteq \mathbb{R}^d$ with $\Lambda(D) < \infty$ where Λ is given by (2.5)
  randomness used:
      $N \sim \mathrm{Pois}(\Lambda(D))$
      i.i.d. samples $X_i \sim \mathbb{1}_D \lambda(\cdot)/\Lambda(D)$ for $i = 1, 2, \ldots, N$
  output:
      a sample from the Poisson process on D with intensity λ
  1: generate $N \sim \mathrm{Pois}(\Lambda(D))$
  2: $\Pi \leftarrow \emptyset$
  3: **for** $i = 1, 2, \ldots, N$ **do**
  4:     generate $X_i \sim \frac{1}{\Lambda(D)} \mathbb{1}_D \lambda(\cdot)$
  5:     $\Pi \leftarrow \Pi \cup \{X_i\}$
  6: **end for**
  7: return Π

The function $\mathbb{1}_D \lambda/\Lambda(D)$ in the algorithm, is given by

$$\frac{\mathbb{1}_D \lambda(x)}{\Lambda(D)} = \begin{cases} \dfrac{\lambda(x)}{\Lambda(D)} & \text{if } x \in D \text{ and} \\ 0 & \text{otherwise.} \end{cases}$$

This function is a probability density whenever $\Lambda(D) > 0$. For $\Lambda(D) = 0$ this expression is undefined but, since in this case we always have $N = 0$, we will never need to generate any X_i when the density is not defined, so there is no problem.

Proposition 2.37   Let $D \subseteq \mathbb{R}^d$ be a set and $\lambda \colon \mathbb{R}^d \to [0, \infty)$ a function such that Λ defined by (2.5) satisfies $\Lambda(D) < \infty$. Then the following statements hold:

(a) The output Π of algorithm 2.36 is a sample of a Poisson process on D with intensity λ.

(b) The number of iterations of the loop in algorithm 2.36 is random with expectation $\Lambda(D)$.

Proof   To show that the output Π is a Poisson process, we have to verify that Π satisfies the two conditions from definition 2.34: let $A \subseteq D$ and $k \in \mathbb{N}$. Then

$$P(|\Pi \cap A| = k) = P\left(\bigcup_{n=k}^{\infty}\{|\Pi \cap A| = k, N = n\}\right)$$

$$= \sum_{n=k}^{\infty} P(N = n)P(|\Pi \cap A| = k|N = n).$$

By construction of Π, each of the X_i independently takes a value in A with probability

$$p = P(X_i \in A) = \frac{1}{\Lambda(D)} \int_A \lambda(x)\,dx = \frac{\Lambda(A)}{\Lambda(D)}.$$

Consequently, the probability that k out of the n values X_i are in A is given by the binomial distribution $B(n, p)$ and we find

$$P\left(|\Pi \cap A| = k\right)$$

$$= \sum_{n=k}^{\infty} e^{-\Lambda(D)}\frac{\Lambda(D)^n}{n!} \cdot \frac{n!}{k!(n-k)!}\left(\frac{\Lambda(A)}{\Lambda(D)}\right)^k \left(\frac{\Lambda(D) - \Lambda(A)}{\Lambda(D)}\right)^{n-k}$$

$$= e^{-\Lambda(D)}\frac{\Lambda(A)^k}{k!} \sum_{n=k}^{\infty} \frac{(\Lambda(D) - \Lambda(A))^{n-k}}{(n-k)!}$$

$$= e^{-\Lambda(D)}\frac{\Lambda(A)^k}{k!}e^{\Lambda(D)-\Lambda(A)}$$

$$= e^{-\Lambda(A)}\frac{\Lambda(A)^k}{k!}.$$

Thus, $|\Pi \cap A|$ is Poisson-distributed with parameter $\Lambda(A)$ as required.

For the second part of the definition, let $A, B \subseteq D$ be disjoint. Similar to the previous calculation, we find

$$P\left(|\Pi \cap A| = k, |\Pi \cap B| = l\right)$$

$$= \sum_{n=k+l}^{\infty} P(N = n)P\left(|\Pi \cap A| = k, |\Pi \cap B| = l \,\Big|\, N = n\right).$$

Since A and B are disjoint, each of the X_i is either in A or in B or in $D \setminus (A \cup B)$. Consequently, $|\Pi \cap A| = k$ and $|\Pi \cap B| = l$ follow a multinomial distribution:

$$P\left(|\Pi \cap A| = k, |\Pi \cap B| = l\right)$$

$$= \sum_{n=k+l}^{\infty} e^{-\Lambda(D)}\frac{\Lambda(D)^n}{n!} \frac{n!}{k!\,l!\,(n-k-l)!}$$

$$\cdot \left(\frac{\Lambda(A)}{\Lambda(D)} \right)^k \left(\frac{\Lambda(B)}{\Lambda(D)} \right)^l \left(\frac{\Lambda(D) - \Lambda(A) - \Lambda(B)}{\Lambda(D)} \right)^{n-k-l}$$

$$= e^{-\Lambda(D)} \frac{\Lambda(A)^k}{k!} \frac{\Lambda(B)^l}{l!} \sum_{n=k+l}^{\infty} \frac{(\Lambda(D) - \Lambda(A) - \Lambda(B))^{n-k-l}}{(n-k-l)!}$$

$$= e^{-\Lambda(D)} \frac{\Lambda(A)^k}{k!} \frac{\Lambda(B)^l}{l!} e^{\Lambda(D) - \Lambda(A) - \Lambda(B)}$$

$$= e^{-\Lambda(A)} \frac{\Lambda(A)^k}{k!} \cdot e^{-\Lambda(B)} \frac{\Lambda(B)^l}{l!}$$

$$= P\left(|\Pi \cap A| = k \right) \cdot P\left(|\Pi \cap B| = l \right),$$

for all $k, l \in \mathbb{N}_0$. Thus, the two random variables $|\Pi \cap A| = k$ and $|\Pi \cap B| = l$ are independent. This completes the proof of the first statement.

The second statement, about computational cost, is clear from the first property in lemma 2.33. $\qquad\square$

Example 2.38   Assume that we want to sample from a Poisson process with constant intensity $\lambda \in \mathbb{R}$ on an interval $D = [a, b] \subseteq \mathbb{R}$. Then we have $\Lambda(D) = \int_a^b \lambda \, dx = \lambda(b - a)$ and thus $\lambda/\Lambda(D) = 1/(b - a)$ is the density of the uniform distribution on $[a, b]$. Consequently, by proposition 2.37, we can generate a sample of a Poisson process on $[a, b]$ by the following procedure:

(a) Generate $N \sim \text{Pois}(\lambda(b - a))$.

(b) Generate $X_1, X_2, \ldots, X_N \sim \mathcal{U}[a, b]$ i.i.d.

(c) Let $\Pi = \{X_1, X_2, \ldots, X_N\}$.

Example 2.39   We can use lemma 2.35 to simulate a Poisson process with a piecewise constant intensity function λ as shown, for example, in Figure 2.3. Let $a = t_0 < t_1 < \cdots < t_n = b$ be given and assume that λ satisfies

$$\lambda(t) = \begin{cases} \lambda_1 & \text{if } t \in [t_0, t_1] \\ \lambda_2 & \text{if } t \in [t_1, t_2] \\ \vdots \\ \lambda_n & \text{if } t \in [t_{n-1}, t_n]. \end{cases}$$

We can use algorithm 2.36 to generate a sample of a Poisson process on each of the intervals $[t_{i-1}, t_i)$. Since λ is constant on these intervals, we can just generate $N_i \sim \text{Pois}(\lambda_i)$ and then let $\Pi_i = \{X_1^{(i)}, \ldots, X_{N_i}^{(i)}\}$ where $X_1^{(i)}, \ldots, X_{N_i}^{(i)} \sim \mathcal{U}[t_{i-1}, t_i)$ are i.i.d. Finally, by lemma 2.35 the set $\Pi = \Pi_1 \cup \Pi_2 \cup \cdots \cup \Pi_n$ is a sample of a Poisson process on $[a, b]$ with intensity λ.

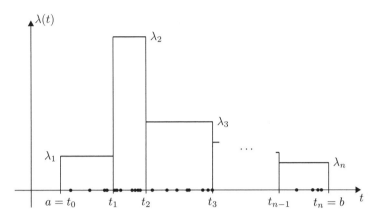

Figure 2.3 A Poisson process on an interval $[a, b] \subseteq \mathbb{R}$ with piecewise constant intensity function λ. The function λ satisfies $\lambda(t) = \lambda_i$ whenever $t \in [t_{i-1}, t_i)$. The marked points on the t-axis form one sample of the corresponding Poisson process.

Example 2.40   Assume that we want to sample a Poisson process on the domain $D = [-6, 6] \times [-4, 4] \subseteq \mathbb{R}^2$ with intensity function

$$\lambda(x) = 100 \exp\left(-\frac{x_1^2 + x_2^2}{2}\right) + 100 \exp\left(-(x_1 - 4)^2 - (x_2 - 2)^2\right)$$

for all $x \in \mathbb{R}^2$. In this case, $\Lambda(D)$ is difficult to determine and we cannot directly apply algorithm 2.36. Instead, we can get samples from this Poisson process by first simulating the Poisson process on $\mathbb{R}^2$ with intensity λ, and the use lemma 2.35 to restrict the resulting process to the given domain D.

By rewriting the intensity-function λ as

$$\lambda(x) = 200\pi \frac{1}{2\pi} \exp\left(-\frac{x_1^2 + x_2^2}{2}\right) + 100\pi \frac{1}{2\pi\frac{1}{2}} \exp\left(-\frac{(x_1 - 4)^2 + (x_2 - 2)^2}{2 \cdot \frac{1}{2}}\right),$$

we see that λ is 300π times the density of the mixture distribution

$$\frac{2}{3}\mathcal{N}(0, I_2) + \frac{1}{3}\mathcal{N}(\mu, I_2/2)$$

where $\mu = (4, 2)$ and I_2 is the identity matrix in $\mathbb{R}^2$. Consequently, we have

$$\Lambda(\mathbb{R}^2) = \int_{\mathbb{R}^2} \lambda(x)\,dx = 200\,\pi + 100\,\pi = 300\,\pi.$$

Thus, if we set $N \sim \text{Pois}(300\,\pi)$ and define $\tilde{\Pi} = \{X_1, \ldots, X_N\}$ where the X_i are independent samples from the mixture distribution with density $\lambda/300\pi$, then $\tilde{\Pi}$ is a Poisson process on $\mathbb{R}^2$ with intensity λ. Finally we define $\Pi = \tilde{\Pi} \cap D$. Then, by lemma 2.35, Π is a Poisson process on D with intensity λ. The result of a simulation using the method described in this example is shown in Figure 2.2.

To conclude this section, we describe a rejection algorithm for converting a Poisson process with intensity λ into a Poisson process with intensity $\tilde{\lambda} \leq \lambda$ by randomly omitting some of the points.

Algorithm 2.41   (thinning method for Poisson processes)
    input:
        intensity functions $\lambda, \tilde{\lambda} \colon \mathbb{R}^d \to \mathbb{R}$ with $\tilde{\lambda} < \lambda$
        a set $D \subseteq \mathbb{R}^d$ with $\Lambda(D) < \infty$ where Λ is given by (2.5)
    randomness used:
        $N \sim \mathrm{Pois}(\Lambda(D))$
        i.i.d. samples $X_i \sim \mathbb{1}_D \lambda(\cdot)/\Lambda(D)$ for $i = 1, 2, \ldots, N$
        $U_i \sim \mathcal{U}[0, 1]$ i.i.d.
    output:
        a sample from the Poisson process on D with intensity $\tilde{\lambda}$
  1: generate $N \sim \mathrm{Pois}(\Lambda(D))$
  2: $\tilde{\Pi} \leftarrow \emptyset$
  3: **for** $i = 1, 2, \ldots, N$ **do**
  4:     generate $X_i \sim \frac{1}{\Lambda(D)} \mathbb{1}_D \lambda(\cdot)$
  5:     generate $U_i \sim \mathcal{U}[0, 1]$
  6:     **if** $U_i \leq \tilde{\lambda}(X_i)/\lambda(X_i)$ **then**
  7:         $\tilde{\Pi} \leftarrow \tilde{\Pi} \cup \{X_i\}$
  8:     **end if**
  9: **end for**
 10: return $\tilde{\Pi}$

Proposition 2.42 shows that this algorithm returns a sample from a Poisson process with intensity $\tilde{\lambda}$. While the result does not depend on the choice of the auxiliary density λ, the choice of λ affects efficiency. Under the constraint $\lambda \geq \tilde{\lambda}$, the intensity λ should be chosen such that the samples X_i in step 4 of the algorithm can be generated efficiently and that $\Lambda(D)$ is as small as possible.

Proposition 2.42   Let Π be a Poisson process on $D \subseteq \mathbb{R}^d$ with intensity $\lambda \colon \mathbb{R}^d \to [0, \infty)$. Let $\tilde{\lambda} \colon \mathbb{R}^d \to [0, \infty)$ such that $\tilde{\lambda}(x) \leq \lambda(x)$ for all $x \in D$ and define a random subset $\tilde{\Pi} \subseteq \Pi$ by randomly including each point $x \in \Pi$ into $\tilde{\Pi}$ with probability $\tilde{\lambda}(x)/\lambda(x)$, independently of each other and of Π. Then $\tilde{\Pi}$ is a Poisson process with intensity function $\tilde{\lambda}$.

Proof   We prove the statement of the proposition by verifying that $\tilde{\Pi}$ satisfies definition 2.34: let $A \subseteq \mathbb{R}^d$. Then, to check the first statement from definition 2.34, we have to show that $|A \cap \tilde{\Pi}|$ is Poisson-distributed with the correct expectation.

We start the proof by first considering the conditional distribution of Π, conditioned on $|A \cap \Pi| = n \in \mathbb{N}_0$. Using the construction of Π given in algorithm 2.36 and proposition 2.37 we see that, under this condition, the set $A \cap \Pi$ consists of n random points $X_1, \ldots, X_n \in A$, independently distributed with density $\lambda/\Lambda(A)$. Depending on its location in A, each point X_i is included in $\tilde{\Pi}$ with probability $\tilde{\lambda}(X_i)/\lambda(X_i)$ and averaging over the possible locations for X_i, we find the probability of X_i being included in $\tilde{\Pi}$ as

$$
p = \int_A \frac{\tilde{\lambda}(x)}{\lambda(x)} \cdot \frac{\lambda(x)}{\Lambda(A)} \, dx = \frac{1}{\Lambda(A)} \int_A \tilde{\lambda}(x) \, dx = \frac{\tilde{\Lambda}(A)}{\Lambda(A)},
$$

where we used the abbreviation

$$
\tilde{\Lambda}(A) = \int_A \tilde{\lambda}(x) \, dx.
$$

Since this probability does not depend on i, the total number of points in $|\tilde{\Pi} \cap A|$, conditioned on $|A \cap \Pi| = n$, is $\mathcal{B}(n, p)$-distributed. Thus we have

$$
\begin{aligned}
P\left(|\tilde{\Pi} \cap A| = k \;\middle|\; |A \cap \Pi| = n\right) \\
= \binom{n}{k} \left(\frac{\tilde{\Lambda}(A)}{\Lambda(A)}\right)^k \left(1 - \frac{\tilde{\Lambda}(A)}{\Lambda(A)}\right)^{n-k} \\
= \binom{n}{k} \frac{\tilde{\Lambda}(A)^k \left(\Lambda(A) - \tilde{\Lambda}(A)\right)^{n-k}}{\Lambda(A)^n}.
\end{aligned} \tag{2.6}
$$

To find the distribution of $|\tilde{\Pi} \cap A|$, we have to average the result from equation (2.6) over the possible values of $|\Pi \cap A|$. Since $|A \cap \Pi| \sim \text{Pois}(\Lambda(A))$, we get

$$
\begin{aligned}
P\left(|\tilde{\Pi} \cap A| = k\right) &= \sum_{n=k}^{\infty} P\left(|\tilde{\Pi} \cap A| = k \;\middle|\; |A \cap \Pi| = n\right) P\left(|A \cap \Pi| = n\right) \\
&= \sum_{n=k}^{\infty} \frac{n!}{k!(n-k)!} \frac{\tilde{\Lambda}(A)^k \left(\Lambda(A) - \tilde{\Lambda}(A)\right)^{n-k}}{\Lambda(A)^n} e^{-\Lambda(A)} \frac{\Lambda(A)^n}{n!} \\
&= \frac{\tilde{\Lambda}(A)^k}{k!} \sum_{n=k}^{\infty} \frac{\left(\Lambda(A) - \tilde{\Lambda}(A)\right)^{n-k}}{(n-k)!} e^{-\Lambda(A)} \\
&= \frac{\tilde{\Lambda}(A)^k}{k!} \sum_{l=0}^{\infty} \frac{\left(\Lambda(A) - \tilde{\Lambda}(A)\right)^l}{l!} e^{-\Lambda(A)} \\
&= \frac{\tilde{\Lambda}(A)^k}{k!} e^{\Lambda(A) - \tilde{\Lambda}(A)} e^{-\Lambda(A)} \\
&= e^{-\tilde{\Lambda}(A)} \frac{\tilde{\Lambda}(A)^k}{k!}
\end{aligned}
$$

for all $k \in \mathbb{N}_0$. Thus $|\tilde{\Pi} \cap A| \sim \text{Pois}(\tilde{\Lambda}(A))$ as required.

For the second condition from definition 2.34, let $A, B \subseteq D$ be disjoint. Then, since Π is a Poisson process, the numbers $|\Pi \cap A|$ and $|\Pi \cap B|$ are independent. Also, given Π, the choices whether to include any of the $x \in \Pi$ into the subset $\tilde{\Pi} \subseteq \Pi$ are independent, and thus $|\tilde{\Pi} \cap A|$ and $|\tilde{\Pi} \cap B|$ are independent. This completes the proof. □

2.5   Summary and further reading

In this chapter we illustrated with the help of examples, how sequences of independent random numbers can be used as building blocks to simulate more complex statistical models. The key aspect of such simulations is that the random components of statistical models are often no longer independent but can feature a complex dependence structure.

Among the models discussed here were Markov chains (Section 2.3) and Poisson processes (Section 2.4). For both classes of processes, we restricted discussion of theoretical aspects to a minimum. More detail can be found in the literature, for example in the books by Norris (1997) for Markov chains and Kingman (1993) for Poisson processes. Space restrictions force us to leave out many popular classes of models, for example we did not discuss the autoregressive and moving average models for time series (see e.g. Chatfield, 2004), but often simulation of such processes is either straightforward or can be performed building on the ideas illustrated in this chapter.

This concludes our study of methods for simulating statistical models. In the following chapters we will learn how the samples by such simulations can be used to study the underlying models. We will return to the question of how to simulate statistical models in a slightly different context in Chapter 6, where we will consider stochastic processes in continuous time.

Exercises

E2.1 Write a program which generates samples from the mixture of $P_1 = \mathcal{N}(1, 0.01^2)$, $P_2 = \mathcal{N}(2, 0.5^2)$ and $P_3 = \mathcal{N}(5, 0.02^2)$ with weights $\theta_1 = 0.1$, $\theta_2 = 0.7$ and $\theta_3 = 0.2$. Generate a plot showing a histogram of $10\,000$ samples of this distribution, together with the density of the mixture distribution.

E2.2 Let $\varepsilon_j \sim \mathcal{N}(0, 1)$ i.i.d. for $j \in \mathbb{N}$ and let $X = (X_j)_{j \in \mathbb{N}}$ be defined by $X_0 = 0$, $X_1 = \varepsilon_1$ and $X_j = X_{j-1} + X_{j-2} + \varepsilon_j$ for all $j \geq 2$. Show that the process X is not a Markov chain.

E2.3 Let X be the Markov chain with transition matrix

$$P = \begin{pmatrix} 2/3 & 1/3 & 0 & 0 \\ 1/10 & 9/10 & 0 & 0 \\ 1/10 & 0 & 9/10 & 0 \\ 1/10 & 0 & 0 & 9/10 \end{pmatrix}$$

and initial distribution $\mu = (1/4, 1/4, 1/4, 1/4)$.

    (a) Write a program which generates a random path $X_0, X_1, X_2, \ldots, X_N$ from this Markov chain.

    (b) Use the program from (a) and Monte Carlo integration to numerically determine the distribution of X_{10} (i.e. you have to estimate the probabilities $P(X_{10} = x)$ for $x = 1, 2, 3, 4$).

    (c) Analytically compute the distribution of X_{10}. You may use a computer to obtain your answer.

E2.4

    (a) Let P be a stochastic matrix. Show that the vector $v = (1, 1, \ldots, 1)$ is an eigenvector of P and determine the corresponding eigenvalue.

    (b) Let v be an eigenvector of a matrix A with eigenvalue λ. Show that for $\alpha \neq 0$ the vector αv is also an eigenvector of A.

    (c) Find a stationary distribution for the stochastic matrix P from exercise E2.3 by computing the eigenvectors of $P^\top$. You can use the R functions t and eigen for this exercise. Some care is needed because the entries of the computed eigenvector may not be positive and may not sum up to 1 (see (b) for an explanation). How can we solve this problem?

E2.5 Write a program which simulates a Poisson process with constant intensity $\lambda > 0$ on an interval $[a, b] \subseteq \mathbb{R}$.

E2.6 Write a program which simulates a Poisson process with piecewise constant intensity function λ. The program should take $t_0 < t_1 < \cdots < t_n$ and $\lambda_1, \ldots, \lambda_n \geq 0$ as inputs (see Figure 2.3 and example 2.39) and should return one sample of the Poisson process.

3

Monte Carlo methods

So far, in the first two chapters of this book, we have learned how to simulate statistical models on a computer. In this and the following two chapters we will discuss how such simulations can be used to study properties of the underlying statistical model. In this chapter we will concentrate on the approach to directly generate a large number of samples from the given model. The idea is then that the samples reflect the statistical behaviour of the model; questions about the model can then be answered by studying statistical properties of the samples. The resulting methods are called *Monte Carlo methods*.

3.1 Studying models via simulation

When studying statistical models, analytical calculations often are only possible under assumptions such as independence of samples, normality of samples or large sample size. For this reason, many problems occurring in 'real life' situations are only approximately covered by the available analytical results. This chapter presents an alternative approach to such problems, based on estimates derived from computer simulations instead of analytical calculations.

The fundamental observation underlying the methods discussed in this and the following chapters is the following: if we can simulate a statistical model on a computer, then we can generate a large set of samples from the model and then we can learn about the behaviour of the model by studying the computer-generated set of samples instead of the model itself. We give three examples for this approach:

- As a consequence of the law of large numbers (see theorem A.8), the expected value of a random variable can be approximated by generating a large number of samples of the random variable and then considering the average value.

An Introduction to Statistical Computing: A Simulation-based Approach, First Edition. Jochen Voss.
© 2014 John Wiley & Sons, Ltd. Published 2014 by John Wiley & Sons, Ltd.

- The probability of an event can be approximated by generating a large number of samples and then considering the proportion of samples where the event occurs.

- The quality of a method for statistical inference can be assessed by repeatedly generating synthetic data with a known distribution and then analysing how well the inference method recovers the (known) properties of the underlying distribution from the synthetic data sets.

Since many interesting questions can be reduced to computing the expectation of some random variable, we will mostly restrict our attention to the problem of computing expectations of the form $\mathbb{E}(f(X))$ where X is a random sample from the system under consideration and f is a real-valued function, determining some quantity of interest in the system. There are several different methods to compute such an expectation:

(a) Sometimes we can find the answer analytically. For example, if the distribution of X has a density φ, we can use the relation

$$\mathbb{E}(f(X)) = \int f(x)\,\varphi(x)\,dx \qquad (3.1)$$

to obtain the value of the expectation (see Section A.3). This method only works if we can solve the resulting integral.

(b) If the integral in (3.1) cannot be solved analytically, we can try to use numerical integration to get an approximation to the value of the integral. When X takes values in a low-dimensional space, this method often works well, but for higher dimensional spaces numerical approximation can become very expensive and the resulting method may no longer be efficient. Since numerical integration is outside the topic of statistical computing, we will not follow this approach here.

(c) The approach we will study in this chapter is called *Monte Carlo estimation* or *Monte Carlo integration*. This technique is based on the strong law of large numbers: if $(X_j)_{j\in\mathbb{N}}$ is a sequence of *i.i.d.* random variables with the same distribution as X, then

$$\mathbb{E}(f(X)) = \lim_{N\to\infty} \frac{1}{N} \sum_{j=1}^{N} f(X_j) \qquad (3.2)$$

with probability 1.

Our aim for this chapter is to study approximations for $\mathbb{E}(f(X))$ based on Equation (3.2). While the exact equality in (3.2) holds only in the limit $N \to \infty$, we can use the approximation with fixed, large N to get the following approximation method.

Definition 3.1   A *Monte Carlo method* for estimating the expectation $\mathbb{E}(f(X))$ is a numerical method based on the approximation

$$\mathbb{E}(f(X)) \approx \frac{1}{N} \sum_{j=1}^{N} f(X_j), \qquad (3.3)$$

where $(X_j)_{j\in\mathbb{N}}$ are i.i.d. with the same distribution as X.

In order to compute a numerical value for this approximation we will need to generate a large number of samples X_j from the model. This can be done using the techniques from Chapters 1 and 2. In the present chapter we will assume that we have solved the problem of generating these samples and we will concentrate on the estimate (3.3) itself: we will consider how the error in the approximation (3.3) depends on the computational cost and we will derive improved variants of the basic approximation (3.3).

Example 3.2   For $f(x) = x$, the Monte Carlo estimate (3.3) reduces to

$$\mathbb{E}(X) \approx \frac{1}{N} \sum_{j=1}^{N} X_j = \bar{X}.$$

This is just the usual estimator for the mean.

Example 3.3   Assume that we want to compute the expectation $\mathbb{E}(\sin(X)^2)$ where $X \sim \mathcal{N}(\mu, \sigma^2)$. Obtaining the exact value analytically will be difficult, but we can easily get an approximation using Monte Carlo estimation: if we choose the number of samples N large and generate independent, $\mathcal{N}(\mu, \sigma)$-distributed random variables $X_1, X_2, \ldots, X_N$, then, by the strong law of large numbers, we have

$$\mathbb{E}(\sin(X)) \approx \frac{1}{N} \sum_{j=1}^{N} \sin(X_j)^2.$$

The right-hand side of this approximation can be easily evaluated numerically, giving an estimate for the required expectation.

While the approximation (3.3) is only stated for the problem of estimating an expectation, the same technique applies to a wider class of problems. To illustrate this, we consider two different problems, which can be reduced to the problem of computing an expectation.

As a first application of Monte Carlo estimates we consider here the problem of estimating probabilities. While computing expectations and computing probabilities at first look like different problems, the latter can be reduced to the former: if X

is a random variable, we have $P(X \in A) = \mathbb{E}(\mathbb{1}_A(X))$; this relation is, for example, explained in Section A.3. Using this equality, we can estimate $P(X \in A)$ by

$$P(X \in A) = \mathbb{E}\,(\mathbb{1}_A(X)) \approx \frac{1}{N} \sum_{j=1}^{N} \mathbb{1}_A(X_j) \qquad (3.4)$$

for sufficiently large N.

Example 3.4   Let $X \sim \mathcal{N}(0, 1)$ and $a \in \mathbb{R}$. Then the probability $p = P(X \leq a)$ cannot be computed analytically in an explicit form, but

$$p_N = \frac{1}{N} \sum_{j=1}^{N} \mathbb{1}_{(-\infty,a]}(X_j)$$

can be used as an approximation for p.

As a second application of Monte Carlo estimation we consider the problem of approximating the value of integrals such as $\int_a^b f(x)\,dx$. Again, this problem at first looks like it is distinct from the problem of computing expectations, but it transpires that the two problems are closely related. To see this, we utilise the relation (3.1) from the beginning of this chapter as follows: let $X, X_j \sim \mathcal{U}[a, b]$ be i.i.d. Then the density of X_j is $\varphi(x) = \frac{1}{b-a}\mathbb{1}_{[a,b]}$. We get

$$\int_a^b f(x)\,dx = (b - a) \int_a^b f(x)\varphi(x)\,dx$$
$$= (b - a)\mathbb{E}(f(X)) \qquad (3.5)$$
$$\approx \frac{b - a}{N} \sum_{j=1}^{N} f(X_j)$$

for sufficiently large N.

Example 3.5   To estimate the integral $\int_0^{2\pi} e^{\kappa \cos(x)}\,dx$ we can generate $X_j \sim \mathcal{U}[0, 2\pi]$ and then use the approximation

$$\int_0^{2\pi} e^{\kappa \cos(x)}\,dx \approx \frac{2\pi}{N} \sum_{j=1}^{N} e^{\kappa \cos(X_j)}.$$

To conclude this section, we give a longer example showing how the methods from Chapter 1 can be used in Monte Carlo estimates.

Example 3.6   Consider a simple Bayesian inference problem, where we want to make inference about $X \sim \text{Exp}(1)$ using a single observation of $Y \sim \mathcal{N}(0, X)$. To solve this problem, we have to find the posterior distribution, that is the conditional distribution of X given the observation $Y = y$.

We can find the density of the conditional distribution of X using Bayes' rule

$$p_{X|Y}(x|y) = \frac{p_{Y|X}(y|x)p_X(x)}{p_Y(y)}$$

as found, for example, using equation (A.5). The value $X \sim \text{Exp}(1)$ has density $p_X(x) = \exp(-x)$ and, given $X = x$, the conditional density of $Y \sim \mathcal{N}(0, X)$ is

$$p_{Y|X}(y|x) = \frac{1}{\sqrt{2\pi x}} \exp\left(-y^2/2x\right).$$

By averaging over the possible values of X we find the unconditional density of Y as

$$p_Y(y) = \int_0^\infty p_{Y|X}(y|\tilde{x}) \, p_X(\tilde{x}) \, d\tilde{x} = \int_0^\infty \frac{1}{\sqrt{2\pi \tilde{x}}} \exp\left(-y^2/2\tilde{x} - \tilde{x}\right) \, d\tilde{x}.$$

Thus, the conditional density of X given $Y = y$ is

$$
\begin{aligned}
p_{X|Y}(x|y) &= \frac{p_{Y|X}(y|x)p_X(x)}{p_Y(y)} \\
&= \frac{\frac{1}{\sqrt{2\pi x}} \exp\left(-y^2/2x - x\right)}{\int_0^\infty \frac{1}{\sqrt{2\pi \tilde{x}}} \exp\left(-y^2/2\tilde{x} - \tilde{x}\right) \, d\tilde{x}} \\
&= \frac{1}{Z_f} f(x)
\end{aligned}
\tag{3.6}
$$

for all $x \geq 0$, where

$$f(x) = \frac{1}{\sqrt{x}} \exp\left(-y^2/2x - x\right)$$

and $Z_f = \int_0^\infty f(\tilde{x}) \, d\tilde{x}$. This is the required density of the conditional distribution of X given Y.

To derive estimates about X using Monte Carlo estimation, we need to be able to generate samples from the posterior distribution of X, that is we need to be able to generate samples from the distribution with the density $p_{X|Y}$ given in equation (3.6). Generating samples from this distribution is complicated by the fact that we do not know the value of the integral Z_f. But, since the rejection sampling algorithm from Section 1.4 can still be applied when the normalising constant Z_f is unknown, we can use algorithm 1.22 for this purpose. Since f is bounded near 0 and since

$f(x) \approx \exp(-x)$ when x is large, we try to use an $\mathrm{Exp}(1)$-distribution for the proposals; the corresponding density is

$$g(x) = \exp(-x)$$

for all $x \geq 0$. For this to work, we need to find a constant $c > 0$ with $f(x) \leq cg(x)$ for all $x \geq 0$. A straightforward calculation (see exercise E1.8) shows that this bound is satisfied for the value

$$c = \frac{1}{|y|} \exp(-1/2)$$

and thus the rejection sampling algorithm 1.22 can be applied to generate the required samples.

Using samples generated by the rejection algorithm, we can now answer questions about the posterior distribution. If we generate samples $X_1, X_2, \ldots, X_N$ for large N, for example for $N = 10\,000$, we can use estimates such as

$$\mathbb{E}(X|Y = y) \approx \frac{1}{N} \sum_{j=1}^{N} X_j = \bar{X}$$

and

$$\mathrm{Var}(X|Y = y) \approx \frac{1}{N-1} \sum_{j=1}^{N} \left(X_j - \bar{X}\right)^2.$$

Without the use of Monte Carlo estimation, the posterior mean $\mathbb{E}(X|Y = y)$ and the posterior variance $\mathrm{Var}(X|Y = y)$ would be difficult to determine.

3.2   Monte Carlo estimates

In this section we study the basic properties of Monte Carlo estimates as introduced in the previous section.

Definition 3.7   Let X be a random variable and f be a function such that $f(X) \in \mathbb{R}$. Then the *Monte Carlo estimate* for $\mathbb{E}(f(X))$ is given by

$$Z_N^{\mathrm{MC}} = \frac{1}{N} \sum_{j=1}^{N} f(X_j), \tag{3.7}$$

where $X_1, \ldots, X_N$ are i.i.d. with the same distribution as X.

Since the estimate Z_N^{MC} is constructed from random samples X_j, it is a random quantity itself. The random variables X_j in (3.7) are sometimes called *i.i.d. copies* of X.

3.2.1   Computing Monte Carlo estimates

We can rewrite equation (3.7) as the following very simple algorithm.

Algorithm 3.8   (Monte Carlo estimate)
  input:
    a function f with values in $\mathbb{R}$
    $N \in \mathbb{N}$
  randomness used:
    i.i.d. copies $(X_j)_{j \in \mathbb{N}}$ of X
  output:
    an estimate Z_N^{MC} for $\mathbb{E}(f(X))$
  1: $s \leftarrow 0$
  2: **for** $j = 1, 2, \ldots, N$ **do**
  3:    generate X_j, with the same distribution as X has
  4:    $s \leftarrow s + f(X_j)$
  5: **end for**
  6: return s/N

Ignoring the negligible computational cost for the initial assignment of the variable s and of the final division by N, the execution time of algorithm 3.8 is proportional to N. On the other hand, by the law of large numbers, we know that the estimate Z_N^{MC} converges to the correct value $\mathbb{E}(f(X))$ only as N increases, and thus the error decreases as N increases. For this reason, choosing the sample size N in algorithm 3.8 involves a trade-off between computational cost and accuracy of the result. We will discuss this trade-off in more detail in the following section.

An alternative to algorithm 3.8 would be to first generate all samples $f(X_1)$, $f(X_2), \ldots, f(X_n)$ and then, in a second step, to compute and return the average of the samples. The advantage of using the temporary variable s in algorithm 3.8 to accumulate the results instead is, that the samples $f(X_j)$ do not need to be stored in the computer's memory, thus greatly reducing the memory requirements of the algorithm: the memory required to execute algorithm 3.8 is independent of N.

Definition 3.7 and algorithm 3.8 do not explicitly state the space the random variables X and X_j take their values in. The reason for this omission is that algorithm 3.8 poses no restrictions on the random variable X except for the requirement that the expectation $\mathbb{E}(f(X))$ must exist. In simple cases, the random variable X could take values in $\mathbb{R}$ or in $\mathbb{R}^d$, but many practical applications require the values of X to have a more complicated structure. As an example, in Section 6.5 we will consider the case where X is a random function. In contrast, we will assume throughout this text that $f(X) \in \mathbb{R}$ so that there are no complications to define the sum in equation (3.7) or to define the expectation $\mathbb{E}(f(X))$.

3.2.2   Monte Carlo error

Since the estimate Z_N^{MC} from definition 3.7 and algorithm 3.8 is random, the Monte Carlo error $Z_N^{MC} - \mathbb{E}(f(X))$ is also random. To quantify the magnitude of this random error, we use the concepts of bias and mean squared error from statistics.

Definition 3.9   The *bias* of an estimator $\hat{\theta} = \hat{\theta}(X)$ for a parameter θ is given by

$$\text{bias}(\hat{\theta}) = \mathbb{E}_\theta\left(\hat{\theta}(X) - \theta\right) = \mathbb{E}_\theta\left(\hat{\theta}(X)\right) - \theta,$$

where the subscript θ in the expectations on the right-hand side indicates that the sample X in the expectation comes from the distribution with true parameter θ.

Since the bias as given in definition 3.9 depends on the value of θ, sometimes the notation $\text{bias}_\theta(\hat{\theta})$ is used to indicate the dependence on θ. While the bias measures how far off the estimate is on average, a small value of the bias does not necessarily indicate a useful estimator: even when the estimator is correct on average, the actual values of the estimator may fluctuate so wildly around θ that they are not useful in practice. For this reason, the size of fluctuations of an estimator is considered.

Definition 3.10   The *standard error* of an estimator $\hat{\theta} = \hat{\theta}(X)$ is given by

$$\text{se}(\hat{\theta}) = \text{stdev}_\theta\left(\hat{\theta}(X)\right),$$

where the subscript θ on the standard deviation indicates that the sample X comes from the distribution with true parameter θ.

Finally, the mean squared error, introduced in the following definition, combines both kinds of error: it measures the fluctuations of the estimator around the true value of the parameter.

Definition 3.11   The *mean squared error* (MSE) of an estimator $\hat{\theta} = \hat{\theta}(X)$ for a parameter θ is given by

$$\text{MSE}(\hat{\theta}) = \mathbb{E}_\theta\left(\left(\hat{\theta}(X) - \theta\right)^2\right).$$

As for the bias, the expressions for the standard error and for the mean squared error depend on the value of θ and sometimes the notations $\text{se}_\theta(\hat{\theta})$ and $\text{MSE}_\theta(\hat{\theta})$ are used to indicate this dependence. The following lemma shows that the mean squared error can be computed from the bias and the standard error.

Lemma 3.12   Let $\hat{\theta} = \hat{\theta}(X_1, \ldots, X_n)$ be an estimator for a parameter $\theta \in \mathbb{R}$. Then the mean squared error of $\hat{\theta}$ satisfies

$$\text{MSE}(\hat{\theta}) = \text{Var}(\hat{\theta}) + \text{bias}(\hat{\theta})^2 = \text{se}(\hat{\theta})^2 + \text{bias}(\hat{\theta})^2.$$

Proof   We have

$$
\begin{aligned}
\text{MSE}(\hat{\theta}) &= \mathbb{E}_\theta \left((\hat{\theta} - \theta)^2 \right) \\
&= \mathbb{E}_\theta(\hat{\theta}^2) - 2\theta \mathbb{E}_\theta(\hat{\theta}) + \theta^2 \\
&= \mathbb{E}_\theta(\hat{\theta}^2) - \mathbb{E}_\theta(\hat{\theta})^2 + \mathbb{E}_\theta(\hat{\theta})^2 - 2\theta \mathbb{E}_\theta(\hat{\theta}) + \theta^2 \\
&= \mathbb{E}_\theta(\hat{\theta}^2) - \mathbb{E}_\theta(\hat{\theta})^2 + \left(\mathbb{E}_\theta(\hat{\theta}) - \theta \right)^2 \\
&= \text{Var}(\hat{\theta}) + \text{bias}(\hat{\theta})^2.
\end{aligned}
$$

This completes the proof. □

Example 3.13   Let $X_1, \ldots, X_N \sim \mathcal{N}(\mu, \sigma^2)$ be i.i.d. with a known variance σ^2 and unknown mean $\mu \in \mathbb{R}$. Then

$$
\hat{\mu}(X) = \frac{1}{n} \sum_{i=1}^n X_i
$$

is an estimator for μ with

$$
\text{bias}(\hat{\mu}) = \mathbb{E}_\mu \left(\frac{1}{n} \sum_{i=1}^n X_i \right) - \mu = \frac{1}{n} \sum_{i=1}^n \mathbb{E}_\mu(X_i) - \mu = \frac{1}{n} \sum_{i=1}^n \mu - \mu = 0
$$

and, using the independence of the X_i,

$$
\text{se}(\hat{\mu}) = \sqrt{ \text{Var}_\mu \left(\frac{1}{n} \sum_{i=1}^n X_i \right) } = \sqrt{ \frac{1}{n^2} \sum_{i=1}^n \text{Var}_\mu(X_i) } = \sqrt{ \frac{n\sigma^2}{n^2} } = \frac{\sigma}{\sqrt{n}}
$$

for all $\mu \in \mathbb{R}$. Finally, using lemma 3.12, we find $\text{MSE}(\hat{\theta}) = (\sigma/\sqrt{n})^2 + 0^2 = \sigma^2/n$.

Proposition 3.14   The Monte Carlo estimate Z_N^{MC} for $E(f(X))$, as computed in algorithm 3.8, has

$$
\text{bias}\left(Z_N^{\text{MC}} \right) = 0
$$

and

$$
\text{MSE}\left(Z_N^{\text{MC}} \right) = \text{Var}\left(Z_N^{\text{MC}} \right) = \frac{1}{N} \text{Var}(f(X)). \tag{3.8}
$$

Proof   The expectation of Z_N^{MC} is given by

$$\mathbb{E}\left(Z_N^{\text{MC}}\right) = \mathbb{E}\left(\frac{1}{N}\sum_{j=1}^{N} f(X_j)\right) = \frac{1}{N}\sum_{j=1}^{N}\mathbb{E}\left(f(X_j)\right) = \mathbb{E}(f(X))$$

and thus we have

$$\text{bias}\left(Z_N^{\text{MC}}\right) = \mathbb{E}\left(Z_N^{\text{MC}}\right) - \mathbb{E}(f(X)) = 0.$$

Since the X_j are independent, the variance of the Monte Carlo estimate can be found as

$$\text{Var}\left(Z_N^{\text{MC}}\right) = \text{Var}\left(\frac{1}{N}\sum_{j=1}^{N} f(X_j)\right) = \frac{1}{N^2}\sum_{j=1}^{N}\text{Var}\left(f(X_j)\right) = \frac{1}{N}\text{Var}(f(X)).$$

Finally, using lemma 3.12, we get

$$\text{MSE}\left(Z_N^{\text{MC}}\right) = \text{Var}\left(Z_N^{\text{MC}}\right) + \text{bias}\left(Z_N^{\text{MC}}\right)^2 = \frac{1}{N}\text{Var}(f(X)) + 0.$$

This completes the proof.   □

Example 3.15   Let $X \sim \mathcal{N}(0, 1)$ and assume that we want to estimate the expectation $\mathbb{E}(\sin(X)^2)$. The Monte Carlo estimate for this expectation is given by

$$Z_N^{\text{MC}} = \sum_{j=1}^{N} \sin(X_j)^2,$$

where $X_1, \ldots, X_N \sim \mathcal{N}(0, 1)$ are independent. From proposition 3.14 we know that Z_N^{MC} is random and that, for every value of N, we have $\mathbb{E}(Z_N^{\text{MC}}) = \mathbb{E}(\sin(X)^2)$. But due to random fluctuations, individual Monte Carlo estimates Z_N^{MC} will not be equal to this value, but instead will fluctuate around this value with variance proportional to $1/N$. This is illustrated in Figure 3.1 which shows the spread of the distribution of Z_N^{MC} for two different values of N. The figure shows histograms of a large number of Monte Carlo estimates with sample size $N = 1000$ (a) and $N = 10\,000$ (b). As expected from proposition 3.14, the distribution of the estimates with $N = 10\,000$ has noticeably smaller variance than the distribution of the estimates with $N = 1000$.

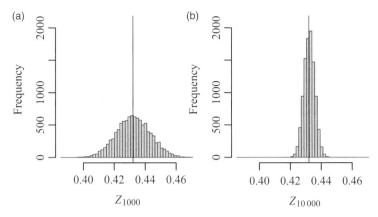

Figure 3.1   Histograms of Monte Carlo estimates for the expectation $\mathbb{E}(\sin(X)^2)$ *where* $X \sim \mathcal{N}(0, 1)$. *The plots clearly show that the estimates with (a)* $N = 1000$ *and (b)* $N = 10\,000$ *are clustered around the same mean. (b) shows that the estimates using* $N = 10\,000$ *samples have much smaller variance, and thus smaller error, than the estimates for* $N = 1000$. *This confirms the result of proposition 3.14.*

Since the mean squared error as considered in proposition 3.14 is a squared quantity, its magnitude is not directly comparable with the magnitude of the estimate Z_N^{MC} itself. For this reason, we occasionally consider the *root-mean-square error*, given by

$$\text{RMSE}\left(Z_N^{MC}\right) = \sqrt{\text{MSE}\left(Z_N^{MC}\right)} = \frac{\text{stdev}(f(X))}{\sqrt{N}}.$$

From this we see that the error of a Monte Carlo estimate decays proportionally to $1/\sqrt{N}$. As a consequence of this result, in practice huge numbers of samples can be required to achieve a reasonable level of error. For example, to increase accuracy by a factor of 10, that is to get one more significant digit of the result, one needs to increase the number of samples, and thus the computational cost of the method, by a factor of 100.

An alternative way to write the results from this section is using the so-called 'big $\mathcal{O}$ notation', as described in the following definition.

Definition 3.16   Given two function $f \colon \mathbb{N} \to \mathbb{R}$ and $g \colon \mathbb{N} \to \mathbb{R}$ we say that f is of *order* $\mathcal{O}(g)$, written symbolically as $f(N) = \mathcal{O}(g(N))$, as $N \to \infty$ if there are constants $N_0 \in \mathbb{N}$ and $c > 0$ such that

$$|f(N)| \leq c\,|g(N)|$$

for all $N \geq N_0$.

Similar definitions exist for other limits, for example for $f(\delta) = \mathcal{O}(g(\delta))$ as $\delta \downarrow 0$. Using this notation we can summarise the result of proposition 3.14 as

$$\text{MSE}\left(Z_N^{\text{MC}}\right) = \mathcal{O}\left(\frac{1}{N}\right)$$

and for the root-mean-square error we get

$$\text{RMSE}\left(Z_N^{\text{MC}}\right) = \mathcal{O}\left(\frac{1}{\sqrt{N}}\right).$$

3.2.3   Choice of sample size

The error bound from proposition 3.14 can be used to guide the choice of sample size N in the Monte Carlo method from algorithm 3.8. The most direct way to use the proposition in this context is to use equation (3.8) to determine the error of the result after a run of algorithm 3.8 has completed. If the error is too large, another run with a larger value of N can be started, until the required precision is reached. The bound of equation (3.8) can only be applied directly, if the sample variance $\text{Var}(f(X))$ is known. If the sample variance is unknown, it can be estimated together during the computation of the Monte Carlo estimate: The estimate Z_N^{MC} has mean squared error

$$\text{MSE}\left(Z_N^{\text{MC}}\right) = \frac{\text{Var}(f(X))}{N} \approx \frac{\hat{\sigma}^2}{N},$$

where

$$\hat{\sigma}^2 = \frac{1}{N-1} \sum_{j=1}^{N} \left(f(X_j) - Z_N^{\text{MC}}\right)^2 \tag{3.9}$$

is the sample variance of the generated values $f(X_1), \ldots, f(X_N)$.

An alternative way to use proposition 3.14 is to determine the required sample size N for a run of algorithm 3.8 in advance: by solving equation (3.8) for N we see that, in order to achieve error $\text{MSE}(Z_N^{\text{MC}}) \leq \varepsilon^2$, the sample size N must satisfy

$$N \geq \frac{\text{Var}(f(X))}{\varepsilon^2}. \tag{3.10}$$

If the value of $\text{Var}(f(X))$ is known, equation (3.10) can be directly used to find an appropriate sample size N for algorithm 3.8. Normally, the variance $\text{Var}(f(X))$ in a Monte Carlo estimate is not explicitly known, but there are various ways to work around this problem. For example, if an upper bound for $\text{Var}(f(X))$ is known, this bound can be used in place of the true variance.

Example 3.17   Assume $\mathrm{Var}(f(X)) = 1$. In order to estimate $\mathbb{E}(f(X))$ so that the error satisfies $\mathrm{MSE}(Z_N^{\mathrm{MC}}) = \varepsilon^2$ for $\varepsilon = 0.01$, we can use a Monte Carlo estimate with

$$N \geq \frac{\mathrm{Var}(f(X))}{\varepsilon^2} = \frac{1}{(0.01)^2} = 10\,000$$

samples.

Example 3.18   Let X be a real-valued random variable and $A \subseteq \mathbb{R}$. As we have seen in equation (3.4), we can estimate $p = P(X \in A)$ by

$$Z_N^{\mathrm{MC}} = \frac{1}{N} \sum_{j=1}^{N} \mathbb{1}_A(X_j),$$

where the X_j are i.i.d. copies of X. The variance of the Monte Carlo samples is

$$\mathrm{Var}\left(\mathbb{1}_A(X)\right) = \mathbb{E}\left(\mathbb{1}_A(X)^2\right) - \mathbb{E}\left(\mathbb{1}_A(X)\right)^2 = p - p^2 = p(1 - p).$$

Thus, from equation (3.10) we know that we can achieve $\mathrm{MSE}(Z_N^{\mathrm{MC}}) \leq \varepsilon^2$ by choosing

$$N \geq \frac{p(1 - p)}{\varepsilon^2}. \tag{3.11}$$

This bound depends on the unknown probability p but, since $p(1 - p) \leq 1/4$ for all $p \in [0, 1]$, choosing

$$N \geq \frac{1}{4\varepsilon^2}$$

is always sufficient to reduce the mean squared error to ε^2.

Another approach to using the bound (3.10) in cases where the variance of the Monte Carlo samples is not known is the following: one can try a two-step procedure where first Monte Carlo integration with a fixed number N_0 of samples (say $N_0 = 10\,000$) is used to estimate $\mathrm{Var}(f(X))$. Then, in a second step, one can use estimates as above to determine the required value of N to achieve a mean squared error of less than ε^2. In this approach, new samples $X_1, \ldots, X_N$ should be generated for the final estimate instead of reusing the initial N_0 samples.

Finally, sequential methods can be used to control the error of a Monte Carlo estimate. In these methods, one generates samples X_j one-by-one and estimates the standard deviation of the generated $f(X_j)$ from time to time. The procedure is continued until the estimated standard deviation falls below a prescribed limit. This approach should be used with care, since a bias can be introduced by the fact that N now depends on the samples used in the estimate.

So far we have seen how the description of the mean squared error from proposition 3.14 can be used to determine the error of a given Monte Carlo method. Another application of the mean squared error is to compare different Monte Carlo estimates for the same quantity. The smaller the variance $\text{Var}(f(X))$ of the samples is, the better is the resulting Monte Carlo method. We will discuss this idea in detail in Section 3.3.

3.2.4   Refined error bounds

We will conclude this section by showing how the central limit theorem can be used to obtain refined bounds for the Monte Carlo error.

Lemma 3.19   Let $\alpha \in (0, 1)$ and $q_\alpha = \Phi^{-1}(1 - \alpha/2)$ where Φ is the CDF of the standard normal distribution $\mathcal{N}(0, 1)$. Furthermore let $\sigma^2 = \text{Var}(f(X))$ and

$$N \geq \frac{q_\alpha^2 \sigma^2}{\varepsilon^2}.$$

Then, approximately for large N, the Monte Carlo estimate Z_N^{MC} for $\mathbb{E}(f(X))$, given by equation (3.7), satisfies

$$P\left(\left|Z_N^{\text{MC}} - \mathbb{E}(f(X))\right| \leq \varepsilon\right) \geq 1 - \alpha.$$

Proof   As an abbreviation, write

$$e_N^{\text{MC}} = Z_N^{\text{MC}} - \mathbb{E}(f(X)) = \frac{1}{N} \sum_{j=1}^{N} f(X_j) - \mathbb{E}(f(X)).$$

Using this notation, the results of proposition 3.14 state that $\mathbb{E}(e_N^{\text{MC}}) = 0$ and $\text{Var}(e_N^{\text{MC}}) = \sigma^2/N$. By the central limit theorem (see theorem A.9), for large values of N we find, approximately,

$$\frac{\sqrt{N}}{\sigma} e_N^{\text{MC}} \sim \mathcal{N}(0, 1)$$

and consequently

$$P\left(|e_N^{\text{MC}}| \leq \varepsilon\right) = P\left(\frac{|e^{\text{MC}}|}{\sigma/\sqrt{N}} \leq \frac{\varepsilon\sqrt{N}}{\sigma}\right)$$

$$\approx \Phi\left(\frac{\varepsilon\sqrt{N}}{\sigma}\right) - \Phi\left(-\frac{\varepsilon\sqrt{N}}{\sigma}\right) = 2\Phi\left(\frac{\varepsilon\sqrt{N}}{\sigma}\right) - 1.$$

Thus we have

$$P\left(|e_N^{\text{MC}}| \leq \varepsilon\right) \geq 1 - \alpha$$

if and only if

$$2\Phi\left(\frac{\varepsilon\sqrt{N}}{\sigma}\right) - 1 \geq 1 - \alpha.$$

The latter inequality is equivalent to

$$\frac{\varepsilon\sqrt{N}}{\sigma} \geq \Phi^{-1}\left(1 - \frac{\alpha}{2}\right) = q_\alpha$$

and solving this inequality for N gives the required lower bound on N.   □

As a special case of lemma 3.19, for $\alpha = 5\%$ we have $q_{0.05} = \Phi^{-1}(0.975) \approx 1.96$ and thus we need

$$N \geq \frac{1.96^2\sigma^2}{\varepsilon^2}$$

samples in order to have an absolute error of at most ε with 95% probability. Comparing this inequality with (3.10), we see that approximately 4 times as many samples are required as for the condition $\mathrm{MSE}(Z_N^{\mathrm{MC}}) \leq \varepsilon^2$.

Example 3.20   Assume $\mathrm{Var}(f(X)) = 1$. In order to estimate $\mathbb{E}(f(X))$ so that the error $|Z_N^{\mathrm{MC}} - E(f(X))|$ is at most $\varepsilon = 0.01$ with probability at least $1 - \alpha = 95\%$, we can use a Monte Carlo estimate with

$$N \geq \frac{1.96^2\mathrm{Var}(f(X))}{\varepsilon^2} = \frac{1.96^2}{(0.01)^2} = 38\,416$$

samples.

An alternative way to express the result of lemma 3.19 is to replace the point estimator Z_N^{MC} by a confidence interval for $\mathbb{E}(f(X))$: we find that Z_N^{MC} (approximately) satisfies

$$P\left(\mathbb{E}(f(X)) \in \left[Z_N^{\mathrm{MC}} - \frac{\sigma q_\alpha}{\sqrt{N}}, Z_N^{\mathrm{MC}} + \frac{\sigma q_\alpha}{\sqrt{N}}\right]\right) \geq 1 - \alpha \qquad (3.12)$$

for sufficiently large N, where q_α and σ^2 are as in lemma 3.19.

If the standard deviation σ in (3.12) is unknown, it can be replaced by the estimate $\hat{\sigma}$ from equation (3.9). A standard result from statistics shows that in this case the value q_α should be replaced by the corresponding quantile of Student's t-distribution with $N - 1$ degrees of freedom. These quantiles converge to q_α as $N \to \infty$ and for the values of N used in a Monte Carlo estimate, q_α can be used without problems.

In all the error estimates of this section, we used the fact that $f(X)$ has finite variance. As $\mathrm{Var}(f(X))$ gets bigger, convergence to the correct result gets slower and slower, and the required number of samples to obtain a given error increases to infinity. Since the strong law of large numbers does not require the random variables to have finite variance, equation (3.2) still holds for $\mathrm{Var}(f(X)) = \infty$, but in this case the convergence in (3.2) will be extremely slow and the resulting method will not be useful in practice.

3.3   Variance reduction methods

As we have seen, the efficiency of Monte Carlo estimation is determined by the variance of the estimate: the higher the variance, the more samples required to obtain a given accuracy. This chapter describes methods to improve efficiency by considering modified Monte Carlo methods. Compared with the basic Monte Carlo method from Section 3.2, the improved methods considered here produce estimates with a lower variance and thus with smaller error.

3.3.1   Importance sampling

The importance sampling method is based on the following argument. Assume that X is a random variable with density φ, that f is a real-valued function and that ψ is another probability density with $\psi(x) > 0$ whenever $f(x)\varphi(x) > 0$. Then we have

$$\mathbb{E}(f(X)) = \int f(x)\,\varphi(x)\,dx = \int f(x)\frac{\varphi(x)}{\psi(x)}\,\psi(x)\,dx,$$

where we define the fraction to be 0 whenever the denominator (and thus the numerator) equals 0. Since ψ is a probability density, the integral on the right can be written as an expectation again: if Y has density ψ, we have

$$\mathbb{E}(f(X)) = \mathbb{E}\left(f(Y)\frac{\varphi(Y)}{\psi(Y)} \right). \tag{3.13}$$

Now we can use a basic Monte Carlo estimate for the expectation on the right-hand side to get the following estimate.

Definition 3.21   Let X be a random variable with density φ and let f be a function such that $f(X) \in \mathbb{R}$. Furthermore, let ψ be another probability density, on the same space as φ. Then the *importance sampling estimate* for $\mathbb{E}(f(X))$ is given by

$$Z_N^{\mathrm{IS}} = \frac{1}{N}\sum_{j=1}^{N} f(Y_j)\frac{\varphi(Y_j)}{\psi(Y_j)} \tag{3.14}$$

where the Y_j are i.i.d. with density ψ.

The estimator Z_N^{IS} can be used as an alternative to the basic Monte Carlo estimator Z_N^{MC} from (3.7). Instead of the arithmetic average used in the basic Monte Carlo method, the importance sampling method uses a weighted average where each sample Y_j is assigned the weight $\varphi(Y_j)/\psi(Y_j)$. We can write the resulting estimation method as the following algorithm.

Algorithm 3.22   (importance sampling)
    input:
        a function f
        the density φ of X
        an auxiliary density ψ
        $N \in \mathbb{N}$
    randomness used:
        an i.i.d. sequence $(Y_j)_{j \in \mathbb{N}}$ with density ψ
    output:
        an estimate Z_N^{IS} for $\mathbb{E}(f(X))$
  1: $s \leftarrow 0$
  2: **for** $j = 1, 2, \ldots, N$ **do**
  3:    generate $Y_j \sim \psi$
  4:    $s \leftarrow s + f(Y_j)\varphi(Y_j)/\psi(Y_j)$
  5: **end for**
  6: return s/N

This method is a generalisation of the basic Monte Carlo method: if we choose $\psi = \varphi$, the two densities in (3.14) cancel and the Y_j are i.i.d. copies of X; for this case, the method is identical to basic Monte Carlo estimation. The usefulness of importance sampling lies in the fact that we can choose the density ψ (and thus the distribution of the Y_j) in order to maximise efficiency.

As for the basic Monte Carlo method, the sample size N can be used to control the balance between error and computational cost. As N increases, the computational cost increases but the error of the method decreases.

Proposition 3.23   The importance sampling estimate Z_N^{IS} from (3.14), as computed by algorithm 3.22, has

$$\text{bias}\left(Z_N^{IS}\right) = 0$$

and

$$\text{MSE}\left(Z_N^{IS}\right) = \frac{1}{N} \text{Var}\left(f(Y)\frac{\varphi(Y)}{\psi(Y)}\right)$$
$$= \frac{1}{N}\left(\text{Var}(f(X)) - \mathbb{E}\left(f(X)^2\left(1 - \frac{\varphi(X)}{\psi(X)}\right)\right)\right). \qquad (3.15)$$

Proof   Using the definition of Z_N^{IS} and equation (3.13) we find

$$\mathbb{E}\left(Z_N^{\text{IS}}\right) = \mathbb{E}\left(\frac{1}{N}\sum_{j=1}^{N} f(Y_j)\frac{\varphi(Y_j)}{\psi(Y_j)}\right) = \mathbb{E}\left(f(Y)\frac{\varphi(Y)}{\psi(Y)}\right) = \mathbb{E}(f(X))$$

and thus

$$\text{bias}\left(Z_N^{\text{MC}}\right) = \mathbb{E}\left(Z_N^{\text{IS}}\right) - \mathbb{E}(f(X)) = 0.$$

For the variance we have

$$\text{Var}\left(Z_N^{\text{IS}}\right) = \text{Var}\left(\frac{1}{N}\sum_{j=1}^{N} f(Y_j)\frac{\varphi(Y_j)}{\psi(Y_j)}\right) = \frac{1}{N}\text{Var}\left(f(Y)\frac{\varphi(Y)}{\psi(Y)}\right). \qquad (3.16)$$

Rewriting the right-hand side of this equation, we get

$$\begin{aligned}
\text{Var}\left(\frac{f(Y)\varphi(Y)}{\psi(Y)}\right) &= \mathbb{E}\left(\frac{f(Y)^2\varphi(Y)^2}{\psi(Y)^2}\right) - \mathbb{E}\left(\frac{f(Y)\varphi(Y)}{\psi(Y)}\right)^2 \\
&= \int \frac{f(y)^2\varphi(y)^2}{\psi(y)^2}\psi(y)\,dy - \mathbb{E}(f(X))^2 \\
&= \int \frac{f(x)^2\varphi(x)}{\psi(x)}\varphi(x)\,dx - \mathbb{E}(f(X))^2,
\end{aligned}$$

where the integration variable is changed from y to x in the last line. Transforming the integral back to an expectation, we get

$$\begin{aligned}
\text{Var}\left(\frac{f(Y)\varphi(Y)}{\psi(Y)}\right) &= \mathbb{E}\left(f^2(X)\frac{\varphi(X)}{\psi(X)}\right) - \mathbb{E}(f(X))^2 \\
&= \mathbb{E}\left(f(X)^2\right) - \mathbb{E}(f(X))^2 - \mathbb{E}\left(f(X)^2\right) + \mathbb{E}\left(f^2(X)\frac{\varphi(X)}{\psi(X)}\right) \\
&= \text{Var}(f(X)) - \mathbb{E}\left(f(X)^2\left(1 - \frac{\varphi(X)}{\psi(X)}\right)\right),
\end{aligned}$$

and thus, substituting this expression into (3.16), we get

$$\begin{aligned}
\text{MSE}\left(Z_N^{\text{IS}}\right) &= \text{Var}\left(Z_N^{\text{IS}}\right) + \text{bias}\left(Z_N^{\text{IS}}\right)^2 \\
&= \frac{1}{N}\text{Var}\left(f(Y)\frac{\varphi(Y)}{\psi(Y)}\right) + 0^2 \\
&= \frac{1}{N}\left(\text{Var}(f(X)) - \mathbb{E}\left(f(X)^2\left(1 - \frac{\varphi(X)}{\psi(X)}\right)\right)\right).
\end{aligned}$$

This completes the proof.   □

By comparing the mean squared error (3.15) for the importance sampling estimate to the mean squared error (3.8) of the basic Monte Carlo estimate, we see that the importance sampling method has smaller error than the basic Monte Carlo method whenever the density ψ satisfies the condition

$$c_\psi = \mathbb{E}\left(f(X)^2\left(1 - \frac{\varphi(X)}{\psi(X)}\right)\right) > 0.$$

The importance sampling method is efficient, if both of the following criteria are satisfied:

(a) The samples Y_j can be generated efficiently.

(b) $\mathrm{Var}(f(Y)\varphi(Y)/\psi(Y))$ is small or, equivalently, the constant c_ψ is large.

To understand how ψ and thus the distribution of Y can be chosen to minimise the variance in the second criterion, we first consider the extreme case where $f\varphi/\psi$ is constant: for this choice of ψ, the variance of the Monte Carlo estimate is 0 and therefore there is no error at all! In this case we have

$$\psi(x) = \frac{1}{a}f(x)\varphi(x) \tag{3.17}$$

for all x where a is the constant value of $f\varphi/\psi$. We can find the value of a by using the fact that ψ is a probability density:

$$a = a\int \psi(x)\,dx = \int f(x)\,\varphi(x)\,dx = \mathbb{E}(f(X)).$$

Therefore, in order to choose ψ as in (3.17) we have to already have solved the problem of computing the expectation $\mathbb{E}(f(X))$ and thus we do not get a useful method for this case. Still, this boundary case offers some guidance: since we get optimal efficiency if ψ is chosen proportional to $f\varphi$, we can expect, at least for $f \geq 0$, that we will get good efficiency if we choose ψ to be approximately proportional to the function $f\varphi$ or even if we choose a distribution such that ψ is big wherever $|f|$ is big.

Example 3.24   Let X be a real-valued random variable and $A \subseteq \mathbb{R}$. Then the importance sampling estimate for the probability $P(X \in A) = \mathbb{E}(\mathbb{1}_A(X))$ is given by

$$Z_N^{\mathrm{IS}} = \frac{1}{N}\sum_{j=1}^{N}\frac{\mathbb{1}_A(Y_j)\varphi(Y_j)}{\psi(Y_j)}$$

where ψ is a probability density, satisfying $\psi(x) > 0$ for all points $x \in A$ with $\varphi(x) > 0$, and $(Y_j)_{j\in\mathbb{N}}$ is a sequence of i.i.d. random variables with density ψ. Then, by proposition 3.23, the mean squared error of the importance sampling estimate for $P(X \in A)$ is given by

$$
\begin{aligned}
\mathrm{MSE}\left(Z_N^{\mathrm{IS}}\right) &= \frac{1}{N}\mathrm{Var}\left(\mathbb{1}_A(X)\right) - \frac{1}{N}\mathbb{E}\left(\mathbb{1}_A(X)\left(1 - \frac{\varphi(X)}{\psi(X)}\right)\right) \\
&= \mathrm{MSE}\left(Z_N^{\mathrm{MC}}\right) - \frac{1}{N}\mathbb{E}\left(\mathbb{1}_A(X)\left(1 - \frac{\varphi(X)}{\psi(X)}\right)\right).
\end{aligned}
$$

We see that the importance sampling method will have a smaller mean squared error than basic Monte Carlo estimation, if we can choose the density ψ such that $\psi > \varphi$ on the set A.

For the choice of the sample size N in algorithm 3.22, the same considerations apply as for basic Monte Carlo estimation: an estimate for the mean squared error $\mathrm{MSE}\left(Z_N^{\mathrm{IS}}\right)$ can be computed together with Z_N^{IS} itself, by using the following relation from proposition 3.23. The estimate Z_N^{IS} has mean squared error

$$
\mathrm{MSE}\left(Z_N^{\mathrm{IS}}\right) = \frac{1}{N}\mathrm{Var}\left(\frac{f(Y)\varphi(Y)}{\psi(Y)}\right) \approx \frac{\hat{\sigma}^2}{N}, \tag{3.18}
$$

where

$$
\hat{\sigma}^2 = \frac{1}{N-1}\sum_{j=1}^{N}\left(\frac{f(Y_j)\varphi(Y_j)}{\psi(Y_j)} - Z_N^{\mathrm{IS}}\right)^2 \tag{3.19}
$$

is the sample variance of the weighted samples generated to compute Z_N^{IS}. As for the basic Monte Carlo method, the relation (3.18) can also be solved for N to determine the value of N required to achieve a given level of error.

3.3.2 Antithetic variables

The *antithetic variables* method (also called *antithetic variates* method) reduces the variance and thus the error of Monte Carlo estimates by using pairwise dependent samples X_j instead of the independent samples used in basic Monte Carlo estimation.

For illustration, we first consider the case $N = 2$: assume that X and X' are identically distributed random variables, which are *not* independent. As for the independent case we have

$$
\mathbb{E}\left(\frac{f(X) + f(X')}{2}\right) = \frac{\mathbb{E}(f(X)) + \mathbb{E}\left(f(X')\right)}{2} = \mathbb{E}(f(X)),
$$

but for the variance we get

$$
\begin{aligned}
\mathrm{Var}\left(\frac{f(X) + f(X')}{2}\right) &= \frac{\mathrm{Var}(f(X)) + 2\,\mathrm{Cov}\left(f(X), f(X')\right) + \mathrm{Var}\left(f(X')\right)}{4} \\
&= \frac{1}{2}\mathrm{Var}(f(X)) + \frac{1}{2}\mathrm{Cov}\left(f(X), f(X')\right).
\end{aligned}
$$

Compared with the expression for the variance in the independent case, an additional covariance term $\frac{1}{2}\text{Cov}(f(X), f(X'))$ is present. The idea of the antithetic variables method is to construct X and X' such that $\text{Cov}(f(X), f(X'))$ is negative, thereby reducing the total variance.

There are many methods available to construct pairs of samples X, X' which have the required properties for use in the antithetic variables method (one possible construction is given below). Once we have found a way to construct such pairs of a given distribution, we can use the following approximation.

Definition 3.25   Let X be a random variable and let f be a function such that $f(X) \in \mathbb{R}$. Furthermore, let X' be another random variable, with the same distribution as X. Then the *antithetic variables estimate* for $\mathbb{E}(f(X))$ with sample size $N \in 2\mathbb{N}$ is given by

$$Z_N^{\text{AV}} = \frac{1}{N} \sum_{k=1}^{N/2} \left(f(X_k) + f(X_k') \right) \tag{3.20}$$

where (X_k, X_k') are i.i.d. copies of (X, X').

In the definition, $2\mathbb{N}$ denotes the set of even integers. Since each term in the sum contributes two terms, $f(X_k)$ and $f(X_k')$, to the estimate Z_N^{AV}, the sum in (3.20) ranges only from 1 to $N/2$. When applying the algorithm, X and X' will be dependent. For fixed k, the pair (X_k, X_k') has the same dependence structure as (X, X'), but X_k and X_k' are independent of X_j and X_j' when $j \neq k$. The pairs (X, X') and (X_k, X_k') are called *antithetic pairs*. We can write the resulting method as the following algorithm.

Algorithm 3.26   (antithetic variables)
　　input:
　　　　a function f
　　　　$N \in \mathbb{N}$ even
　　randomness used:
　　　　i.i.d. copies (X_k, X_k') of (X, X')
　　output:
　　　　the estimate Z_N^{AV} for $\mathbb{E}(f(X))$
　　1: $s \leftarrow 0$
　　2: **for** $k = 1, 2, \ldots, N/2$ **do**
　　3:　　generate (X_k, X_k')
　　4:　　$s \leftarrow s + f(X_k) + f(X_k')$
　　5: **end for**
　　6: return s/N

As for the Monte Carlo methods discussed so far, the running time of algorithm 3.26 is proportional to N. The loop in line 2 of the algorithm has only $N/2$ iterations, compared with N iterations for the basic Monte Carlo method, but in each iteration two terms are added to the cumulative sum s and thus the final result is still the sum of N terms. For this reason, the computational cost of the antithetic variables method is usually very similar to the computational cost of the corresponding basic Monte Carlo algorithm.

Proposition 3.27   Let X and X' be two random variables with the same distribution and let $\rho = \mathrm{Corr}(f(X), f(X'))$. Then the antithetic variables estimate Z_N^{AV} for $\mathbb{E}(f(X))$ satisfies

$$\mathrm{bias}\left(Z_N^{AV}\right) = 0$$

and

$$\mathrm{MSE}(Z_N^{AV}) = \frac{1}{N}\mathrm{Var}(f(X))(1+\rho). \tag{3.21}$$

Proof   The expectation of Z_N^{AV} is given by

$$\mathbb{E}\left(Z_N^{AV}\right) = \frac{1}{N}\sum_{k=1}^{N/2}\mathbb{E}\left(f(X_k) + f(X_k')\right) = \frac{1}{N}\cdot\frac{N}{2}\cdot 2\mathbb{E}(f(X)) = \mathbb{E}(f(X)).$$

Thus the estimate Z_N^{AV} is unbiased and the mean squared error of Z_N^{AV} coincides with the variance.

For the variance we can use the fact that the pairs (X_j, X_j') and (X_k, X_k') are independent for $j \neq k$. This gives

$$\mathrm{Var}\left(Z_N^{AV}\right) = \frac{1}{N^2}\sum_{j,k=1}^{N/2}\mathrm{Cov}\left(\left(f(X_j) + f(X_j')\right), \left(f(X_k) + f(X_k')\right)\right)$$

$$= \frac{1}{N^2}\sum_{k=1}^{N/2}\left(\mathrm{Var}\left(f(X_k)\right) + \mathrm{Var}\left(f(X_k')\right) + 2\,\mathrm{Cov}\left(f(X_k), f(X_k')\right)\right)$$

$$= \frac{1}{N^2}\cdot\frac{N}{2}\cdot\left(\mathrm{Var}(f(X)) + \mathrm{Var}\left(f(X')\right) + 2\,\mathrm{Cov}\left(f(X), f(X')\right)\right)$$

$$= \frac{1}{N}\left(\mathrm{Var}(f(X)) + \mathrm{Cov}\left(f(X), f(X')\right)\right).$$

Finally, since

$$\rho = \mathrm{Corr}\left(f(X), f(X')\right) = \frac{\mathrm{Cov}\left(f(X), f(X')\right)}{\sqrt{\mathrm{Var}(f(X))\mathrm{Var}\left(f(X')\right)}} = \frac{\mathrm{Cov}\left(f(X), f(X')\right)}{\mathrm{Var}(f(X))},$$

we find

$$\mathrm{MSE}\left(e_N^{AV}\right) = \mathrm{Var}\left(e_N^{AV}\right) = \frac{1}{N}\left(\mathrm{Var}(f(X)) + \mathrm{Var}(f(X))\rho\right).$$

This completes the proof. □

Comparing equation (3.8) and equation (3.21) for the mean squared error of the basic Monte Carlo estimate and of the antithetic variables estimate, respectively, we see that the antithetic variables method has smaller mean squared error than the basic Monte Carlo method if and only if $\rho < 0$.

In order to apply the antithetic variables method, we need to construct the pairs (X, X') of samples such that both values have the correct distribution but, at the same time, $f(X)$ and $f(X')$ are negatively correlated. There is no generic method to construct such antithetic pairs; here will restrict ourselves to discussing ideas which can help to construct antithetic pairs in specific cases.

A first idea for constructing antithetic pairs, if the distribution of X is symmetric, is to use $X' = -X$. The correlation $\mathrm{Corr}(f(X), f(X'))$ depends on the function f, but in many cases we have $\mathrm{Corr}(f(X), f(-X)) < 0$. This idea is illustrated in the following example and, in a much more complex situation, in Section 6.5.2.

Example 3.28 Let $X \sim \mathcal{N}(0, 1)$ and consider the problem of estimating the probability $p = P(X \in [1, 3]) = \mathbb{E}(\mathbb{1}_{[1,3]}(X))$. Since the distribution of X is symmetric, we can try to use (X, X') with $X' = -X$ as an antithetic pair. For this choice we find

$$
\begin{aligned}
\mathrm{Cov}\left(\mathbb{1}_{[1,3]}(X), \mathbb{1}_{[1,3]}(-X)\right) \\
= \mathbb{E}\left(\mathbb{1}_{[1,3]}(X)\mathbb{1}_{[1,3]}(-X)\right) - \mathbb{E}\left(\mathbb{1}_{[1,3]}(X)\right) \cdot \mathbb{E}\left(\mathbb{1}_{[1,3]}(-X)\right) \\
= 0 - p \cdot p \\
= -p^2,
\end{aligned}
$$

since the two values $\mathbb{1}_{[1,3]}(X)$ and $\mathbb{1}_{[1,3]}(-X)$ cannot be non-zero simultaneously. As in example 3.18 we find

$$
\mathrm{Var}\left(\mathbb{1}_{[1,3]}(X)\right) = \mathrm{Var}\left(\mathbb{1}_{[1,3]}(-X)\right) = p(1 - p)
$$

and thus

$$
\begin{aligned}
\rho &= \mathrm{Corr}\left(\mathbb{1}_{[1,3]}(X), \mathbb{1}_{[1,3]}(X')\right) \\
&= \frac{\mathrm{Cov}\left(\mathbb{1}_{[1,3]}(X), \mathbb{1}_{[1,3]}(X')\right)}{\sqrt{\mathrm{Var}\left(\mathbb{1}_{[1,3]}(X)\right)\mathrm{Var}\left(\mathbb{1}_{[1,3]}(X')\right)}} \\
&= -\frac{p}{1 - p}.
\end{aligned}
$$

Thus, by the result of proposition 3.27, the mean squared error of the antithetic variables estimate Z_N^{AV} for this example, compared with the error for the basic Monte Carlo method, is

$$\text{MSE}\left(Z_N^{AV}\right) = (1 + \rho)\text{MSE}\left(Z_N^{MC}\right) = \left(1 - \frac{p}{1 - p}\right)\text{MSE}\left(Z_N^{MC}\right).$$

By computing an estimate for p we find $p \approx 0.16$ and $\rho = -p/(1 - p) \approx -0.19$. Thus, in this example the mean squared error of Z_N^{AV} is only about 81% of the error for Z_N^{MC}.

Another method for generating antithetic pairs can be applied if X is one-dimensional and if we can generate samples of X using the inverse transform method from Section 1.3: let F be the distribution function of X and let $U \sim \mathcal{U}[0, 1]$. Then $1 - U \sim \mathcal{U}[0, 1]$ and we can use $X = F^{-1}(U)$ and $X' = F^{-1}(1 - U)$ to generate an antithetic pair. Since the inverse F^{-1} of the distribution function is monotonically increasing, X increases as U increases while X' decreases as U increases. For this reason we expect X and X' to be negatively correlated.

Lemma 3.29   Let $g: \mathbb{R} \to \mathbb{R}$ be monotonically increasing and $U \sim \mathcal{U}[0, 1]$. Then

$$\text{Cov}\left(g(U), g(1 - U)\right) \leq 0.$$

Proof   The proof uses a trick, which is based on the idea of introducing a new random variable $V \sim \mathcal{U}[0, 1]$, independent of U. We distinguish two cases: if $U \leq V$ we have $g(U) \leq g(V)$ and $g(1 - U) \geq g(1 - V)$. Otherwise, if $U > V$, we have $g(U) \geq g(V)$ and $g(1 - U) \leq g(1 - V)$. Thus, in both cases we have

$$(g(U) - g(V))(g(1 - U) - g(1 - V)) \leq 0$$

and consequently

$$
\begin{aligned}
\text{Cov} & \left(g(U), g(1 - U)\right) \\
&= \mathbb{E}\left(g(U)g(1 - U)\right) - \mathbb{E}\left(g(U)\right)\mathbb{E}\left(g(1 - U)\right) \\
&= \frac{1}{2}\mathbb{E}\left(g(U)g(1 - U) + g(V)g(1 - V) - g(U)g(1 - V) - g(V)g(1 - U)\right) \\
&= \frac{1}{2}\mathbb{E}\left((g(U) - g(V))(g(1 - U) - g(1 - V))\right) \leq 0.
\end{aligned}
$$

This completes the proof. $\square$

If the function f is monotonically increasing, we can apply the lemma to $g(u) = f(F^{-1}(u))$. In this case we have

$$f(X) = f(F^{-1}(U)) = g(U)$$

and

$$f(X') = f\left(F^{-1}(1-U)\right) = g(1-U).$$

The lemma allows then to conclude $\text{Cov}(f(X_1), f(X_2)) \leq 0$. While lemma 3.29 only guarantees that the variance for the antithetic variables method is smaller or equal to the variance of standard Monte Carlo estimation, in practice often a significant reduction of variance is observed.

3.3.3   Control variates

The control variates method is another method to reduce the variance of Monte Carlo estimates for expectations of the form $\mathbb{E}(f(X))$. The method is based on the following idea: if we can find a 'simpler' function $g \approx f$ such that $\mathbb{E}(g(X))$ can be computed analytically, then we can use our knowledge of $\mathbb{E}(g(X))$ to assist with the estimation of $\mathbb{E}(f(X))$. In the control variates methods, this is done by rewriting the expectation of interest as

$$\mathbb{E}(f(X)) = \mathbb{E}\left(f(X) - g(X)\right) + \mathbb{E}(g(X)).$$

Since we know $\mathbb{E}(g(X))$, the Monte Carlo estimation can now be restricted to the term $\mathbb{E}(f(X) - g(X))$ and since $f(X) \approx g(X)$, the random quantity $f(X) - g(X)$ has smaller variance and thus smaller Monte Carlo error than $f(X)$ has on its own. In this context, the random variable $g(X)$ is called a *control variate* for $f(X)$.

Definition 3.30   Let g be a function such that $\mathbb{E}(g(X))$ is known. Then the *control variates estimate* for $\mathbb{E}(f(X))$ is given by

$$Z_N^{CV} = \frac{1}{N} \sum_{j=1}^{N} \left(f(X_j) - g(X_j)\right) + \mathbb{E}(g(X))$$

where the X_j are i.i.d. copies of X.

The algorithm for computing the estimate Z_N^{CV} is a trivial modification of the basic Monte Carlo algorithm.

Algorithm 3.31   (control variates)
    input:
        a function f
        a function $g \approx f$ such that $\mathbb{E}(g(X))$ is known
        $N \in \mathbb{N}$
    randomness used:
        a sequence $(X_j)_{j \in \mathbb{N}}$ of i.i.d. copies of X

output:
    the estimate Z_N^{CV} for $\mathbb{E}(f(X))$
1: $s \leftarrow 0$
2: **for** $j = 1, 2, \ldots, N$ **do**
3:    generate X_j with the same distribution as X has
4:    $s \leftarrow s + f(X_j) - g(X_j)$
5: **end for**
6: return $s/N + \mathbb{E}(g(X))$

As for the other Monte Carlo algorithms, the sample size N controls the trade-off between error and computational cost. The computational cost of this algorithm increases with N, but at the same time the error decreases when N gets larger.

Proposition 3.32   The control variates estimate Z_N^{CV} for $\mathbb{E}(f(X))$ satisfies

$$\text{bias}\left(Z_N^{CV}\right) = 0$$

and

$$\text{MSE}\left(Z_N^{CV}\right) = \frac{1}{N}\text{Var}\left(f(X) - g(X)\right).$$

Proof   This result is a direct consequence of proposition 3.14. We have

$$\mathbb{E}\left(Z_N^{CV}\right) = \mathbb{E}\left(\frac{1}{N}\sum_{j=1}^{N}\left(f(X_j) - g(X_j)\right) + \mathbb{E}(g(X))\right)$$
$$= \mathbb{E}\left(f(X) - g(X)\right) + \mathbb{E}(g(X))$$
$$= \mathbb{E}(f(X))$$

and thus

$$\text{bias}\left(Z_N^{CV}\right) = \mathbb{E}\left(Z_N^{CV}\right) - \mathbb{E}(f(X)) = 0.$$

Using proposition 3.14 again, we find the mean squared error as

$$\text{MSE}\left(Z_N^{CV}\right) = \text{Var}\left(\frac{1}{N}\sum_{j=1}^{N}\left(f(X_j) - g(X_j)\right) + \mathbb{E}(g(X))\right)$$
$$= \text{Var}\left(\frac{1}{N}\sum_{j=1}^{N}\left(f(X_j) - g(X_j)\right)\right)$$
$$= \frac{1}{N}\text{Var}\left(f(X) - g(X)\right).$$

This completes the proof.   □

The control variates method described above is a special case of a more general method. Using a correlated control variate Y, every random variable Z can be transformed into a new random variable $\tilde{Z}$ with the same mean but smaller variance. This technique is described in the following lemma.

Lemma 3.33   Let Z be a random variable with $\mathbb{E}(Z) = \mu$ and $\mathrm{Var}(Z) = \sigma^2$. Furthermore, let Y be a random variable with $\mathrm{Corr}(Y, Z) = \rho$ and define

$$\tilde{Z} = Z - \frac{\mathrm{Cov}(Y, Z)}{\mathrm{Var}(Y)}\,(Y - \mathbb{E}(Y)).$$

Then the random variable $\tilde{Z}$ satisfies $\mathbb{E}(\tilde{Z}) = \mu$ and

$$\mathrm{Var}(\tilde{Z}) = \left(1 - \rho^2\right)\sigma^2 \leq \sigma^2.$$

Proof   For $c \in \mathbb{R}$ define

$$Z_c = Z - c\,(Y - \mathbb{E}(Y)).$$

Then the random variable Z_c has expectation

$$\mathbb{E}(Z_c) = \mathbb{E}(Z) - c\,(\mathbb{E}(Y) - \mathbb{E}(Y)) = \mathbb{E}(Z)$$

and the variance of Z_c is given by

$$\mathrm{Var}(Z_c) = \mathrm{Var}(Z) - 2c\,\mathrm{Cov}(Y, Z) + c^2\mathrm{Var}(Y). \tag{3.22}$$

The value $\tilde{Z}$ in the statement satisfies $\tilde{Z} = Z_c$ for $c = \mathrm{Cov}(Y, Z)/\mathrm{Var}(Y)$ and thus we get

$$\begin{aligned}
\mathrm{Var}(\tilde{Z}) &= \mathrm{Var}(Z) - 2\frac{\mathrm{Cov}(Y, Z)}{\mathrm{Var}(Y)}\mathrm{Cov}(Y, Z) + \frac{\mathrm{Cov}(Y, Z)^2}{\mathrm{Var}(Y)^2}\mathrm{Var}(Y) \\
&= \mathrm{Var}(Z) - \frac{\mathrm{Cov}(Y, Z)^2}{\mathrm{Var}(Y)} \\
&= \left(1 - \mathrm{Corr}(Y, Z)^2\right)\mathrm{Var}(Z).
\end{aligned}$$

This completes the proof.   □

Lemma 3.33 can, for example, be used to construct improved versions of an unbiased estimator: if Z is an unbiased estimator for a parameter θ, then

$$\mathrm{bias}(\tilde{Z}) = \mathbb{E}(\tilde{Z}) - \theta = \mathbb{E}(Z) - \theta = 0$$

and

$$\mathrm{MSE}(\tilde{Z}) = \mathrm{Var}(\tilde{Z}) < \mathrm{Var}(Z) = \mathrm{MSE}(Z),$$

that is $\tilde{Z}$ is then also an unbiased estimator for θ, but has smaller mean squared error than Z.

It is easy to check that the value of c used in (3.22) is optimal: the c which minimises the variance satisfies the condition

$$0 = \frac{d}{dc}\mathrm{Var}(Z_c) = -2\,\mathrm{Cov}(Y, Z) + 2c\,\mathrm{Var}(Y)$$

and thus the optimal value of c is given by

$$c^* = \frac{\mathrm{Cov}(Y, Z)}{\mathrm{Var}(Y)}.$$

In practice, the exact covariance between Z and Y is often unknown. From equation (3.22) we know that it suffices to have

$$2c\,\mathrm{Cov}(Y, Z) > c^2\,\mathrm{Var}(Y)$$

for variance reduction to occur. It is easy to check that this condition is satisfied for all values of c between 0 and $2\,\mathrm{Cov}(Z, Y)/\mathrm{Var}(Y)$. Therefore we can replace the estimator $\tilde{Z}$ from lemma 3.33 by

$$Z_c = Z - c\,(Y - \mathbb{E}(Y))$$

where c is only an approximation to $\mathrm{Cov}(Y, Z)/\mathrm{Var}(Y)$, and still expect the estimator Z_c to have smaller variance than Z.

3.4   Applications to statistical inference

In this section we will use a series of examples to illustrate how Monte Carlo methods can be used to study methods from statistical inference. We will consider point estimates in Section 3.4.1, confidence intervals in Section 3.4.2 and hypothesis tests in Section 3.4.3.

In statistical inference problems, we have observed data $x = (x_1, \ldots, x_n)$ and our aim is to decide which statistical model, typically chosen from a family of models under consideration, the observed data could have been generated by. More specifically, we consider a family $(P_\theta)_{\theta \in \Theta}$ of probability distributions, where θ is the parameter vector of the models and Θ is the set of all possible parameter values, and we assume that the observed data are a sample of a random variable $X = (X_1, \ldots, X_n) \sim P_\theta$, for an unknown parameter value $\theta \in \Theta$. In this context, the random variable X is called the *random sample*.

3.4.1   Point estimators

A *point estimator* (or an *estimator* in short) for the parameter θ is any function of the random sample X with values in Θ. Typically, we write $\hat{\theta} = \hat{\theta}(X) = \hat{\theta}(X_1, \ldots, X_n)$ to denote an estimator for a parameter θ. The value of the estimator for the observed data x, that is $\hat{\theta}(x)$ is called a *point estimate* (or an *estimate*) for θ.

While the definition of an estimator does not refer to the 'true' value θ, useful estimators will have the property that $\hat{\theta}$ is close to θ. How close the estimate is to the exact value determines the quality of an estimator. This is measured by quantities like the bias and the standard error. In simple cases, for example when the data consist of independent, normally distributed values or in the limit $n \to \infty$, it is possible to determine the bias and standard error of estimators analytically. In more complicated cases, exact expressions for the bias and the standard error are no longer available. Here we will illustrate how Monte Carlo estimation can be used in these cases, to obtain numerical approximations for the bias and the standard error of an estimator.

3.4.1.1   Monte Carlo estimates of the bias

From definition 3.9 we know that the bias of an estimator $\hat{\theta} = \hat{\theta}(X)$ for a parameter θ is given by

$$\text{bias}_\theta(\hat{\theta}) = \mathbb{E}\left(\hat{\theta}(X)\right) - \theta$$

for all $\theta \in \Theta$. For given values of θ, we can use basic Monte Carlo estimation to approximate the expection $\mathbb{E}(\hat{\theta}(X))$. The resulting Monte Carlo estimate for the bias is given by

$$\widehat{\text{bias}}_\theta(\hat{\theta}) = \frac{1}{N} \sum_{j=1}^{N} \hat{\theta}(X^{(j)}) - \theta,$$

where the samples $X^{(j)} = (X_1^{(j)}, \ldots, X_n^{(j)})$ are i.i.d. copies of X, using the given parameter value θ.

While this procedure is a simple application of Monte Carlo estimation as described in proposition 3.14, some care is needed to not confuse the two conceptual levels characterised by the size n of samples, and the number N of samples used in the Monte Carlo estimate: while $\hat{\theta}$ can be computed from a sample of size n, we need to generate N samples, that is $n \cdot N$ random values, to compute the estimate $\hat{\mu}$. Similarly, while $\hat{\theta}$ is an estimator for θ, the value $\widehat{\text{bias}}_\theta(\hat{\theta})$ is an *estimate for the bias of an estimator*.

Of course, the true value of the parameter θ is normally not known, but we can systematically compute $\widehat{\text{bias}}(\theta)$ for a range of different θ to get, for example, an approximate upper bound for the bias of an estimator. This approach is illustrated in the following example.

Example 3.34   Let $\rho \in [-1, 1]$ and $X, \eta \sim \mathcal{N}(0, 1)$ and define

$$Y = \rho X + \sqrt{1 - \rho^2}\, \eta.$$

Then

$$\begin{aligned}
\mathrm{Cov}(X, Y) &= \mathrm{Cov}(X, \rho X + \sqrt{1 - \rho^2}\, \eta) \\
&= \rho \mathrm{Cov}(X, X) + \sqrt{1 - \rho^2}\mathrm{Cov}(X, \eta) \\
&= \rho \mathrm{Var}(X) \\
&= \rho,
\end{aligned}$$

since X and η are independent. Using independence again, we find

$$\begin{aligned}
\mathrm{Var}(Y) &= \mathrm{Var}\left(\rho X + \sqrt{1 - \rho^2}\, \eta\right) \\
&= \rho^2 \mathrm{Var}(X) + (1 - \rho^2)\mathrm{Var}(\eta) \\
&= \rho^2 + 1 - \rho^2 \\
&= 1.
\end{aligned}$$

Consequently,

$$\mathrm{Corr}(X, Y) = \frac{\mathrm{Cov}(X, Y)}{\sqrt{\mathrm{Var}(X)\mathrm{Var}(Y)}} = \frac{\rho}{\sqrt{1 \cdot 1}} = \rho.$$

The correlation between any two random variables X and Y can be estimated by the *sample correlation*

$$\hat{\rho}(X, Y) = \frac{\sum_{i=1}^{n}(X_i - \bar{X})(Y_i - \bar{Y})}{\sqrt{\sum_{i=1}^{n}(X_i - \bar{X})^2 \sum_{i=1}^{n}(Y_i - \bar{Y})^2}} \tag{3.23}$$

where $\bar{X} = \frac{1}{n}\sum_{i=1}^{n} X_i$, $\bar{Y} = \frac{1}{n}\sum_{i=1}^{n} Y_i$ and (X_i, Y_i), $i = 1, 2, \ldots, n$ is a sequence of i.i.d. copies of (X, Y). Thus, we can use $\hat{\rho}(X, Y)$ as an estimator for ρ. Our aim in this example is to determine the bias of this estimator.

For given ρ, the estimator $\widehat{\mathrm{bias}}_\rho(\hat{\rho})$ for the bias of the estimator $\hat{\rho}$ from equation (3.23) can be computed using the following steps:

1: $s \leftarrow 0$
2: **for** $j = 1, 2, \ldots, N$ **do**
3:    generate $X_1^{(j)}, \ldots, X_n^{(j)} \sim \mathcal{N}(0, 1)$
4:    generate $\eta_1^{(j)}, \ldots, \eta_n^{(j)} \sim \mathcal{N}(0, 1)$
5:    let $Y_i^{(j)} = \rho X_i^{(j)} + \sqrt{1 - \rho^2}\, \eta_i^{(j)}$ for $i = 1, 2, \ldots, n$
6:    compute $\hat{\rho}^{(j)} = \hat{\rho}(X^{(j)}, Y^{(j)})$ using (3.23)

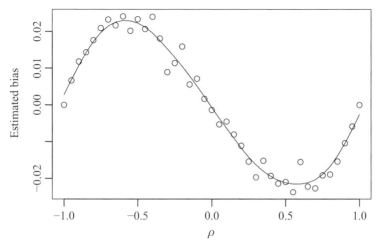

Figure 3.2   The estimated bias of the sample correlation (3.23) when estimating the correlation ρ. The circles denote Monte Carlo estimates for the bias, for different values of ρ, obtained using the method described in example 3.34. The solid line is obtained by fitting a smooth curve to the Monte Carlo estimates.

7:   $s \leftarrow s + \hat{\rho}^{(j)}$
8: **end for**
9: return $s/N - \rho$

Finally, this code can be run repeatedly for different values of $\rho \in [-1, +1]$ to get the dependence of the bias on the parameter ρ. The result of one run of this procedure is shown in Figure 3.2.

3.4.1.2   Monte Carlo estimates of the standard error

From definition 3.10 we know that the standard error of an estimator $\hat{\theta} = \hat{\theta}(X)$ is given by

$$\text{se}_\theta(\hat{\theta}) = \text{stdev}_\theta\left(\hat{\theta}(X)\right) = \sqrt{\text{Var}_\theta\left(\hat{\theta}(X)\right)}$$

for all $\theta \in \Theta$. While this expression does not exactly have the form $\mathbb{E}(f(x))$ considered for Monte Carlo estimates in this chapter, we can still us a Monte Carlo approach to estimating the standard error: for given θ, we can estimate $\text{se}_\theta(\hat{\theta})$ as

$$\widehat{\text{se}}_\theta(\hat{\theta}) = \sqrt{\frac{1}{N-1} \sum_{j=1}^{N} \left(\hat{\theta}(X^{(j)}) - \overline{\hat{\theta}^{(\cdot)}}\right)^2},$$

where

$$\overline{\hat{\theta}^{(\cdot)}} = \frac{1}{N} \sum_{j=1}^{N} \hat{\theta}(X^{(j)})$$

and $X^{(j)}$ for $j = 1, 2, \ldots, N$ are i.i.d. copies of X and in the formula for $\widehat{\mathrm{se}}_\theta(\hat{\theta})$ we use the standard estimate for the variance.

3.4.2   Confidence intervals

If the set Θ of all possible parameter values is one-dimensional, that is if $\Theta \subseteq \mathbb{R}$, we can draw inference about the unknown parameter θ using confidence intervals. Confidence intervals serve a similar purpose as point estimators but, instead of returning just one 'plausible' value of the parameter, they determine a range of possible parameter values, chosen large enough so that the true parameter value lies inside the range with high probability.

Definition 3.35   A *confidence interval* with confidence coefficient $1 - \alpha$ for a parameter θ is a random interval $[U, V] \subset \mathbb{R}$ where $U = U(X)$ and $V = V(X)$ are functions of the random sample $X = (X_1, \ldots, X_n)$, such that

$$P_\theta \left(\theta \in [U(X), V(X)] \right) \geq 1 - \alpha \tag{3.24}$$

for all $\theta \in \Theta$. The subscript θ on the probability P indicates that the random sample $X = (X_1, \ldots, X_n)$, for the purpose of computing the probability in (3.24), is distributed according to the distribution with parameter θ.

It is important to note that in equation (3.24) the interval $[U, V]$ is random, since it depends on the random sample X, but the value θ is not. The usefulness of confidence intervals lies in the fact that equation (3.24) holds for all possible values of θ simultaneously. Thus, even without knowing the true value of θ, we can be certain that the relation (3.24) holds and for given data x we can use $[U(x), V(x)]$ as an interval estimate for θ.

In many cases, a confidence interval for a parameter θ can be constructed by considering a point estimator $\hat{\theta}(X)$ for θ and then choosing the boundaries of the confidence interval as $U(X) = \hat{\theta}(X) - \varepsilon$ and $V(X) = \hat{\theta}(X) + \varepsilon$ for an appropriate value $\varepsilon > 0$. In this case, the condition (3.24) can be written as

$$P_\theta \left(\hat{\theta} - \theta \in [-\varepsilon, \varepsilon] \right) \geq 1 - \alpha \tag{3.25}$$

and, if the distribution of $\hat{\theta} - \theta$ is known, this relation can be used to choose the value ε. This approach is illustrated in the following example.

Example 3.36   Let $X_1, \ldots, X_N \sim \mathcal{N}(\mu, \sigma^2)$ be i.i.d. with a known variance σ^2 and assume that we want to construct the confidence interval for the unknown mean μ. As the centre of the confidence interval we choose the estimator

$$\hat{\mu} = \hat{\mu}(X) = \sum_{i=1}^{n} X_i / n$$

from example 3.13. As a sum of independent, normally distributed random variables, $\hat{\mu}$ is itself normally distributed, and computing the mean and variance as in example 3.13, we find $\hat{\mu} - \mu \sim \mathcal{N}(0, \sigma^2/n)$. Denoting the CDF of the standard normal distribution by Φ, we get

$$P_\mu \left(\hat{\theta} - \theta \in [-\varepsilon, \varepsilon] \right) = P_\mu \left(\frac{\hat{\theta} - \theta}{\sigma/\sqrt{n}} \in \left[-\frac{\varepsilon}{\sigma/\sqrt{n}}, \frac{\varepsilon}{\sigma/\sqrt{n}} \right] \right)$$

$$= \Phi \left(\frac{\varepsilon \sqrt{n}}{\sigma} \right) - \Phi \left(-\frac{\varepsilon \sqrt{n}}{\sigma} \right)$$

$$= 2\Phi \left(\frac{\varepsilon \sqrt{n}}{\sigma} \right) - 1.$$

In order for the condition (3.25) to be satisfied, we need to choose ε such that

$$2\Phi \left(\frac{\varepsilon \sqrt{n}}{\sigma} \right) - 1 \geq 1 - \alpha$$

or, equivalently,

$$\varepsilon \geq \frac{\sigma}{\sqrt{n}} \cdot \Phi^{-1} \left(1 - \frac{\alpha}{2} \right) = \frac{\sigma q_\alpha}{\sqrt{n}}$$

holds, where we use $q_\alpha = \Phi^{-1}(1 - \alpha/2)$. Using the smallest valid choice of ε, we find

$$I(X) = \left[\hat{\mu}(X) - \frac{q_\alpha \sigma}{\sqrt{n}}, \hat{\mu}(X) + \frac{q_\alpha \sigma}{\sqrt{n}} \right]$$

as a confidence interval for μ. Commonly used values for q_α are $q_{0.1} \approx 1.64$, $q_{0.05} \approx 1.96$ and $q_{0.01} \approx 2.58$, corresponding to confidence coefficients of 90%, 95% and 99%, respectively.

Example 3.37   If, in the situation of example 3.36, the variance σ is not known, the interval

$$I(X) = \left[\hat{\mu}(X) - \frac{p_{n,\alpha} \hat{\sigma}}{\sqrt{n}}, \hat{\mu}(X) + \frac{p_{n,\alpha} \hat{\sigma}}{\sqrt{n}} \right], \tag{3.26}$$

where $\hat{\sigma}$ is given by

$$\hat{\sigma}^2 = \frac{1}{n-1} \sum_{i=1}^{n} (X_i - \hat{\mu}(X))^2,$$

can be used as a confidence interval for the mean. A standard result from statistics shows that $p_{n,\alpha}$ can be chosen as the $(1 - \alpha/2)$-quantile of Student's t-distribution with $n - 1$ degrees of freedom, in order for this confidence interval to be exact.

If the X_i are not normally distributed, theoretical analysis becomes difficult and, in particular for small n, often only approximate confidence intervals can be derived. Monte Carlo estimates can be used, both to assist with the construction of confidence intervals and to assess the confidence coefficient of a given confidence interval. The underlying idea for both of these approaches is, following equation (3.4), to approximate the expectation in (3.24) as

$$P_\theta (\theta \in [U, V]) \approx \frac{1}{N} \sum_{j=1}^{N} \mathbb{1}_{[U^{(j)}, V^{(j)}]}(\theta), \tag{3.27}$$

where $U^{(j)} = U(X^{(j)})$, $V^{(j)} = V(X^{(j)})$, and the vectors $X^{(j)} = (X_1^{(j)}, \ldots, X_n^{(j)})$ for $j = 1, 2, \ldots, N$ are i.i.d. copies of $X = (X_1, \ldots, X_n)$. Since equation (3.24) is required to hold for all $\theta \in \Theta$, we have to compute the value of the approximation (3.27) for a representative sample of parameter values θ.

Example 3.38   If the data X do not consist of independent, normally distributed samples, the confidence interval for the mean given by (3.26) in example 3.37 is no longer exact. In such a situation, we can use Monte Carlo estimation to estimate the resulting confidence coefficient.

Here, we consider the case where $X_1, \ldots, X_n$ are independent and Poisson-distributed with parameter λ. Since the mean of the $\text{Pois}(\lambda)$ distribution is λ, we need to estimate the probability $P_\lambda(U \le \lambda \le V)$, where U and V are the boundaries of the confidence interval (3.26). A Monte Carlo estimate for this probability, for given λ, can be obtained by the following algorithm.

```
1: k ← 0
2: for j = 1, 2, ..., N do
3:    generate X₁, ..., Xₙ ~ Pois(λ)
4:    μ̂ ← ∑ᴺᵢ₌₁ Xᵢ/n
5:    σ̂ ← √(∑ⁿᵢ₌₁(Xᵢ − μ̂)²/(n − 1))
6:    U ← μ̂ − pₙ,ₐσ̂/√n
7:    V ← μ̂ + pₙ,ₐσ̂/√n
8:    if U ≤ λ ≤ V then
9:       k ← k + 1
```

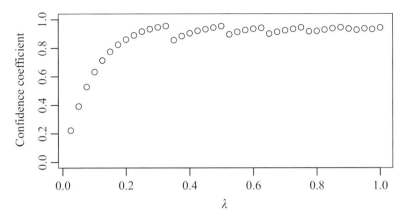

Figure 3.3   Estimated confidence coefficients for the confidence interval (3.26) with
$\alpha = 5\%$, *when the data are Pois(λ) distributed, for different values of λ. The con-*
fidence coefficients are estimated using Monte Carlo estimates, as described in
example 3.38. The plot shows that for Poisson-distributed data, in particular when λ
is small, the confidence coefficient is much smaller than 95%. This indicates that the
confidence interval (3.26), designed for normally distributed data, is too small for
data following a Poisson distribution.

10:   **end if**
11: **end for**
12: return k/N

The above algorithm determines the probability that the confidence interval covers
the mean for fixed λ. Running this algorithm repeatedly, we can systematically
estimate $P_\lambda(U \leq \lambda \leq V)$ for a range of λ. The result of such an estimation, for
$\alpha = 5\%$, is shown in Figure 3.3. The figure clearly shows that the interval given in
(3.26) is not a 95% confidence interval for Poisson-distributed values. In particular, for
small parameter values, the probability of the interval covering the actual parameter
value is very small. Thus, to derive an acceptable confidence interval for the parameter
of a Poisson distribution, the interval (3.26) needs to be enlarged. A more detailed
Monte Carlo analysis could be used to derive approximate bounds for such improved
confidence intervals.

3.4.3   Hypothesis tests

In hypothesis testing, inference about an unknown parameter is restricted to the
question of whether or not the parameter satisfies a given 'hypothesis' H_0. Such a
hypothesis about the parameter θ could, for example, be a statement like '$\theta = 0$' or
'$\theta > 0$'. The alternative hypothesis, that is the hypothesis that θ does not satisfy H_0,
is denoted by H_1.

While the dichotomy between H_0 and H_1 is symmetric, it transpires that in most situations any given statistical test can only determine for *one* of the hypotheses whether it is likely to be true, whereas the other hypothesis can only be shown to be likely to be wrong. Traditionally the names are chosen such that H_0 is the hypothesis which can only be disproved (called the *null hypothesis*) and H_1 is the hypothesis which can be proved. The two possible outcomes of a statistical test are then 'H_0 has been rejected' and 'H_0 has not been rejected'.

Example 3.39   Assume that $X_1, \ldots, X_n \sim \mathcal{N}(\mu, \sigma^2)$ are i.i.d. with unknown mean μ. Furthermore assume that, given one instance $x_1, \ldots, x_n \in \mathbb{R}$ of the random sample $X_1, \ldots, X_n$, we want to test the hypothesis that $\mu = 0$. In this case, for example if all the x_i are concentrated far away from 0, it is possible to conclude that the observations x_i are not compatible with the hypothesis $\mu = 0$. In contrast, even if the observations are clustered around 0, we will never be able to exclude the possibility that μ has a value which is only very close to but not identical with 0. Thus, for this example, we should choose H_0 to be the hypothesis $\mu = 0$ and we should choose H_1 to be the alternative hypothesis $\mu \neq 0$.

Since the null-hypothesis H_0 is a statement about the parameter θ, we can describe the hypothesis H_0 by a subset Θ_0 of the set Θ of all possible parameter values where H_0 holds.

Definition 3.40   A *statistical hypothesis test* (or *statistical test*) of size $\alpha \in (0, 1)$ for the hypothesis $H_0 = \{\theta \in \Theta_0\}$ for $\Theta_0 \subseteq \Theta$ is given by a function $T = T(X)$ of the random sample $X = (X_1, \ldots, X_n)$ together with a set C, such that

$$P_\theta\big(T(X) \in C\big) \leq \alpha \qquad (3.28)$$

for all $\theta \in \Theta_0$. The test *rejects* the hypothesis H_0 if and only if $T(X) \in C$. The function T is called the *test statistic* and the set C is called the *critical region* of the test.

A statistical test can fail in two different ways: if $\theta \in \Theta_0$, the random sample X could take a value such that $T(X) \in C$. In this case, the hypothesis H_0 is wrongly rejected despite being true. This is called a *type I error*. From equation (3.28) we know $P_\theta(T(X) \in C) \leq \alpha$ for all $\theta \in \Theta_0$ and thus the probability of type I errors is bounded from above by the size α of the test. Conversely, if $\theta \notin \Theta_0$, it can happen that $T(X) \notin C$. In this case, the test fails to reject the hypothesis H_0, despite it being wrong. The probability of the corresponding type II error is given by $P_\theta(T(X) \notin C)$ for all $\theta \in \Theta \setminus \Theta_0$. This probability depends on θ and can only be expected to be small if the parameter value θ is 'sufficiently far away' from Θ_0.

The rôle of the parameter α is to allow for rare, untypical behaviour of the random sample X. If H_0 holds and X shows typical behaviour, the null hypothesis H_0 will not be wrongly rejected. But in rare cases, with probability less than α, the random behaviour of X will be such that H_0 is wrongly rejected, that is such that a type I error occurs. By choosing the value of α, a trade-off between accuracy and sensitivity can be made: smaller values of α lead to reduced probability of type I errors, but at the same time the test will become less and less likely to reject H_0 and the probability of type II errors increases.

In many cases the relation (3.28) can only be proven (or only holds) for large values of n while for small values of n the real significance level of the test will not be known exactly.

Example 3.41   The *skewness*

$$\gamma = \mathbb{E}\left(\left(\frac{X - \mu}{\sigma}\right)^3\right) = \frac{\mathbb{E}\left((X - \mu)^3\right)}{\sigma^3}$$

of a random variable with mean μ and standard deviation σ can be estimated by

$$\hat{\gamma}_n = \frac{\frac{1}{n}\sum_{i=1}^{n}(X_i - \bar{X})^3}{\left(\frac{1}{n}\sum_{i=1}^{n}(X_i - \bar{X})^2\right)^{3/2}},$$

where $X_1, X_2, \ldots, X_n$ are i.i.d. copies of X and $\bar{X}$ is the average of the X_i. If $X \sim \mathcal{N}(\mu, \sigma^2)$, then $\gamma = 0$ and one can show that

$$\sqrt{\frac{n}{\sigma}}\hat{\gamma}_n \longrightarrow \mathcal{N}(0, 1) \tag{3.29}$$

as $n \to \infty$.

Assume that we want to construct a test for the null hypothesis H_0 that X is normally distributed with variance σ^2. As a consequence of (3.29), for large n, we can use the test statistic $T = \sqrt{n/\sigma}\,|\hat{\gamma}_n|$ and the critical region

$$C = (1.96, \infty) \subseteq \mathbb{R}$$

to construct a test of size $\alpha = 5\%$: We reject H_0 if $T \in C$, that is if $|\hat{\gamma}_n| \geq 1.96\sqrt{\sigma/n}$.

One problem with the test constructed in the preceding example is, that the convergence of the distribution of $\sqrt{n/\sigma}\hat{\gamma}_n$ to $\mathcal{N}(0, 1)$ is very slow. For small or moderate n the probability of wrongly rejecting H_0 (type I error) may be bigger than α.

We can use Monte Carlo estimation to estimate the probability of type I errors of statistical tests. For simple hypotheses (i.e. if H_0 specifies the distribution of the test statistic completely), this can be done as follows:

(a) For $j = 1, 2, \ldots, N$, generate samples $(X_1^{(j)}, \ldots, X_n^{(j)})$ according to the distribution given by the null hypothesis.

(b) Compute $T^{(j)} = T(X_1^{(j)}, \ldots, X_n^{(j)})$ for $j = 1, 2, \ldots, N$.

(c) Check for which percentage of samples H_0 is (wrongly) rejected:

$$P\,(T \in C) \approx \frac{1}{N} \sum_{j=1}^{N} \mathbb{1}_C(T^{(j)}).$$

3.5   Summary and further reading

In this chapter we have learned how Monte Carlo methods can be used to study statistical models via simulation. We have given special consideration to the relation between computational cost and accuracy of the results: the error of Monte Carlo estimates decays proportional to $1/\sqrt{N}$ where N is the number of samples used. We have also studied several 'variance reduction methods' which allow to reduce the constant of proportionality in this relation, thus allowing for more efficient estimates. Some information about Monte Carlo methods and variance reduction can be found in Ripley (1987). A more extensive discussion of Monte Carlo methods and variance reduction is contained in Chapters 3–5 of Robert and Casella (2004). The historical origins of the Monte Carlo method are described in Metropolis (1987).

Finally, we illustrated the methods introduced in this chapter by studying problems in statistical inference using Monte Carlo estimates. While we only touched the basics of statistical inference, there are many textbooks about statistics available, ranging from application oriented texts to more theoretical treatments. Details about statistical inference can, for example, be found in the books Garthwaite *et al.* (2002) and Casella and Berger (2001).

Exercises

E3.1   Assume $X \sim \text{Exp}(1)$ and $Y \sim \mathcal{N}(0, X)$, that is Y is normally distributed with a random variance. Use Monte Carlo estimation to estimate $\mathbb{E}(X|Y = 4)$ and $\text{Var}(X|Y = 4)$.

E3.2   Write a program which, for given N, computes the Monte Carlo estimate from algorithm 3.8 for the expectation $z = \mathbb{E}(\sin(X)^2)$. For $N = 1000$, use this program to repeatedly generate estimates for z and then use these estimates to assess the accuracy of the Monte Carlo method for the given problem. Repeat this experiment for $N = 10\,000$ and use the results to illustrate that the resulting estimates for z are more accurate than the estimates obtained for $N = 1000$.

E3.3   Let $X \sim \mathcal{N}(0, 1)$. Use Monte Carlo estimation to obtain an estimate for $\mathbb{E}(\cos(X))$ to three digits of accuracy.

E3.4   Let $X \sim \mathcal{N}(0, 1)$ and $a \geq 0$. Consider the estimates

$$p_N = \frac{1}{N} \sum_{j=1}^{N} \mathbb{1}_{(-\infty, a]}(X_j)$$

and

$$\tilde{p}_N = \frac{1}{2} + \frac{a}{\sqrt{2\pi N}} \sum_{j=1}^{N} e^{-a^2 U_j^2/2}$$

for $p = P(X \leq a)$, where $X_j \sim \mathcal{N}(0, 1)$ and $U_j \sim \mathcal{U}[0, 1]$ for $j \in \mathbb{N}$ are i.i.d.

(a) Using equation (3.5) or otherwise, show that $\tilde{p}_N \to p$ as $N \to \infty$.

(b) For $a = 1$, perform a numerical experiment to determine which of the two estimates has smaller mean squared error.

E3.5 The naïve approach to compute the Monte Carlo sample variance $\hat{\sigma}^2$ from equation (3.9) is to store all generated samples $f(X_j)$ in memory and to compute the sample variance (3.9) only after all samples are generated. A disadvantage of this approach is that this requires an amount of memory which is proportional to N. Describe a way to compute $\hat{\sigma}^2$ which requires only a constant (i.e. independent of N) amount of memory.

E3.6 Let $X \sim \mathcal{N}(0, 1)$ and $A = [3, 4]$ and consider importance sampling estimates for the probability $P(X \in A)$, using samples Y_j from the following sample distributions:

- $Y_j \sim \mathcal{N}(1, 1)$;
- $Y_j \sim \mathcal{N}(2, 1)$;
- $Y_j \sim \mathcal{N}(3.5, 1)$;
- $Y_j \sim \mathrm{Exp}(1) + 3$.

Each of these four distributions gives rise to a different importance sampling method. Our aim is to compare the resulting estimates.

(a) For each of the four methods, determine the sample variance $\hat{\sigma}^2$ for the weighted samples, as given in (3.19). Which of these four methods gives the best results?

(b) Determine a good estimate for $P(X \in A)$ and discuss the accuracy of your estimate.

(c) For each of the four methods, approximately how many samples Y_j are required to reduce the error of the estimate of $P(X \in A)$ to 1%?

E3.7 In this question we study the Monte Carlo estimate $\widehat{\mathrm{bias}}_\rho(\hat{\rho})$ for the bias of the estimator $\hat{\rho}$ from (3.23) for the correlation $\rho = \mathrm{Corr}(X, Y)$.

(a) Implement a function which computes, for given $n \in \mathbb{N}$ and $\rho \in [-1, 1]$, the estimate $\widehat{\mathrm{bias}}_\rho(\hat{\rho})$ as in example 3.34. Comment on your choice of the sample size N for the Monte Carlo estimate.

(b) For $n = 10$, create a plot which shows the bias of $r(X, Y)$ as a function of ρ.

(c) Use your results to give an (approximately) unbiased estimate of the correlation for the following data.

i	1	2	3	4	5	...
X_i	0.218	0.0826	0.091	0.095	−0.826	...
Y_i	0.369	0.715	−1.027	−1.499	1.291	...

i	...	6	7	8	9	10
X_i	...	0.208	0.600	−0.058	0.602	0.620
Y_i	...	−0.213	−2.400	1.064	−0.367	−1.490

E3.8 Write a program to compute the Monte Carlo estimate (3.27) for the confidence coefficient of the confidence interval (3.26). For $n = 10$ and $\alpha = 5\%$, using the value $p_{10,0.05} = 2.262157$, estimate the confidence coefficient when $X_1, \ldots, X_{10} \sim \text{Pois}(\lambda)$ for a range of λ and create a plot of the results. Is (3.26) a useful confidence interval for Poisson-distributed data?

4

Markov Chain Monte Carlo methods

Monte Carlo methods, as discussed in Chapter 3, use a sequence $(X_j)_{j \in \mathbb{N}}$ of i.i.d. samples as a tool to explore the behaviour of a statistical model. Large numbers of samples are required and thus these methods depend on our ability to generate the samples from a given distribution efficiently. Sometimes, when generating these samples is difficult, it can be easier to replace the i.i.d. sequence $(X_j)_{j \in \mathbb{N}}$ by a Markov chain instead. The resulting methods are called Markov Chain Monte Carlo methods.

Definition 4.1 A Markov Chain Monte Carlo (MCMC) method for estimating the expectation $\mathbb{E}(f(X))$ is a numerical method based on the approximation

$$\mathbb{E}(f(X)) \approx \frac{1}{N} \sum_{j=1}^{N} f(X_j), \tag{4.1}$$

where $(X_j)_{j \in \mathbb{N}}$ is a Markov chain with the distribution of X as its stationary distribution.

MCMC methods form a generalisation of the Monte Carlo methods discussed in the previous chapter. Where Monte Carlo methods use independent samples, MCMC methods use samples which can have direct dependence on the immediately preceding value. Since the dependence structure of samples is restricted in this way, the samples in MCMC methods are in some sense still 'close to being independent'.

Compared with the situation of Monte Carlo methods, the theory behind MCMC methods is more challenging but, at the same time, the resulting methods are often

An Introduction to Statistical Computing: A Simulation-based Approach, First Edition. Jochen Voss.
© 2014 John Wiley & Sons, Ltd. Published 2014 by John Wiley & Sons, Ltd.

easier to apply since no extra knowledge about random number generation is required for MCMC methods.

4.1   The Metropolis–Hastings method

In order to apply MCMC methods as in definition 4.1, we need to be able to construct Markov chains with a prescribed stationary distribution. The Metropolis–Hastings method, presented in this section, is a popular method to solve this problem. The resulting algorithm is similar to the rejection sampling algorithm: starting from a nearly arbitrary Markov chain, the Metropolis–Hastings algorithm generates a new Markov chain with the required stationary distribution by 'rejecting' some of the state transitions of the original Markov chain.

Here we concentrate solely on the problem of constructing a Markov chain with a given stationary distribution. Use of the resulting Markov chain in Monte Carlo estimates will be discussed in Section 4.2. The discussion in this and the following sections requires basic knowledge about Markov chains; we refer to Section 2.3 for a short introduction.

4.1.1   Continuous state space

In its general form, the Metropolis–Hastings method can be used on nearly arbitrary state spaces. In order to avoid technical complications, we do not give the general form of the algorithm here but, instead, consider the most important special cases separately. In this section, we will discuss the case where the state space is $S = \mathbb{R}^d$. The following section considers the case of finite or countable state space.

Algorithm 4.2   (Metropolis–Hastings method for continuous state space)
    input:
        a probability density π (the *target density*)
        a transition density $p\colon S \times S \to [0, \infty)$
        $X_0 \in \{x \in S \mid \pi(x) > 0\}$
    randomness used:
        independent samples Y_j from the transition density p (the *proposals*)
        $U_j \sim \mathcal{U}[0, 1]$ i.i.d.
    output:
        a sample of a Markov chain X with stationary density π.
    As an abbreviation we define a function $\alpha\colon S \times S \to [0, 1]$ by

$$\alpha(x, y) = \min\left(\frac{\pi(y)p(y, x)}{\pi(x)p(x, y)}, 1\right)$$

    for all $x, y \in S$ with $\pi(x)p(x, y) > 0$.
    1: **for** $j = 1, 2, 3, \ldots$ **do**
    2:    generate Y_j with density $p(X_{j-1}, \cdot)$
    3:    generate $U_j \sim \mathcal{U}[0, 1]$

```
4:   if U_j ≤ α(X_{j-1}, Y_j) then
5:       X_j ← Y_j
6:   else
7:       X_j ← X_{j-1}
8:   end if
9:   output X_j
10: end for
```

The acceptance probability $\alpha(x, y)$ is only defined for elements $x, y \in S$ which satisfy $\pi(x)p(x, y) > 0$. At a first glance it seems that the algorithm could fail by hitting a proposal Y_j such that $\alpha(X_{j-1}, Y_j)$ is undefined, but it transpires that this cannot happen: Assume that we have already found $X_0, X_1, \ldots, X_{j-1}$. Then we have $\pi(X_{j-1}) > 0$ (it would not have been accepted otherwise in a previous step) and, given X_{j-1}, the proposal Y_j satisfies $p(X_{j-1}, Y_j) > 0$ with probability 1 (it would not have been proposed otherwise). Consequently, $\alpha(X_{j-1}, Y_j)$ is defined and we can compute the next value X_j.

The purpose of algorithm 4.2 is to generate a Markov chain with stationary density π, that is ideally we want to achieve $X_j \sim \pi$ for all $j \in \mathbb{N}$. The following results shows that, assuming we have $X_0 \sim \pi$, the algorithm achieves this aim. In practice, if we could generate samples with density π, we would be able to use basic Monte Carlo estimation and there would be no need to employ MCMC methods. Thus, realistically we will not exactly have $X_0 \sim \pi$ but the results of Section 4.2 will show that, under mild additional assumptions, the Markov chains generated by algorithm 4.2 can be used as the basis for MCMC methods, even if the algorithm is not started with $X_0 \sim \pi$.

Proposition 4.3 The process $(X_j)_{j \in \mathbb{N}}$ constructed in the Metropolis–Hastings algorithm 4.2 is a Markov chain with stationary density π.

Proof Assume that X_{j-1} has density π. Then we have

$$
\begin{aligned}
P(X_j \in A) &= E\left(\mathbb{1}_A(X_j)\right) \\
&= \int_S \pi(x) \int_S p(x, y)\left(\alpha(x, y)\mathbb{1}_A(y) + (1 - \alpha(x, y))\,\mathbb{1}_A(x)\right) dy\, dx \\
&= \int_S \pi(x) \int_S p(x, y)\,\alpha(x, y)\mathbb{1}_A(y)\, dy\, dx \\
&\quad + \int_S \pi(x) \int_S p(x, y)(1 - \alpha(x, y))\,\mathbb{1}_A(x)\, dy\, dx.
\end{aligned}
$$

From the definition of the acceptance probability α we know

$$
\pi(x)p(x, y)\alpha(x, y) = \min\left(\pi(x)p(x, y), \pi(y)p(y, x)\right) = \pi(y)p(y, x)\alpha(y, x)
$$

and thus

$$P(X_j \in A) = \int_S \int_S \pi(y)p(y,x)\alpha(y,x)\,\mathbb{1}_A(y)\,dy\,dx$$
$$+ \int_S \pi(x) \int_S p(x,y)(1-\alpha(x,y))\,\mathbb{1}_A(x)\,dy\,dx.$$

Finally, interchanging the variables x and y in the first double integral, we get

$$P(X_j \in A) = \int_S \int_S \pi(x)p(x,y)\alpha(x,y)\mathbb{1}_A(x)\,dx\,dy$$
$$+ \int_S \pi(x) \int_S p(x,y)(1-\alpha(x,y))\,\mathbb{1}_A(x)\,dy\,dx$$
$$= \int_S \pi(x) \int_S p(x,y)(\alpha(x,y)\mathbb{1}_A(x)+(1-\alpha(x,y))\,\mathbb{1}_A(x))\,dy\,dx$$
$$= \int_S \pi(x) \int_S p(x,y)\mathbb{1}_A(x)\,dy\,dx$$
$$= \int_S \pi(x)\mathbb{1}_A(x)\,dx$$
$$= P(X_{j-1} \in A)$$

for all sets A. This shows that the distributions of X_{j-1} and X_j are the same and thus that π is a stationary density of the Markov chain. □

The result of proposition 4.3 can be slightly improved: it transpires that the Markov chain constructed by algorithm 4.2, when started with initial distribution π, is balanced in the sense that

$$P(X_j \in A, X_{j-1} \in B) = P(X_j \in B, X_{j-1} \in A) \qquad (4.2)$$

for all sets $A, B \subseteq \mathbb{R}^d$. The condition (4.2), for $B = \mathbb{R}^d$, implies

$$P(X_j \in A) = P(X_j \in A, X_{j-1} \in \mathbb{R}^d)$$
$$= P(X_j \in \mathbb{R}^d, X_{j-1} \in A)$$
$$= P(X_{j-1} \in A)$$

for all $j \in \mathbb{N}$ and by induction we find $P(X_j \in A) = P(X_0 \in A) = \pi(A)$. Thus, condition (4.2) for $X_0 \sim \pi$ implies the statement of proposition 4.3. A process which satisfies condition (4.2) for $X_j \sim \pi$ is called π-reversible. Since we do not require reversibility in the following, we restrict proofs of reversibility to the simpler, discrete case and only prove stationarity as in proposition 4.3 for the continuous case.

One big advantage of the Metropolis–Hastings algorithm is that, as for the rejection sampling algorithm, the target density π only needs to be known up to a constant: the only place where π occurs in the algorithm is in the acceptance probability α; if we only know the product $c \cdot \pi$ but not the value of the constant c, we can still evaluate the function α, since

$$\alpha(x, y) = \min\left(\frac{\pi(y)p(y, x)}{\pi(x)p(x, y)}, 1\right) = \min\left(\frac{c\,\pi(y)p(y, x)}{c\,\pi(x)p(x, y)}, 1\right).$$

This property of the Metropolis–Hastings algorithm can, for example, be used in situations such as the one from Section 4.3 where we need to sample from the density

$$p(\theta \mid x) = \frac{p(x \mid \theta)p(\theta)}{\int p\left(x \mid \tilde{\theta}\right) p(\tilde{\theta})\, d\tilde{\theta}}$$

with an unknown normalisation constant $\int p\left(x \mid \tilde{\theta}\right) p(\tilde{\theta})\, d\tilde{\theta}$.

4.1.2   Discrete state space

In this section, we state the Metropolis–Hastings algorithm for discrete state spaces. The algorithm for this case is obtained from algorithm 4.2 by replacing densities with probability weights. Since the situation of discrete state is less technically challenging, here we prove a slightly better result than we did for the continuous case in proposition 4.3.

Algorithm 4.4   (Metropolis–Hastings method for discrete state space)
    input:
      a probability vector $\pi \in \mathbb{R}^S$ (the *target distribution*)
      a transition matrix $P = (p_{xy})_{x,y \in S}$
      $X_0 \in S$ with $\pi_{X_0} > 0$
    randomness used:
      independent samples Y_j from the transition matrix P (the *proposals*)
      $U_j \sim \mathcal{U}[0, 1]$ i.i.d.
    output:
      a sample of a Markov chain X with stationary distribution π.
    As an abbreviation we define a function $\alpha\colon S \times S \to [0, 1]$ by

$$\alpha(x, y) = \min\left(\frac{\pi_y\, p_{yx}}{\pi_x\, p_{xy}}, 1\right) \tag{4.3}$$

for all $x, y \in S$ with $\pi_x\, p_{xy} > 0$.
  1: **for** $j = 1, 2, 3, \ldots$ **do**
  2:    generate Y_j with $P(Y_j = y) = p_{X_{j-1},y}$ for all $y \in S$
  3:    generate $U_j \sim \mathcal{U}[0, 1]$
  4:    **if** $U_j \leq \alpha(X_{j-1}, Y_j)$ **then**

5:     $X_j \leftarrow Y_j$
6:   **else**
7:     $X_j \leftarrow X_{j-1}$
8:   **end if**
9:   output X_j
10: **end for**

While we could follow the structure of Section 4.1.1 and prove a result in analogy to proposition 4.3, we will prove a slightly better result in this section, by showing that the Markov chain constructed by algorithm 4.4 is reversible in the sense of the following definition.

Definition 4.5   A time-homogeneous Markov chain X with state space S and transition matrix $P = (p_{xy})_{x,y \in S}$ satisfies the *detailed balance condition*, if there is a probability vector $\pi \in \mathbb{R}^S$ with

$$\pi_x p_{xy} = \pi_y p_{yx}$$

for all $x, y \in S$. In this case we say that the Markov chain X is π-*reversible*.

The following lemma shows that being π-reversible is indeed a stronger condition for a Markov chain than having stationary distribution π.

Lemma 4.6   Let X be a π-reversible Markov chain. Then π is a stationary distribution of X.

Proof   If X satisfies the detailed balance condition for a probability vector π, we have

$$\sum_{x \in S} \pi_x p_{xy} = \sum_{x \in S} \pi_y p_{yx} = \pi_y \sum_{x \in S} p_{yx} = \pi_y$$

for all $y \in S$. Thus, π satisfies the condition from equation (2.4) and consequently is a stationary distribution.   □

Proposition 4.7   The process $(X_j)_{j \in \mathbb{N}}$ constructed in the Metropolis–Hastings algorithm 4.4 is a π-reversible Markov chain. In particular, X has stationary distribution π.

Proof   Since X_j in the algorithm depends only on X_{j-1} and on the additional, independent randomness from Y_j and U_j, the process $(X_j)_{j \in \mathbb{N}}$ is a Markov chain. Let $Q = (q_{xy})_{x,y \in S}$ be the transition matrix of this Markov chain, that is

$$q_{xy} = P(X_j = y \mid X_{j-1} = x)$$

for all $x, y \in S$. Then we have to show that Q satisfies the detailed balance condition

$$\pi_x q_{xy} = \pi_y q_{yx} \tag{4.4}$$

for all $x, y \in S$.

In the case $x = y$, the relation (4.4) is trivially true; thus we can assume $x \neq y$. For the process X to jump from x to y, the proposal Y_n must equal y and then the proposal must be accepted. Since these two events are independent, we find

$$q_{xy} = p_{xy} \cdot \alpha(x, y)$$

and thus

$$
\begin{aligned}
\pi_x q_{xy} &= \pi_x p_{xy} \min\left(\frac{\pi_y p_{yx}}{\pi_x p_{xy}}, 1\right) \\
&= \min\left(\pi_y p_{yx}, \pi_x p_{xy}\right) \\
&= \pi_y p_{yx} \min\left(1, \frac{\pi_x p_{xy}}{\pi_y p_{yx}}\right) \\
&= \pi_y q_{yx}
\end{aligned}
$$

for all $x, y \in S$ with $\pi_x p_{xy} > 0$ and $\pi_y p_{yx} > 0$. There are various cases with $\pi_x p_{xy} = 0$ or $\pi_y p_{yx} = 0$; using the that fact the $p_{xy} = 0$ implies $q_{xy} = 0$ (since transitions from x to y are never proposed) it is easy to check that in all of these cases $\pi_x q_{xy} = 0 = \pi_y q_{yx}$ holds. Thus, equation (4.4) is satisfied for all $x, y \in S$, the process X is π-reversible and, by lemma 4.6, π is a stationary distribution of X.   □

Example 4.8 Let $\pi_x = 2^{-|x|}/3$ for all $x \in \mathbb{Z}$. Then

$$
\sum_{x \in \mathbb{Z}} \pi_x = \frac{1}{3}\left(\cdots + \frac{1}{4} + \frac{1}{2} + 1 + \frac{1}{2} + \frac{1}{4} + \cdots\right)
$$

$$
= \frac{1}{3}\left(1 + 2\sum_{x=1}^{\infty} 2^{-x}\right) = 1,
$$

that is the infinite vector π is a probability vector on $S = \mathbb{Z}$. Using algorithm 4.4, we can easily find a Markov chain which has π as a stationary distribution: for the algorithm, we can choose the transition matrix P for the proposals. For this example we consider

$$
P(Y_j = x + 1 \mid X_{j-1} = x) = P(Y_j = x - 1 \mid X_{j-1} = x) = \frac{1}{2}
$$

for all $x \in S$ and all $n \in \mathbb{N}$. This corresponds to $p_{xy} = 1/2$ if $y = x + 1$ or $y = x - 1$ and $p_{xy} = 0$ otherwise. From this we can compute the acceptance probabilities $\alpha(x, y)$:

$$\alpha(x, y) = \min\left(\frac{\pi_y p_{yx}}{\pi_x p_{xy}}, 1\right) = \min\left(\frac{\frac{2^{-|y|}}{3} p_{yx}}{\frac{2^{-|x|}}{3} p_{xy}}, 1\right) = \min\left(2^{|x|-|y|}\frac{p_{yx}}{p_{xy}}, 1\right).$$

In the Metropolis–Hastings algorithm, the function $\alpha(x, y)$ is only evaluated for $x = y + 1$ and $x = y - 1$. In either of these cases we have $p_{xy} = 1/2$ and thus

$$\alpha(x, y) = \begin{cases} 2^{|x|-|y|} & \text{if } |y| > |x| \text{ and} \\ 1 & \text{otherwise.} \end{cases}$$

Finally, substituting this transition matrix P and the corresponding function α into algorithm 4.4 we get the following:

1: **for** $j = 1, 2, 3, \ldots$ **do**
2:     Let $Y_j \leftarrow X_{j-1} + \varepsilon_j$ where $\varepsilon_j \sim \mathcal{U}\{-1, 1\}$.
3:     generate $U_j \sim \mathcal{U}[0, 1]$
4:     **if** $U_j \le 2^{|X_{j-1}|-|Y_j|}$ **then**
5:         $X_j \leftarrow Y_j$
6:     **else**
7:         $X_j \leftarrow X_{j-1}$
8:     **end if**
9: **end for**

By proposition 4.7, the resulting process X is a Markov chain with stationary distribution π.

4.1.3   Random walk Metropolis sampling

The random walk Metropolis algorithm discussed in this section is an important special case of the Metropolis–Hastings algorithm. It can be considered both for the continuous and for the discrete case; here we restrict ourselves to the continuous case and refer to example 4.8 for an illustration of the corresponding discrete case.

The Metropolis–Hastings method for the case $p(x, y) = p(y, x)$ is called the *Metropolis method*. In this case, the expression for the acceptance probability α simplifies to

$$\alpha(x, y) = \min\left(\frac{\pi(y)p(y, x)}{\pi(x)p(x, y)}, 1\right) = \min\left(\frac{\pi(y)}{\pi(x)}, 1\right)$$

for all $x, y \in S$ with $\pi(x) > 0$ (or π_y / π_x for discrete state space). The condition $p(x, y) = p(y, x)$ is, for example, satisfied when the proposals Y_j are constructed as

$$Y_j = X_{j-1} + \varepsilon_j$$

where the ε_j are i.i.d. with a symmetric distribution (i.e. ε_j has the same distribution as $-\varepsilon_j$). We only state the version of the resulting algorithm for continuous state space S, the discrete version is found by using a probability vector instead of a density for the target distribution.

Algorithm 4.9 (random walk Metropolis)
    input:
        a probability density $\pi\colon S \to [0, \infty)$ (the *target density*)
        $X_0 \in S$ with $\pi(X_0) > 0$
    randomness used:
        an i.i.d. sequence $(\varepsilon_j)_{j \in \mathbb{N}}$ with a symmetric distribution (the *increments*)
        $U_j \sim \mathcal{U}[0, 1]$ i.i.d.
    output:
        a sample of a Markov chain X with stationary density π.
    As an abbreviation we define a function $\alpha\colon S \times S \to [0, 1]$ by

$$\alpha(x, y) = \min\left(\frac{\pi(y)}{\pi(x)}, 1\right) \tag{4.5}$$

for all $x, y \in S$ with $\pi(x) > 0$.
  1: **for** $j = 1, 2, 3, \ldots$ **do**
  2:    generate ε_j
  3:    let $Y_j \leftarrow X_{j-1} + \varepsilon_j$
  4:    generate $U_j \sim \mathcal{U}[0, 1]$
  5:    **if** $U_j \leq \alpha(X_{j-1}, Y_j)$ **then**
  6:      $X_j \leftarrow Y_j$
  7:    **else**
  8:      $X_j \leftarrow X_{j-1}$
  9:    **end if**
 10: **end for**

Since algorithm 4.9 is a special case of the general Metropolis–Hastings algorithm 4.2, we can use proposition 4.3 to see that the output of algorithm 4.9 is a Markov chain with stationary distribution π, independently of the choice of increments ε_j. The results of Section 4.2 can be used to prove convergence of MCMC methods based on algorithm 4.9. For state space $S = \mathbb{R}^d$, the most common choice of increments is $\varepsilon_j \sim \mathcal{N}(0, \sigma^2)$; convergence of the MCMC methods resulting from this choice of increments is proved in example 4.24.

The proposal variance σ^2 can be chosen to maximise efficiency of the method: if σ^2 is small, we have $Y_j \approx X_{j-1}$, that is $\pi(Y_j) \approx \pi(X_{j-1})$ and consequently

$\alpha(X_{j-1}, Y_j) \approx 1$. In this case, almost all proposals are accepted, but the algorithm moves slowly since the increments in each step are small. On the other hand, if σ^2 is large, the proposals Y_j are widely dispersed and often $\pi(Y_j)$ will be small. In this case, many proposals are rejected; the process X does not move very often but when it does, the increment is typically large. The optimal choice of σ^2 will be between these two extremes. This is illustrated in Figure 4.1.

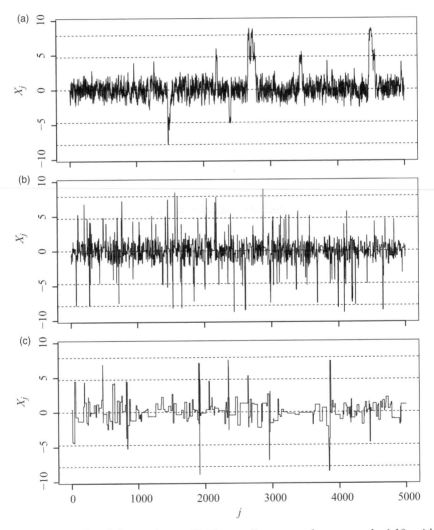

Figure 4.1   Paths of the random walk Metropolis process from example 4.10, with $\mathcal{N}(0, \sigma^2)$-distributed increments, for different values of the proposal variance σ: (a) $\sigma = 1$; (b) $\sigma = 6$; (c) $\sigma = 36$.

Example 4.10   We can use the random walk Metropolis algorithm to generate samples from the density

$$
\pi(x) = \frac{1}{Z} \cdot \begin{cases} \frac{\sin(x)^2}{x^2} & \text{if } x \in [-3\pi, 3\pi] \text{ and} \\ 0 & \text{otherwise,} \end{cases} \tag{4.6}
$$

where Z is the normalisation constant which makes π a probability density. For this target distribution, the acceptance probabilities $\alpha(x, y)$ from equation (4.5) are found as

$$
\begin{aligned}
\alpha(x, y) &= \min\left(\frac{\pi(y)}{\pi(x)}, 1 \right) \\
&= \min\left(\frac{\frac{1}{Z} \cdot \frac{\sin(y)^2}{y^2} \mathbb{1}_{[-3\pi,3\pi]}(y)}{\frac{1}{Z} \cdot \frac{\sin(x)^2}{x^2} \mathbb{1}_{[-3\pi,3\pi]}(x)}, 1 \right) \\
&= \min\left(\left(\frac{x \sin(y)}{y \sin(x)} \right)^2 \mathbb{1}_{[-3\pi,3\pi]}(y), 1 \right)
\end{aligned}
$$

for all $x, y \in \mathbb{R}$ with $\pi(x) > 0$.

In this section we have only argued that the given target density π is a stationary density for the Markov chain generated by the random walk Metropolis algorithm. This result is a consequence of proposition 4.3 and does not make any assumptions on π or on the increments ε_j. Later, when using the generated Markov chain as part of a Monte Carlo method in Section 4.2, we will require additional assumptions to guarantee that the average in the estimate (4.1) converges to the correct value, even if the Markov chain is not started in stationarity. This will require that the increments ε_j are large enough for the process to move freely within the region of value typically taken by the target distribution. For this reason, the random walk Metropolis–Hastings method will often be inefficient if π has several isolated maxima, and the method works best if the target density has just one connected region of high probability.

4.1.4   The independence sampler

Another special case of the Metropolis–Hastings algorithm is obtained by choosing the proposals Y_j independently of X_{j-1}, that is by using a transition density of the form $p(x, y) = p(y)$.

Algorithm 4.11   (independence sampler)
    input:
        a probability density $\pi \in \mathbb{R}^S$ (the *target density*)
        $X_0 \in S$

randomness used:

an i.i.d. sequence $(Y_j)_{j \in \mathbb{N}}$ with density p (the *proposals*)

$U_j \sim \mathcal{U}[0, 1]$ i.i.d.

output:

a sample of a Markov chain X with stationary density π.

As an abbreviation we define a function $\alpha: S \times S \to [0, 1]$ by

$$\alpha(x, y) = \min\left(\frac{\pi(y)p(x)}{\pi(x)p(y)}, 1\right)$$

for all $x, y \in S$ with $\pi(x) > 0$.

1: **for** $j = 1, 2, 3, \ldots$ **do**
2:     generate $Y_j \sim p$
3:     generate $U_j \sim \mathcal{U}[0, 1]$
4:     **if** $U_j \leq \alpha(X_{j-1}, Y_j)$ **then**
5:         $X_j \leftarrow Y_j$
6:     **else**
7:         $X_j \leftarrow X_{j-1}$
8:     **end if**
9: **end for**

While the proposals Y_j in this algorithm are independent, the X_j are still dependent, since the acceptance probability $\alpha(X_{j-1}, Y_j)$ for X_j depends on the value of X_{j-1}. The algorithm is a special case of the general Metropolis–Hastings method from algorithm 4.2 and by proposition 4.3 the generated Markov chain has stationary density π.

The resulting method resembles the rejection sampling method from algorithm 1.22. When we use rejection sampling to generate (independent) samples with density π from proposals with density p, we have the condition $\pi(x)/p(x) \leq c$ for all x and we accept proposals, if the condition

$$cp(Y_j)U_j \leq \pi(Y_j)$$

is satisfied. In contrast, the independence sampler generates dependent samples, but does not require boundedness of $\pi(x)/p(x)$. For the independence sampler, proposals are accepted whenever

$$\frac{\pi(X_j)}{p(X_j)}p(Y_j)U_j \leq \pi(Y_j)$$

holds.

4.1.5   Metropolis–Hastings with different move types

As a simple generalisation of the Metropolis–Hastings algorithm, we can split the transition mechanism into different *move types*. This generalisation allows more flexibility in the design of MCMC algorithms. There are two different situations where this approach can be employed. First, sometimes an existing MCMC method can be improved by adding additional move types which allow the algorithm to explore the state space faster. An example of this approach can be found in example 4.14. Secondly, some MCMC methods are constructed from the ground up as a combination of different move types, each of which serves a different purpose. We will see examples of this approach in Sections 4.4 and 4.5.

A finite or countable set M is used to denote the possible types of move. Instead of directly generating a proposal Y_j with density $p(X_{j-1}, \cdot)$ as in the basic Metropolis–Hastings algorithm, we first randomly choose a move type $m \in M$ and then generate Y_j with a move-dependent density $p_m(X_{j-1}, \cdot)$. Let $\gamma_m(x)$ denote the probability of choosing move m when the current state is $X_{j-1} = x$. Then, given the value of X_{j-1}, the distribution of the proposal Y_j is given by

$$P(Y_j \in A \mid X_{j-1}) = \sum_{m \in M} \gamma_m(X_{j-1}) \int_A p_m(X_{j-1}, y)\, dy.$$

The resulting algorithm follows.

Algorithm 4.12   (Metropolis–Hastings method with different move types)
  input:
    a probability density π (the *target density*)
    $(\gamma_m(x))_{m \in M, x \in S}$ with $\sum_{m \in M} \gamma_m(x) = 1$ for all $x \in S$
    transition densities $p_m \colon S \times S \to [0, \infty)$ for $m \in M$
    $X_0 \in \{x \in S \mid \pi(x) > 0\}$
  randomness used:
    independent $m_j \in M$ with $P(m_j = m) = \gamma_m(x)$ for all $m \in M$
    independent Y_j from the transition densities p_m (the *proposals*)
    $U_j \sim \mathcal{U}[0, 1]$ i.i.d.
  output:
    a sample of a Markov chain X with stationary density π.
  As an abbreviation we define $\alpha_m \colon S \times S \to [0, 1]$ by

$$\cdot\alpha_m(x, y) = \min \left(\frac{\pi(y)\gamma_m(y)p_m(y, x)}{\pi(x)\gamma_m(x)p_m(x, y)}, 1 \right) \tag{4.7}$$

  for all $x, y \in S$ and all $m \in M$.
   1: **for** $j = 1, 2, 3, \ldots$ **do**
   2:    generate $m_j \in M$ with $P(m_j = m) = \gamma_m(X_{j-1})$ for all $m \in M$.
   3:    generate Y_j with density $p_{m_j}(X_{j-1}, \cdot)$
   4:    generate $U_j \sim \mathcal{U}[0, 1]$
   5:    **if** $U_j \leq \alpha_{m_j}(X_{j-1}, Y_j)$ **then**
   6:       $X_j \leftarrow Y_j$
   7:    **else**
   8:       $X_j \leftarrow X_{j-1}$
   9:    **end if**
  10: **end for**

Proposition 4.13   The process $(X_j)_{j \in \mathbb{N}}$ constructed by algorithm 4.12 is a Markov chain with stationary density π.

Proof   Assume that X_{j-1} has density π. Then we have

$$P(X_j \in A) = \int_S \pi(x) \sum_{m \in M} \gamma_m(x) \int_S p_m(x, y)$$

$$(\alpha_m(x, y) \mathbb{1}_A(y) + (1 - \alpha_m(x, y)) \mathbb{1}_A(x)) \, dy \, dx$$

$$= \sum_{m \in M} \int_S \int_S \pi(x) \gamma_m(x) p_m(x, y)$$

$$(\alpha_m(x, y) \mathbb{1}_A(y) + (1 - \alpha_m(x, y)) \mathbb{1}_A(x)) \, dy \, dx.$$

From the definition of the acceptance probabilities α_m we get

$$\pi(x) \gamma_m(x) p_m(x, y) \alpha_m(x, y)$$
$$= \min \left(\pi(x) \gamma_m(x) p_m(x, y), \pi(y) \gamma_m(y) p_m(y, x) \right)$$
$$= \pi(y) \gamma_m(y) p_m(y, x) \alpha_m(y, x)$$

for all $x, y \in S$ and all $m \in M$. Thus, following the same steps as in the proof of proposition 4.3, we get

$$P(X_j \in A) = \sum_{m \in M} \int_S \pi(x) \gamma_m(x) \int_S p_m(x, y)$$

$$(\alpha_m(x, y) \mathbb{1}_A(x) + (1 - \alpha_m(x, y)) \mathbb{1}_A(x)) \, dy \, dx$$

$$= \int_S \pi(x) \sum_{m \in M} \gamma_m(x) \int_S p_m(x, y) \, \mathbb{1}_A(x) \, dy \, dx$$

$$= \int_S \pi(x) \, \mathbb{1}_A(x) \, dx$$

$$= P(X_{j-1} \in A)$$

for all sets A. This shows that the distributions of X_{j-1} and X_j are the same and thus π is a stationary density of the Markov chain.   □

Since the proof of proposition 4.13 considers each transition $X_{j-1} \to X_j$ separately, the statement stays true if the probabilities $\gamma_m(x)$ for choosing move m depend on the time j as well as on the location x. In particular, it is possible to choose the moves solely based on time, for example by applying a pair of moves alternatingly or, more generally, by having move types $M = \{0, 1, \ldots, k-1\}$ and then always choosing move type $(j \bmod k)$ at time j. Since proposition 4.13 still applies, π is still a stationary density even if the move probabilities are time dependent. It transpires that, even if every move type on its own would result in a reversible Markov chain, the combined Markov chain is no longer necessarily reversible.

Example 4.14   Assume that the target distribution π is known to have several disjoint regions $A_1, \ldots, A_n$ of high probability and that π is very small outside these regions. Such a situation could, for example, occur when π is a mixture distribution (see definition 2.6). In this situation we can augment a random walk Metropolis algorithm tuned for moving *inside* these regions by adding a separate move type for moving *between* different regions. A very simple example of this approach is as follows: let $M = \{1, 2\}$. For moves of type $m = 1$ construct proposals as

$$Y_j = X_{j-1} + \varepsilon_j,$$

where $\varepsilon_j \sim \mathcal{N}(0, \sigma_1^2)$ and σ_1 is comparable with the diameter of the regions A_i. For moves of type $m = 2$ construct proposals as

$$Y_j = X_{j-1} + \eta_j,$$

where $\eta_j \sim \mathcal{N}(0, \sigma_2^2)$ and σ_2 is comparable with the distances between the regions A_i.

Some subtleties occur in the context of the Metropolis–Hastings method with different move types: a common choice for move types is to construct moves which only change a single coordinate of the state $X_{j-1} \in \mathbb{R}^d$. In such cases, the values Y_j are concentrated on a lower dimensional subset of $\mathbb{R}^d$ and the distribution of Y_j cannot be described by a density on $\mathbb{R}^d$. Thus, the densities $p_m(x, \cdot)$ from the definition of α_m do not exist in this situation. Nevertheless, inspection of the proof of proposition 4.13 reveals that a slight generalisation of algorithm 4.12 still works in this case. The resulting generalisation is described in the following lemma.

Lemma 4.15   Assume that for each move $m \in M$ in algorithm 4.12 there is a set $C_m \subseteq \mathbb{R}^d \times \mathbb{R}^d$ such that

$$(x, y) \in C_m \iff (y, x) \in C_m \tag{4.8}$$

for all $x, y \in \mathbb{R}^d$ and a density $\varphi_m \colon C_m \to [0, \infty)$ such that, if a proposal Y_j is constructed from a previous state $X_{j-1} \sim \pi$, we have

$$P(Y_j \in B, X_{j-1} \in A) = \int_{C_m} \mathbb{1}_A(x) \mathbb{1}_B(y) \, \varphi_m(x, y) \, dy \, dx \tag{4.9}$$

for all $A, B \subseteq \mathbb{R}^d$. Consider algorithm 4.12, but using acceptance probabilities

$$\alpha_m(x, y) = \min\left(\frac{\gamma_m(y) \varphi_m(y, x)}{\gamma_m(x) \varphi_m(x, y)}, 1 \right) \tag{4.10}$$

instead of the expression from (4.7). The process generated by this algorithm is a Markov chain with stationary density π.

While the pair (x, y) in (4.9) lies in the $2d$-dimensional space $\mathbb{R}^d \times \mathbb{R}^d$, the integral in (4.9) is an integral over the subset C_m only and in many applications we have $\dim(C_m) < 2d$. For example, if C_m is $(d + 1)$-dimensional, the integral will only be a $(d + 1)$-dimensional integral, even if the state space S of the Markov chain X_n has dimension $d > 1$ and thus $2d > d + 1$. Similarly, if C_m contains isolated points, the integral is interpreted as a sum over these points.

Proof   (of lemma 4.15, outline only). We first observe that (4.10) can be obtained from (4.7) by replacing $\pi(y)p_m(y, x)/\pi(x)p_m(x, y)$ with $\varphi_m(y, x)/\varphi_m(x, y)$. Assume that $X_{j-1} \sim \pi$. If the conditional distribution of Y_j given $X_j = x$ for move type m has a density $p_m(x, \cdot)$ then, by (4.9), we have $\varphi_m(x, y) = \pi(x)p_m(x, y)$ and we are in the situation of algorithm 4.12. In this case we get $X_j \sim \pi$ from proposition 4.13.

If the conditional distribution of Y_j has no density, as discussed above, a more advanced concept of integration is required to make sense of the integrals in (4.9) and we omit the required technical details. Once the meaning of the integrals with density $\varphi_m(x, y)$ has been clarified, a proof can be constructed by following the lines of the proof of proposition 4.13 while replacing occurrences of $\pi(x)p_m(x, y)$ with $\varphi_m(x, y)$ and using the modified definition of $\alpha(x, y)$.   □

Example 4.16   If the move $m \in M$ maps the current state $X_{j-1} \in \mathbb{R}^d$ to the proposal $Y_j \in \mathbb{R}^d$, constructed such that the component $Y_{j,1}$ is random with density $q : \mathbb{R} \to [0, \infty)$ and $(Y_{j,2}, \ldots, Y_{j,n}) = (X_{j-1,2}, \ldots, X_{j-1,n})$, then we can apply lemma 4.15 with

$$C = \left\{ (x, y) \in \mathbb{R}^d \times \mathbb{R}^d \mid x_2 = y_2, \ldots, x_d = y_d \right\} \cong \mathbb{R}^d \times \mathbb{R}.$$

Since we have

$$P(Y_j \in B, X_{j-1} \in A) = \int_{\mathbb{R}^d} \pi(x) \mathbb{1}_A(x) \int_{\mathbb{R}} q(y_1) \mathbb{1}_B(y_1, x_2, \ldots, x_d) \, dy_1 \, dx,$$

the pair (X_{j-1}, Y_j) has density $\varphi(x, y) = \pi(x)q(y_1)$ on the $(d + 1)$-dimensional set C and the required acceptance probability for move m is

$$\alpha_m(x, y) = \min \left(\frac{\gamma_m(y)\varphi(y, x)}{\gamma_m(x)\varphi(x, y)}, 1 \right) = \min \left(\frac{\gamma_m(y)\pi(y)q(x_1)}{\gamma_m(x)\pi(x)q(y_1)}, 1 \right).$$

The resulting moves only change the first component of the state vector, but similar moves can be introduced to change any of the remaining components.

Example 4.17   If the move $m \in M$ maps the current state $X \in \mathbb{R}^d$ to the proposal $Y = X + \varepsilon$, where the distribution of $\varepsilon \in \mathbb{R}^d$ is symmetric, that is where $P(\varepsilon \in A) = P(-\varepsilon \in A)$, then lemma 4.15 can be applied with

$$C = \left\{ (x, y) \in \mathbb{R}^d \times \mathbb{R}^d \mid x - y \in \mathrm{supp}(\varepsilon_j) \right\} \cong \mathbb{R}^d \times \mathrm{supp}(\varepsilon).$$

Here $\mathrm{supp}(\varepsilon)$ is the support of the distribution of ε and, since ε is symmetric, the support of ε satisfies $z \in \mathrm{supp}(\varepsilon)$ if and only if $-z \in \mathrm{supp}(\varepsilon)$. Thus condition (4.8) is satisfied. Also, the probabilities for going from x to $y = x + \varepsilon$ and for going from y to $x = y + (-\varepsilon)$ are the same and thus, for this example, we have $\varphi(y, x)/\varphi(x, y) = \pi(y)/\pi(x)$. Consequently, the acceptance probability for this move type is

$$\alpha_m(x, y) = \min \left(\frac{\gamma_m(y)\varphi(y, x)}{\gamma_m(x)\varphi(x, y)}, 1 \right) = \min \left(\frac{\gamma_m(y)\pi(y)}{\gamma_m(x)\pi(x)}, 1 \right).$$

In the case $\dim(\mathrm{supp}(\varepsilon)) < d$, this result is a generalisation of the random walk Metropolis sampling method from Section 4.1.3.

4.2   Convergence of Markov Chain Monte Carlo methods

MCMC methods are based on the following definition.

Definition 4.18   Let X be a random variable and f be a function such that $f(X) \in \mathbb{R}$. Then the *MCMC estimate* for $\mathbb{E}(f(X))$ is given by

$$Z_N^{\mathrm{MCMC}} = \frac{1}{N} \sum_{j=1}^{N} f(X_j) \tag{4.11}$$

where $(X_j)_{j \in \mathbb{N}}$ is a Markov chain with the distribution of X as a stationary distribution.

We have discussed methods for constructing such Markov chains in Section 4.1. In this section we will discuss conditions for the estimate Z_N^{MCMC} to converge to the expectation $\mathbb{E}(f(X))$ and we will also discuss factors affecting the error of this estimate.

4.2.1   Theoretical results

Whether or not the MCMC estimate Z_N^{MCMC} from (4.11) converges to the exact value $E(f(X))$ as $N \to \infty$, depends on the Markov chain X: if the Markov chain does not reach all parts of the state space, or if it moves too slowly, the estimator Z_N^{MCMC} will only correspond to an average of f over the visited part of the state space instead of to the full expectation $\mathbb{E}(f(X))$. Thus, for the approximation (4.1) to work, the Markov chain $(X_j)_{j \in \mathbb{N}}$ must move through the state space quickly enough. In this section we will discuss criteria for the Markov chain to explore the state space fast enough.

Definition 4.19 and definition 4.20, as well as the result from theorem 4.21, use the concept of 'σ-finite measures' from mathematical analysis. A measure μ on S assigns a 'size' $\mu(A)$ to subsets $A \subseteq S$. In order to avoid technical complications, we do not give the formal definition of a 'σ-finite measure' here but instead give a few examples:

- If $S = \mathbb{R}^d$, then the d-dimensional volume, that is $\mu(A) = |A|$ for all $A \subseteq S$, is a σ-finite measure on S.

- If S is finite or countable, the number of elements in a subset, that is $\mu(A) = \#A$ for all $A \subseteq S$, is a σ-finite measure on S.

- Every probability measure is a σ-finite measure on S, that is if X is a random variable we can use $\mu(A) = P(X \in A)$. In particular, the stationary distribution π of a Markov chain X is a σ-finite measure.

Definition 4.19   Let $(X_j)_{j \in \mathbb{N}}$ be a Markov chain with state space S and let μ be a σ-finite measure on S with $\mu(S) > 0$. Then the Markov chain $(X_j)_{j \in \mathbb{N}}$ is called μ-irreducible, if for every $A \subseteq S$ with $\mu(A) > 0$ and for every $x \in S$ there is a $j \in \mathbb{N}$ such that

$$P(X_j \in A \mid X_0 = x) > 0.$$

This definition formalises the idea that there are regions A in the state space which can always be reached by the process, independently of where it is started, thus preventing situations where the state space consists of different parts between which no transitions are possible. The following extension of the concept of μ-irreducibility makes sure that the times between returns to A are small enough so that infinitely many visits to A happen for every individual sample path of the Markov chain.

Definition 4.20   Let $(X_j)_{j \in \mathbb{N}}$ be a Markov chain with state space S. The Markov chain $(X_j)_{j \in \mathbb{N}}$ is called *Harris recurrent*, if there exists a σ-finite measure μ on S with $\mu(S) > 0$ such that the following conditions hold:

(a) $(X_j)_{j \in \mathbb{N}}$ is μ-irreducible.

(b) For every $A \subseteq S$ with $\mu(A) > 0$ and for every $x \in A$ we have

$$P\left(\sum_{j=1}^{\infty} \mathbb{1}_A(X_j) = \infty \,\middle|\, X_0 = x \right) > 0.$$

Using the concept of Harris recurrence we can state the main result of this section, guaranteeing convergence of the estimate Z_N^{MCMC} from (4.11) to the required limit.

Theorem 4.21   (law of large numbers for Markov chains) Let $(X_j)_{j \in \mathbb{N}}$ be a time homogeneous, Harris recurrent Markov chain on a state space S with stationary distribution π and let $X \sim \pi$. Let $f \colon S \to \mathbb{R}$ be a function such that the expectation $\mathbb{E}(f(X))$ exists. Then, for every initial distribution, we have

$$\lim_{N \to \infty} \frac{1}{N} \sum_{j=1}^{N} f(X_j) = \mathbb{E}(f(X))$$

with probability 1.

Proofs of this result can, for example, be found in the books by Robert and Casella (2004, theorem 6.63) and Meyn and Tweedie (2009, theorem 17.0.1).

In order to apply theorem 4.21 to the Markov chains generated by the Metropolis–Hastings algorithms 4.2 and 4.4, we need to find conditions for the resulting Markov chains to be Harris recurrent. While irreducibility as described in definition 4.19 is often relatively easy to check, the second condition in the definition 4.20 of Harris recurrence is less obvious. In many cases, verifying the conditions of theorem 4.21 for Markov chains generated by the Metropolis–Hastings algorithm is greatly simplified by the following result (Robert and Casella, 2004, lemma 7.3).

Lemma 4.22   Let $(X_j)_{j \in \mathbb{N}}$ be the Markov chain generated by the Metropolis–Hastings algorithms 4.2 or 4.4, with stationary distribution π. Assume that $(X_j)_{j \in \mathbb{N}}$ is π-irreducible. Then $(X_j)_{j \in \mathbb{N}}$ is Harris recurrent.

This lemma, together with theorem 4.21 allows the convergence of the Metropolis–Hastings method to be proved in many situations. For reference, we state the combined result as a corollary.

Corollary 4.23   Let $(X_j)_{j \in \mathbb{N}}$ be the Markov chain generated by the Metropolis–Hastings algorithms 4.2 or 4.4, with state space S, initial value $X_0 \in S$ (either random or deterministic) and stationary distribution π. Assume that $(X_j)_{j \in \mathbb{N}}$ is π-irreducible. Let $X \sim \pi$, let $f \colon S \to \mathbb{R}$ be a function such that the expectation $\mathbb{E}(f(X))$ exists and let Z_N^{MCMC} be defined by (4.11). Then we have

$$\lim_{N \to \infty} Z_N^{\text{MCMC}} = \mathbb{E}(f(X))$$

with probability 1.

Proof   The statement is a direct consequence of lemma 4.22 and theorem 4.21.   □

The usefulness of corollary 4.23 is illustrated by the following example.

Example 4.24   Assume that the target distribution is given by a density π on $S = \mathbb{R}^d$. Let $(X_j)_{j \in \mathbb{N}}$ be generated by the random walk Metropolis method from algorithm 4.9 with increments $\varepsilon_j \sim \mathcal{N}(0, \sigma^2)$. Assume that $A \subseteq \mathbb{R}^d$ with

$$\pi(A) = \int_A \pi(x)\,dx > 0$$

and $x \in \mathbb{R}^d$. The proposal Y_1 satisfies

$$P(Y_1 \in A \mid X_0 = x) = \int_A \psi_{0,\sigma^2}(y - x)\,dy,$$

where ψ_{0,σ^2} is the density of the $\mathcal{N}(0, \sigma^2)$-distribution. Similarly, the probability p of the proposal Y_1 falling in the set A and being accepted, conditional on $X_0 = x$, is given by

$$p = \int_A \alpha(x, y)\,\psi_{0,\sigma^2}(y - x)\,dy = \int_A \min\left(\frac{\pi(y)}{\pi(x)}, 1\right) \psi_{0,\sigma^2}(y - x)\,dy. \qquad (4.12)$$

Our aim is to show $p > 0$, thus proving π-irreducibility of the generated Markov chain in one time step.

Assume first that $\pi(x) > 0$. To show that $p > 0$ in this case, we will use the fact that for every function $h \geq 0$ we have $\int_A h(x)\,dx > 0$ if and only if $\left|\{y \in A \mid h(y) > 0\}\right| > 0$ where $|\cdot|$ denotes the d-dimensional volume. Thus, since $\int_A \pi(y)\,dy > 0$, we have

$$\left|\{y \in A \mid \pi(y) > 0\}\right| > 0.$$

Since $\pi(x) > 0$ and since the density ψ_{0,σ^2} is strictly positive, we find

$$\left|\left\{y \in A \mid \min\left(\frac{\pi(y)}{\pi(x)}, 1\right) \psi_{0,\sigma^2}(y - x) > 0\right\}\right| > 0$$

and thus, using (4.12), we get $p > 0$. For the case $\pi(x) = 0$, the value $\alpha(x, y)$ and thus the next step in algorithm 4.9 is not defined. There are two ways to deal with this case. Either we can restrict the state space S to only include points $x \in \mathbb{R}^d$ with $\pi(x) > 0$. In this case, corollary 4.23 guarantees convergence of the random walk Metropolis algorithm for all initial points X_0 with $\pi(X_0) > 0$. Alternatively, we can modify the algorithm to always accept the proposal whenever $\pi(x) = 0$

causes $\alpha(x, y)$ to be undefined. For the modified algorithm we get $p > 0$ for all initial values $x \in \mathbb{R}^d$. In this case, corollary 4.23 gives convergence of Z_N^{MCMC} for all $X_0 \in \mathbb{R}^d$.

The result from theorem 4.21 only states convergence of the *averages* Z_N^{MCMC} to the limit $\mathbb{E}(f(X))$, but does not give convergence of the *distributions* of the X_j to the distribution of X. In fact, it transpires that for convergence of the distributions to hold, the additional assumption of 'aperiodicity' of $(X_j)_{j \in \mathbb{N}}$ is required. In practice, the Markov chains generated by the Metropolis–Hastings method are normally aperiodic and thus, in situations where theorem 4.21 applies we normally also have convergence of the distribution of X_j to π as $j \to \infty$. For discussion of this topic we refer again to the books by Robert and Casella (2004, theorem 6.63) and Meyn and Tweedie (2009, theorem 17.0.1).

When applying the results from this section, it is important to keep in mind that theorem 4.21 and corollary 4.23 only make statements about the behaviour of the estimator Z_N^{MCMC} in the limit as $N \to \infty$. The results do not quantify the speed of convergence and convergence can sometimes be extremely slow (we will see an example of this at the end of Section 4.4.2).

4.2.2 Practical considerations

In the previous section we have seen that the Markov chain constructed for an MCMC method must be irreducible in order for the MCMC estimate to converge to the correct value $\mathbb{E}(f(X))$. In this section we will discuss different aspects of the speed of this convergence. As before, we consider the mean squared error

$$\mathrm{MSE}(Z_N^{\mathrm{MCMC}}) = \mathrm{Var}(Z_N^{\mathrm{MCMC}}) + \mathrm{bias}(Z_N^{\mathrm{MCMC}})^2, \qquad (4.13)$$

where the estimator Z_N^{MCMC} is given by equation (4.11) and we use lemma 3.12 to get the representation (4.13) for the mean squared error.

4.2.2.1 Convergence to stationarity

Since π is a stationary density of the Markov chain X, if we start the Markov chain with $X_0 \sim \pi$, we have $X_j \sim \pi$ for every $j \in \mathbb{N}$. While this is true in theory, in practice we usually cannot easily generate samples from the density π (otherwise we would be using Monte Carlo estimation instead of MCMC) and thus we need to start the Markov chain with a different distribution or with a fixed value. Consequently, for finite j, the distribution of X_j will not have density π and normally we will have $\mathbb{E}(f(X_j)) \neq \mathbb{E}(f(X))$. Thus, the argument from proposition 3.14 will no longer apply and we will have $\mathrm{bias}(Z_N^{\mathrm{MCMC}}) \neq 0$.

On the other hand, assuming aperiodicity of X, the distribution of X_j will converge to the correct distribution as $j \to \infty$ and thus we can normally expect

$\text{bias}(Z_N^{\text{MCMC}}) \to 0$ as $N \to \infty$. A typical approach to reduce the effect of bias is to omit the first samples from the Monte Carlo estimate in order to give the Markov chain time to get close to stationarity. In practice one often uses

$$\mathbb{E}(f(X)) \approx \frac{1}{N-M} \sum_{j=M+1}^{N} f(X_j) \tag{4.14}$$

instead of the estimate Z_N^{MCMC} from (4.11). The value M is chosen large enough that the process can be assumed close to stationarity, but small enough that enough samples remain for the estimate. Even if the samples $X_1, X_2, \ldots, X_M$ are not directly used in the estimate (4.14), the corresponding values still need to be computed in order to get the following values $X_{M+1}, X_{M+2}, \ldots$ In the context of the estimate (4.14), the time interval $1, 2, \ldots, M$ is called the *burn-in period*.

A very simple illustration of the effect of a burn-in period can be found in Figure 4.2. In the situation depicted there, the first values of the process do not resemble typical samples from the target distribution. The samples become useful only after the region of high probability is reached. The same kind of behaviour, a directional motion towards a region of high probability followed by fluctuations inside this region, can sometimes be observed in more realistic applications of MCMC and can the guide the choice of burn-in period.

From the results in Section 4.2.1 we know that a burn-in period is not required for convergence of the method. The main effect of a burn-in period is to remove

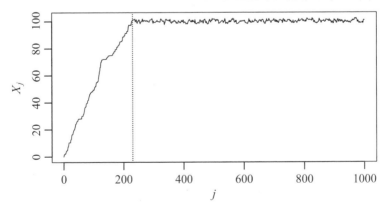

Figure 4.2   Path of a Metropolis–Hastings Markov chain, started far away from equilibrium: the displayed path corresponds to a random walk Metropolis method with target distribution $\mathcal{N}(100, 1)$, started at $X_0 = 0$. The first values of X_j, until the vertical line, do not behave like a sample from the target distribution. A good choice of the burn-in period for this Markov chain would be the time interval before the vertical line.

samples which would be extremely unlikely under the target distribution from the computation, thus reducing the sample size N required to get good estimates. An alternative approach to the use of a burn-in phase is to start the Markov chain with a 'typical value' for the target distribution, for example at the maximum of the target density π.

4.2.2.2 Effective sample size

The mean squared error given in equation (4.13) is determined by the bias (considered above) as well as the variance of the estimator. Since X forms a Markov chain, the samples X_j are not independent, and we have

$$
\begin{aligned}
\text{Var}\left(Z_N^{\text{MCMC}}\right) &= \text{Var}\left(\frac{1}{N}\sum_{j=1}^{N} f(X_j)\right) \\
&= \text{Cov}\left(\frac{1}{N}\sum_{j=1}^{N} f(X_j), \frac{1}{N}\sum_{l=1}^{N} f(X_l)\right) \\
&= \frac{1}{N^2}\sum_{j,l=1}^{N} \text{Cov}\left(f(X_j), f(X_l)\right).
\end{aligned}
\tag{4.15}
$$

If we assume that the Markov chain is close to stationarity, then we have $\text{Var}(X_j) \approx \text{Var}(X)$, that is the variance of X_j is approximately constant in time. Similarly, the joint distribution of X_j and X_l (approximately) only depends on the lag $k = l - j$ but not on the individual values of j and l. Thus, the covariance between $f(X_j)$ and $f(X_l)$ only (approximately) depends on $l - j$ and we get

$$
\text{Cov}\left(f(X_j), f(X_l)\right) \approx \text{Cov}\left(f(X_0), f(X_{l-j})\right).
$$

Similarly, if the Markov chain is close to stationarity, the correlation between $f(X_j)$ and $f(X_l)$ satisfies

$$
\begin{aligned}
\text{Corr}\left(f(X_j), f(X_l)\right) &= \frac{\text{Cov}\left(f(X_j), f(X_l)\right)}{\sqrt{\text{Var}\left(f(X_j)\right)\text{Var}\left(f(X_l)\right)}} \\
&\approx \frac{\text{Cov}\left(f(X_j), f(X_l)\right)}{\text{Var}\left(f(X)\right)}
\end{aligned}
\tag{4.16}
$$

and (approximately) only depends on $l - j$. This correlation is called the lag-$(l - j)$ autocorrelation of the Markov chain X.

Combining equation (4.15) and equation (4.16) and using the symmetry of the covariance we find

$$\text{Var}\left(Z_N^{\text{MCMC}}\right) = \frac{1}{N^2} \sum_{j,l=1}^{N} \text{Cov}\left(f(X_j), f(X_l)\right)$$

$$\approx \frac{1}{N} \text{Var}\left(f(X)\right) + 2\frac{1}{N^2} \sum_{j=1}^{N} \sum_{l=j+1}^{N} \text{Cov}\left(f(X_j), f(X_l)\right)$$

$$\approx \frac{1}{N} \text{Var}\left(f(X)\right) \left(1 + 2\frac{1}{N} \sum_{j=1}^{N} \sum_{l=j+1}^{N} \text{Corr}\left(f(X_0), f(X_{l-j})\right)\right)$$

$$\approx \frac{1}{N} \text{Var}\left(f(X)\right) \left(1 + 2\frac{1}{N} \sum_{j=1}^{N} \sum_{k=1}^{N-j} \text{Corr}\left(f(X_0), f(X_k)\right)\right).$$

Markov chains used in MCMC methods (under the assumption of aperiodicity) have the property that the distribution of a Markov chain is guaranteed to converge to the same distribution π for every initial distribution. Thus, the Markov chain 'forgets' the initial distribution over time and we expect the lag-k autocorrelations $\rho_k = \text{Corr}\left(f(X_0), f(X_k)\right)$ to converge to 0 as $k \to \infty$ (see Figure 4.3 for illustration). For sufficiently large N we have

$$\sum_{k=1}^{N-j} \text{Corr}\left(f(X_0), f(X_k)\right) \approx \sum_{k=1}^{\infty} \text{Corr}\left(f(X_0), f(X_k)\right)$$

and, since the right-hand side no longer depends on j we find

$$\text{Var}\left(Z_N^{\text{MCMC}}\right) \approx \frac{1}{N} \text{Var}\left(f(X)\right) \left(1 + 2\frac{1}{N} \sum_{j=1}^{N} \sum_{k=1}^{\infty} \text{Corr}\left(f(X_0), f(X_k)\right)\right)$$

$$= \frac{1}{N} \text{Var}\left(f(X)\right) \left(1 + 2 \sum_{k=1}^{\infty} \text{Corr}\left(f(X_0), f(X_k)\right)\right). \tag{4.17}$$

Comparing $\text{Var}\left(Z_N^{\text{MCMC}}\right)$ from equation (4.17) to the corresponding variance $\text{Var}(e^{\text{MC}})$ from (3.8) for the basic Monte Carlo estimate we see that the variance, and thus the error, of the MCMC estimate differs by the presence of the additional correlation term on the right-hand side of (4.17). Typically this term is positive and thus MCMC methods are less efficient than a basic Monte Carlo method would be.

While the above derivation is heuristic, an exact result such as (4.17) can be derived under conditions very similar to the conditions required for theorem 4.21.

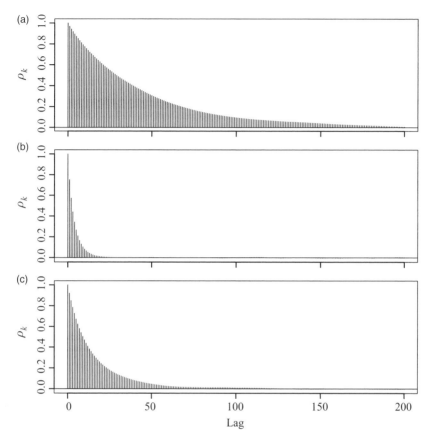

Figure 4.3   The autocorrelation functions for the Markov chain from example 4.10 and example 4.25 for different values of σ: (a) σ = 1; (b) σ = 6; (c) σ = 36. The corresponding estimation problem is to find $\mathbb{E}(X^2)$. The values of σ considered here are the same as for the paths shown in Figure 4.1. The autocorrelation function in (b) decays fastest, indicating that this is the most efficient of the three MCMC methods considered here.

Such a result can, for example, be found as part of theorem 17.0.1 in the book by Meyn and Tweedie (2009).

In analogy to equation (3.8) we write

$$\text{Var}\left(Z_N^{\text{MCMC}}\right) \approx \frac{1}{N_{\text{eff}}} \text{Var}\left(f(X)\right)$$

where

$$N_{\text{eff}} = \frac{N}{1 + 2\sum_{k=1}^{\infty} \text{Corr}\left(f(X_0),\, f(X_k)\right)}$$

is called the *effective sample size*. The effective sample size does not only depend on N and on the Markov chain X, but it also depends on the function f used in the estimation problem. The value N_{eff} can be used to quantify the increase in computational cost caused by the dependence of samples in an MCMC method. Compared with a Monte Carlo method for the same problem, the number of samples generated needs to be increased by a factor N/N_{eff} to achieve a comparable level of error. The empirical sample size can be used to compare the efficiency of different MCMC methods: the larger N_{eff}/N is, the more efficient the resulting method.

Example 4.25   In example 4.10 and the corresponding Figure 4.1 we considered a family of random walk Metropolis methods for the target distribution π given by equation (4.6). In the methods considered there, we can choose the variance σ of the increments of the random walk. To compare the efficiency of the random walk Metropolis algorithm for different values of σ, we can consider the effective sample sizes N_{eff}.

The autocorrelations and thus the effective sample size depend on the specific estimation problem, that is on the choice of the function f in (4.1). Here we consider the problem of estimating the variance of X. Since π is symmetric, we have $\mathbb{E}(X) = 0$ and thus $\text{Var}(X) = \mathbb{E}(X^2)$. Thus, we consider $f(x) = x^2$.

The autocorrelations $\text{Corr}\left(f(X_j), f(X_{j+k})\right)$ for this choice of f are shown in Figure 4.3. Using these autocorrelations, we find the following effective samples sizes.

	$\sigma = 1$	$\sigma = 6$	$\sigma = 36$
N_{eff}	$0.0119 \cdot N$	$0.1317 \cdot N$	$0.0337 \cdot N$
N/N_{eff}	84.0	7.6	29.7

Thus, even ignoring the additional error introduced by bias, the MCMC method from example 4.10 needs for $\sigma = 1$ approximately 84 times more samples than a basic Monte Carlo method with the same error would need. The efficiency can be greatly improved by tuning the parameter σ, for example for $\sigma = 6$ the MCMC method only needs approximately 7.6 times more samples than a basic Monte Carlo method. Thus we see that $\sigma = 6$ is a much better choice for the given problem than either $\sigma = 1$ or $\sigma = 36$.

4.2.2.3   Acceptance rates

By comparing Figure 4.1 and Figure 4.3 we see that the effective sample size in example 4.25 is influenced by two different effects. In Figure 4.1(a) and Figure 4.3(a), where the variance σ^2 of the increments is smallest, the effective sample size is large because the small increments allow the process only slow movement through the state space; reaching different regions of state space takes a long time and thus samples close in time have a tendency to also be close in space. In contrast, the constant

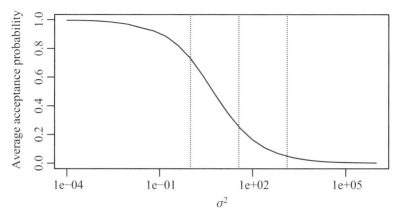

Figure 4.4 The average acceptance probability $\mathbb{E}(\alpha(X_{j-1}, Y_j))$ *of the random walk Metropolis sampler from example 4.10 and example 4.25, as a function of* σ^2. *The three vertical lines (from left to right) correspond to the values* $\sigma^2 = 1^2$, $\sigma^2 = 6^2$, *and* $\sigma^2 = 36^2$ *from Figure 4.3.*

stretches of the path in Figure 4.1(c) indicate that the large σ^2 considered there leads to very low acceptance rates and this, in turn causes the small effective sample size for large values of σ.

The observation that the acceptance rate affects the efficiency of MCMC methods can be used as a simple, heuristic tuning criterion: in a random walk Metropolis method, we expect the average acceptance rate to decrease as the variance σ^2 of the increments increases. (This decrease is not necessarily monotonically, though.) The variance σ^2 should be chosen as large as possible under the constraint that the average acceptance probability should not be too small. To illustrate this, Figure 4.4 shows the average acceptance rate for the random walk Metropolis methods considered in example 4.10 and example 4.25, as a function of σ^2.

4.2.2.4 Convergence diagnostics

Even in cases where convergence of an MCMC method is formally proved convergence to equilibrium can be extremely slow, resulting in inaccurate and misleading estimates. Situations where this problem occurs include:

- In multimodal distributions the Markov chain can get 'stuck' in a mode for a very long time. This will be the case if transitions between modes require a very unlikely sequence of steps, for example if the states between the modes are very unlikely and if direct transitions between the modes are never proposed.

  If the positions of modes of the target distribution are not known in advance, this problem can be very difficult to detect since, before the first transition between modes happens, the path of the Markov chain 'looks stationary'. If the problem is not detected, the resulting estimate Z_N^{MCMC} will estimate the conditional expectation $\mathbb{E}\left(f(X) \mid X \in A\right)$ where A is the mode in question, instead of the full expectation $\mathbb{E}\left(f(X)\right)$.

One way to resolve this issue is to use an MCMC method with different move types (see Section 4.1.5) and to introduce special move types which perform transitions between modes.

- In high-dimensional situations it takes a long time for a Markov chain to explore all of the space. In cases where most of the mass of the target distribution is concentrated in a small region of the state space, the Markov chain may take an extremely long time to find this region or even may never reach the relevant region of space at all.

  This kind of problem can be recognised by the fact that the Markov chain moves through space 'aimlessly', possibly behaving like a random walk, or diverging to infinity. If enough information about the target distribution is available, this problem can be resolved by starting the Markov chain in the region of space where the target distribution is concentrated.

To recognise problems caused by slow convergence, so called *convergence diagnostics* can be used. Here we restrict ourselves to a few very simple cases. First, it is always a good idea to plot some one-dimensional functions of the state of the Markov chain over time. If the Markov chain is one-dimensional, this could be the state X_j itself. In the case of multidimensional Markov chains this could be, for example, the coordinate with the highest variance. The resulting plots (see for example Figure 4.1 and Figure 4.2) allow to visually assess whether the process is close to stationarity.

To detect the presence of different modes, one can run several copies of the Markov chain, using different starting values. The idea here is then to assess whether all of these Markov chains move into the same mode of the distribution, or whether they end up in different modes. This question can either be answered visually by using plots as above, or by analytical methods, for example based on the idea of comparing the sample variance of the samples from each instance of the Markov chain to the sample variance of all generated samples combined. A commonly used convergence diagnostic based on this idea is introduced in Gelman and Rubin (1992).

Convergence diagnostics can be used both to determine a good burn-in period and to determine whether a given value of N is large enough to give useful estimates.

4.2.2.5   Numerical errors

A final consideration when using MCMC methods in practice is the effect of rounding error caused by the number representation in the computer. While stochastic methods are often quite robust in this regard, the effect of rounding errors often causes problems when evaluating the acceptance probability $\alpha(X_{j-1}, Y_j)$ in the Metropolis–Hastings algorithm.

As an example we consider here the case of algorithm 4.2, where the acceptance probability has the form

$$\alpha(x, y) = \min\left(\frac{\pi(y)p(y, x)}{\pi(x)p(x, y)}, 1\right). \tag{4.18}$$

Due to the restrictions of number representation in the computer, the numerical values for probabilities $\pi(x)$ and $p(x, y)$ in the program will not exactly coincide with the exact values and problems can occur in cases where π gets very small. To avoid the resulting problems, equation (4.18) can be rewritten as

$$\alpha(x, y) = \min\left(\exp\left(\log\frac{\pi(y)p(y, x)}{\pi(x)p(x, y)}\right), 1\right) \tag{4.19}$$
$$= \min\left(\exp\left(\log\pi(y) + \log p(y, x) - \log\pi(x) - \log p(x, y)\right), 1\right).$$

This representation avoids numerical problems by performing all arithmetic on the logarithmic terms, which are of much smaller magnitude, and only exponentiating the result after all possible cancellations have taken place.

4.3 Applications to Bayesian inference

In a Bayesian model, one assumes that the parameter θ of a statistical model is itself random. We consider the following setting:

- Data $x = (x_1, \ldots, x_n)$ is given, where x forms a sample of i.i.d. random variables $X_1, \ldots, X_n \in \mathbb{R}^d$. Here we assume that the distribution of the X_i has a density $\varphi_{X|\theta}(x \mid \theta)$, so that the distribution of $X = (X_1, \ldots, X_n)$ is given by

$$p_{X|\theta}(x \mid \theta) = \prod_{i=1}^{n} \varphi_{X|\theta}(x_i \mid \theta) \tag{4.20}$$

  for all $x = (x_1, \ldots, x_n) \in (\mathbb{R}^d)^n$ and all $\theta \in \mathbb{R}^p$.

- The distribution of the data depends on an unknown parameter $\theta \in \mathbb{R}^p$ where θ is assumed to be random with density $p_\theta(\theta)$. The distribution p_θ of θ is called the *prior distribution*.

Thus, the data are assumed to be generated by a two-step procedure where first θ is chosen randomly and then, given the value of θ, samples $X_1, \ldots, X_n \sim \varphi_{X|\theta}(\cdot \mid \theta)$ are generated to obtain the observations $x_1, \ldots, x_n$.

Here, symbols such as $p_{X|\theta}$ denote densities and the subscript indicates that this is the conditional density of X, given the value of θ. The variables used as arguments of these densities are mostly denoted by the corresponding lower case letters, that is $p_X(x)$ is the density of X at the point $x \in (\mathbb{R}^d)^n$. By a slight abuse of notation, we use lower case θ both for the random variable and for the corresponding values.

Our aim is to gather as much information as possible about the unknown para-meter θ from given data x. Typically, the data x do not completely determine the value of θ. Instead, since θ is assumed to be random, the solution of this parameter estimation problem will be given by the conditional distribution of θ, given the observation $X = x$. This distribution is called the *posterior distribution*; it depends both on the data x and on the prior distribution for θ. By Bayes' rule (see Section A.2), we find the density of the posterior distribution as

$$p_{\theta|X}(\theta \mid x) = \frac{p_{X|\theta}(x \mid \theta)p_\theta(\theta)}{\int p_{X|\theta}(x \mid t)p_\theta(t)\,dt} = \frac{1}{Z}\,p_{X|\theta}(x \mid \theta)p_\theta(\theta) \qquad (4.21)$$

for all $\theta \in \mathbb{R}^p$, where $Z = \int p_{X|\theta}(x \mid t)p_\theta(t)\,dt$ is the normalisation constant. The value x in equation (4.21) is the given data and thus a constant and we will consider the given expression as a function of θ.

Denote the true value of θ which was used to generate the observations x by θ^*. If sufficient data are given and if the dependence of the distribution of the X_i on the parameter θ is sensitive enough, the posterior distribution $p_{\theta|X}$ will be concentrated around the true value θ^* and the more data are given the more concentrated the posterior will be. Given the posterior distribution from equation (4.21), we can obtain results such as:

- An estimates for the unknown parameter value can be found by either taking the conditional expectation $\mathbb{E}(\theta \mid X = x)$ or by considering the mode or median of the posterior distribution.

- The uncertainty remaining in the estimate can be quantified by considering the conditional variance $\mathrm{Var}(\theta \mid X = x)$ or the corresponding standard deviation.

- More specific questions about the possible behaviour of the parameter, given the available observations, can be obtained by studying the posterior distribution. For example the probability that the parameter is inside a region $A \subseteq \mathbb{R}^p$ can be found as $P(\theta \in A \mid X = x)$.

While the posterior distribution is known 'explicitly' from equation (4.21), eval-uating expectations or probabilities with respect to the posterior distribution can be challenging in practice. Reasons for this include the fact that often the value of the normalisation constant Z in (4.21) cannot be found, and also the general problems involved in evaluating expectations for distributions with complicated densities.

Following the approach described in Section 3.1, a widely used strategy is to employ Monte Carlo and MCMC methods to evaluate expectations with respect to the posterior distribution. This allows us to evaluate expectations of the form $\mathbb{E}(f(\theta) \mid X = x)$, if we are able to generate samples from the distribution with density

$$\pi(\theta) = p_{\theta|X}(\theta \mid x) = \frac{1}{Z}\,p_{X|\theta}(x \mid \theta)p_\theta(\theta) \qquad (4.22)$$

for all $\theta \in \mathbb{R}^p$, where x is the given data. In some simple situations, for example in example 3.6, it is possible to generate the required samples using rejection sampling, but often it is cumbersome to find a feasible proposal distribution and thus samples from the density (4.22) are more commonly generated using MCMC methods. This is the approach we will consider here.

Since the posterior distribution is typically concentrated around the true parameter value θ^*, we do not expect any problems arising from multimodality and we can use the random walk Metropolis algorithm described in Section 4.1.3. For this algorithm, the probability of accepting a proposal $\tilde{\theta}$ if the previous state was θ is given by

$$\alpha(\theta, \tilde{\theta}) = \min\left(\frac{\pi(\tilde{\theta})}{\pi(\theta)}, 1\right) = \min\left(\frac{p_{X|\theta}(x \mid \tilde{\theta})p_\theta(\tilde{\theta})}{p_{X|\theta}(x \mid \theta)p_\theta(\theta)}, 1\right) \qquad (4.23)$$

where x is the given observations and where we use the fact that the normalisation constant Z from (4.22) in the numerator and denominator cancels. The right-hand side in (4.23) can be further expanded using the definition of $p_{X|\theta}$ from equation (4.20). Since, for large n, the product in (4.20) can be very small, we also use the trick from equation (4.19) to avoid potential problems caused by rounding errors. The resulting expression is

$$\frac{p_{X|\theta}(x \mid \tilde{\theta})p_\theta(\tilde{\theta})}{p_{X|\theta}(x \mid \theta)p_\theta(\theta)} = \frac{\prod_{i=1}^{n} \varphi_{X|\theta}(x_i \mid \tilde{\theta})p_\theta(\tilde{\theta})}{\prod_{i=1}^{n} \varphi_{X|\theta}(x_i \mid \theta)p_\theta(\theta)}$$

$$= \exp\left(\sum_{i=1}^{n} \log \varphi_{X|\theta}(x_i \mid \tilde{\theta}) + \log p_\theta(\tilde{\theta}) \qquad (4.24)\right.$$

$$\left. - \sum_{i=1}^{n} \log \varphi_{X|\theta}(x_i \mid \theta) - \log p_\theta(\theta)\right).$$

Substituting this expression into (4.23) gives the form in which $\alpha(\theta, \tilde{\theta})$ should be used in an implementation of algorithm 4.9.

Example 4.26   For the prior let $\mu \sim \mathcal{U}[-10, 10]$ and $\sigma \sim \text{Exp}(1)$, independently, and set $\theta = (\mu, \sigma) \in \mathbb{R}^2$. Assume that the data consist of i.i.d. values $X_1, \ldots, X_n \sim \mathcal{N}(\mu, \sigma^2)$ and that we have observed values $x = (x_1, \ldots, x_n)$ for the data. In this case, the prior density is

$$p_{\mu,\sigma}(\mu, \sigma) = \frac{1}{20}\mathbb{1}_{[-10,10]}(\mu) \cdot \exp(-\sigma)\mathbb{1}_{[0,\infty)}(\sigma)$$

and the conditional density of the observations given the parameters is

$$p_{X|\mu,\sigma}(x \mid \mu, \sigma) = \prod_{i=1}^{n} \frac{1}{\sqrt{2\pi\sigma^2}} \exp\left(-\frac{(x_i - \mu)^2}{2\sigma^2}\right)$$

$$= \frac{1}{(2\pi)^{n/2}} \cdot \frac{1}{\sigma^n} \exp\left(-\frac{1}{2\sigma^2}\sum_{i=1}^{n}(x_i - \mu)^2\right).$$

The constants $1/20$ and $1/(2\pi)^{n/2}$ cancel in the numerator and denominator of (4.23) and using (4.24) we find

$$
\frac{p_{X|\mu,\sigma}(x \mid \tilde{\mu}, \tilde{\sigma}) p_{\mu,\sigma}(\tilde{\mu}, \tilde{\sigma})}{p_{X|\mu,\sigma}(x \mid \mu, \sigma) p_{\mu,\sigma}(\mu, \sigma)}
$$

$$
= \mathbb{1}_{[-10,10]}(\tilde{\mu}) \cdot \mathbb{1}_{[0,\infty)}(\tilde{\sigma}) \cdot \frac{\sigma^n}{\tilde{\sigma}^n}
$$

$$
\cdot \exp\left(-\frac{1}{2\tilde{\sigma}^2} \sum_{i=1}^{n} (x_i - \tilde{\mu})^2 - \tilde{\sigma} + \frac{1}{2\sigma^2} \sum_{i=1}^{n} (x_i - \mu)^2 + \sigma \right)
$$

for all $\tilde{\mu}, \tilde{\sigma} \in \mathbb{R}$ and all $\mu \in [-10, 10]$ and $\sigma \geq 0$. Here we can assume that μ and σ are inside the valid parameter range (and thus can omit the corresponding indicator functions in the denominator), because $\alpha(\mu, \sigma, \tilde{\mu}, \tilde{\sigma})$ only needs to be defined for (μ, σ) with $\pi(\mu, \sigma) > 0$. Finally, substituting these expressions into equation (4.23) we get the acceptance probability for the random walk Metropolis algorithm for this example:

$$
\alpha(\mu, \sigma, \tilde{\mu}, \tilde{\sigma})
$$

$$
= \min\left(\mathbb{1}_{[-10,10]}(\tilde{\mu}) \cdot \mathbb{1}_{[0,\infty)}(\tilde{\sigma}) \cdot \frac{\sigma^n}{\tilde{\sigma}^n} \right.
$$

$$
\left. \cdot \exp\left(-\frac{1}{2\tilde{\sigma}^2} \sum_{i=1}^{n} (x_i - \tilde{\mu})^2 - \tilde{\sigma} + \frac{1}{2\sigma^2} \sum_{i=1}^{n} (x_i - \mu)^2 + \sigma \right), 1 \right)
$$

In order to apply the random walk Metropolis algorithm as given in algorithm 4.9, we have to choose a distribution for the increments ε_j. This choice affects the efficiency of the resulting MCMC method. For simplicity we choose here $\varepsilon_j \sim \mathcal{N}(0, \sigma^2 I_p)$ where I_p is the p-dimensional identity matrix. The parameter σ^2 can be used to tune the method for efficiency as described in Section 4.2.2. This leads to the following algorithm:

1: choose a θ_0 with $\pi(\theta_0) > 0$
2: **for** $j = 1, 2, 3, \ldots$ **do**
3:     generate $\varepsilon_j \sim \mathcal{N}(0, \sigma^2 I_p)$
4:     $\tilde{\theta}_j \leftarrow \theta_{j-1} + \varepsilon_j$
5:     compute $a = \alpha(\theta_{j-1}, \tilde{\theta}_j)$ using (4.23) and (4.24)
6:     generate $U_j \sim \mathcal{U}[0, 1]$
7:     **if** $U_j \leq a$ **then**
8:         $\theta_j \leftarrow \tilde{\theta}_j$
9:     **else**
10:        $\theta_j \leftarrow \theta_{j-1}$
11:    **end if**
12: **end for**

From proposition 4.3 we know that the output of this algorithm is a Markov chain with stationary distribution $\pi = p_{\theta|X}(\cdot \,|\, x)$, that is that it generates samples from the posterior distribution in our Bayesian parameter estimation problem. Using corollary 4.23 and example 4.24 we the find that

$$\mathbb{E}\left(f(\theta)\,\big|\, X = x\right) = \lim_{N \to \infty} \frac{1}{N} \sum_{j=1}^{N} f\left(\theta_j\right).$$

This convergence holds for every initial value θ_0 with $\pi(\theta_0) > 0$. Thus we have solved the given estimation problem.

While the MCMC method works for every initial value θ_0, we have seen in Section 4.2.2 that convergence of the results can be improved if θ_0 is chosen inside a region where the target distribution π is large. If a classical point estimate $\hat{\theta}_n = \hat{\theta}_n(x_1, \ldots, x_n)$ for θ is available, $\theta_0 = \hat{\theta}_n$ will often be a good choice.

Many generalisations of the methodology presented in this section are possible: if some of the parameters or the observations are discrete instead of continuous, very similar methods can be derived by replacing the corresponding densities with probability weights. Also, MCMC methods can be used for Bayesian inference where the models have a more complex structure than in the situation of the present section. Examples of such situations can be found in Sections 4.4.2 (mixture model with known number of components) and 4.5.2 (mixture model with unknown number of components).

4.4   The Gibbs sampler

The Gibbs sampler is an MCMC method which can be used if the state space has a product structure. It is based on the idea of updating the different components of the state one at a time, thus potentially reducing a high-dimensional sampling problem to a sequence of more manageable, low-dimensional sampling problems.

4.4.1   Description of the method

The Gibbs sampler is applicable in situations where the state space S can be written as a finite product of spaces, that is where we have

$$S = S_1 \times S_2 \times \cdots \times S_n.$$

Elements of this space are vectors $x = (x_i)_{i=1,2,\ldots,n}$ where $x_i \in S_i$ for every i. Situations where the Gibbs sampler can be applied include:

- In Bayesian parameter estimation problems, n is typically small, say 2 or 3, and often the spaces S_i have different dimensions. We will study examples of this type in Section 4.4.2.

- In applications from statistical mechanics or image processing, n is typically large but all spaces S_i are the same. In this case we write

$$S = C^I$$

where $I = \{1, 2, \ldots, n\}$ and C is the state space for the individual components. We will see examples of such situations in Section 4.4.3.

As with other MCMC methods, our aim is to construct a Markov chain with a given distribution π on S as its stationary distribution. In algorithm 4.27 we use the conditional distribution of a single component X_i of $X \in S$, conditioned on the values of all other components and we denote this distribution by $\pi_{X_i|X_{-i}}(\cdot \,|\, x_1, \ldots, x_{i-1}, x_{i+1}, \ldots, x_n)$. Here, X_{-i} stands for X with the component X_i left out. In the case of discrete state space S_i, the conditional distribution of X_i is given by probability weights $\pi_{X_i|X_{-i}}$ whereas, if S_i is continuous, $\pi_{X_i|X_{-i}}$ is normally given as a probability density. To simplify notation, in this section we write the time of the Markov chain as an upper index, so that $X_i^{(j)}$ corresponds to the value of component $i \in I$ of the Markov chain X at time $j \in \mathbb{N}_0$. Using this notation, the general Gibbs sampler algorithm takes the following form.

Algorithm 4.27   (Gibbs sampler)
   input:
      a distribution $\pi = \pi_{X_1,\ldots,X_n}$
      initial state $X_0 \in S = S_1 \times S_2 \times \cdots \times S_n$
   randomness used:
      samples from the distributions $\pi_{X_i|X_{-i}}(\cdot \,|\, x_1, \ldots, x_{i-1}, x_{i+1}, \ldots, x_n)$
   output:
      a sample of a Markov chain $(X^{(j)})_{j \in \mathbb{N}}$ with stationary distribution π
   1: **for** $j = 1, 2, 3, \ldots$ **do**
   2:    let $m \leftarrow (j - 1 \bmod n) + 1$
   3:    generate

$$\xi^{(j)} \sim \pi_{X_m|X_{-m}}(\cdot \,|\, X_1^{(j-1)}, \ldots, X_{m-1}^{(j-1)}, X_{m+1}^{(j-1)}, \ldots, X_n^{(j-1)})$$

   4:    define $X^{(j)} \in S$ by

$$X_i^{(j)} = \begin{cases} X_i^{(j-1)} & \text{if } i \neq m \text{ and} \\ \xi^{(j)} & \text{otherwise} \end{cases}$$

      for all $i = 1, 2, \ldots, n$.
   5: **end for**

The index $m = (j - 1 \mod n) + 1$ constructed in step 2 of algorithm 4.27 periodically cycles through the components $1, 2, \ldots, n$ as j increases. Alternative versions of the Gibbs sampler algorithm can be constructed where the index $m \in \{1, \ldots, n\}$ is chosen randomly in each iteration of the loop in algorithm 4.27 instead.

Proposition 4.28 The process $(X^{(j)})_{j \in \mathbb{N}}$ constructed by algorithm 4.27 is a Markov chain with stationary density π.

Proof Since in each step of the algorithm $X^{(j)}$ is constructed from $X^{(j-1)}$ and additional randomness, the process $X = (X^{(j)})_{j \in \mathbb{N}}$ is a Markov chain starting in X_0. Assume $X_{j-1} \sim \pi$. Then

$$X_{\neg m}^{(j-1)} = (X_1^{(j-1)}, \ldots, X_{m-1}^{(j-1)}, X_{m+1}^{(j-1)}, \ldots, X_n^{(j-1)}) \sim \pi_{\neg m}.$$

The construction of $X_m^{(j)} = \xi^{(j)}$ in line 3 of the algorithm depends on the values of the components of $X_{\neg m}^{(j-1)}$. Assuming the distribution of the components $X_i^{(j)}$ has a density, we can integrate over all possible values of $X_{\neg m}^{(j-1)}$ to find

$$P(X_1^{(j)} \in A_1, \ldots, X_n^{(j)} \in A_n)$$
$$= \int \int \mathbb{1}_{A_1 \times \cdots \times A_n}(x_1, \ldots, x_n)$$
$$\cdot \pi_{X_m | X_{\neg m}}(x_m \mid x_{\neg m}) \, dx_m$$
$$\cdot \pi_{X_{\neg m}}(x_{\neg m}) \, dx_1 \cdots dx_{m-1} \, dx_{m+1} \cdots dx_n$$
$$= \int \int \mathbb{1}_{A_1 \times \cdots \times A_n}(x_1, \ldots, x_n) \cdot \pi_{X_1, \ldots, X_m}(x_1, \ldots, x_m) \, dx_1 \cdots dx_n.$$

This shows that $(X_1^{(j)}, \ldots, X_n^{(j)})$ has density $\pi = \pi_{X_1, \ldots, X_m}$ as required. If the distribution of the $X_i^{(j)}$ is given by probability weights instead of densities, the same result can be obtained by replacing the integrals over densities with sums over probability weights. □

Example 4.29 Let $A \subseteq \mathbb{R}^2$ be a set with area $|A| < \infty$ and let X be uniformly distributed on A, that is $X \sim \pi$ where the density π is given by

$$\pi(x) = \frac{1}{|A|} \mathbb{1}_A(x)$$

for all $x \in \mathbb{R}^2$. Writing the state space $\mathbb{R}^2$ as $\mathbb{R} \times \mathbb{R}$, we can apply the Gibbs sampler from algorithm 4.27 to generate samples from the distribution of X. Using equation (A.5), we can find the conditional density of X_1 given $X_2 = x_2 \in \mathbb{R}$ as

$$
\begin{aligned}
\pi_{X_1 | \neg X_1}(x_1 \mid x_2) &= \pi_{X_1 | X_2}(x_1 \mid x_2) \\
&= \frac{\pi(x_1, x_2)}{\int_{\mathbb{R}} \pi(\tilde{x}_1, x_2) \, d\tilde{x}_1} \\
&= \frac{\frac{1}{|A|} \mathbb{1}_A(x_1, x_2)}{\int_{\mathbb{R}} \frac{1}{|A|} \mathbb{1}_A(\tilde{x}_1, x_2) \, d\tilde{x}_1} \\
&= \frac{1}{\left| \{ \tilde{x}_1 \mid (\tilde{x}_1, x_2) \in A \} \right|} \mathbb{1}_{\{ \tilde{x}_1 | (\tilde{x}_1, x_2) \in A \}}(x_1)
\end{aligned}
$$

for all $x_1 \in \mathbb{R}$. Thus, $\pi_{X_1 | \neg X_1}$ is the density of the uniform distribution on the set $\{ \tilde{x}_1 \mid (\tilde{x}_1, x_2) \in A \}$, that is on the 'slice' of A where the second coordinate is fixed to x_2. Similarly, the density $\pi_{X_2 | \neg X_2} = \pi_{X_2 | X_1}$ transpires to be the density of the uniform distribution on the 'slice' $\{ \tilde{x}_2 \mid (x_1, \tilde{x}_2) \in A \}$, where the first coordinate is fixed to the value x_1. Thus, the Gibbs sampler for π takes the following form:

1: **for** $j = 1, 2, 3, \ldots$ **do**
2:     **if** j is odd **then**
3:         generate $\xi^{(j)} \sim \mathcal{U}\{ x_1 \mid (x_1, X_2^{(j-1)}) \in A \}$
4:         $X^{(j)} \leftarrow \left(\xi^{(j)}, X_2^{(j-1)} \right)$
5:     **else**
6:         generate $\xi^{(j)} \sim \mathcal{U}\{ x_2 \mid (X_1^{(j-1)}, x_2) \in A \}$
7:         $X^{(j)} \leftarrow \left(X_1^{(j-1)}, \xi^{(j)} \right)$
8:     **end if**
9: **end for**

From Lemma 1.33 we know that we can generate samples X with density f by sampling (X, Y) from the uniform distribution on the set

$$
A = \left\{ (x, y) \in \mathbb{R} \times [0, \infty) \mid 0 \leq y < f(x) \right\} \subseteq \mathbb{R}^2.
$$

Such samples can be obtained using the Gibbs sampler as explained in the first part of this example. The resulting method and its higher dimensional variants are known as the *slice sampler*.

Despite the fact that the Gibbs sampler from algorithm 4.27 does not contain a rejection mechanism, the algorithm is a special case of the Metropolis–Hastings algorithm with different move types as introduced in Section 4.1.5. To see this, consider the Metropolis–Hastings algorithm with move types $M = \{ 0, 1, \ldots, n - 1 \}$.

Assume that the algorithm, at time j, always performs a move of type $m = j \bmod n$ by constructing the proposal Y from the current state X as

$$Y_i = \begin{cases} X_i & \text{for } i \neq m \text{ and} \\ \xi & \text{for } i = m \end{cases}$$

for $i = 1, 2, \ldots, n$, where

$$\xi \sim \pi_{X_m \mid X_{\neg m}}(\cdot \mid X_1, \ldots, X_{m-1}, X_{m+1}, \ldots, X_n).$$

The random variable Y, given X, is concentrated on the subspace $\{X_1\} \times \cdots \times \{X_{m-1}\} \times S_m \times \{X_{m+1}\} \times \cdots \times \{X_n\}$. Thus, Y has no density but instead we are in the situation of lemma 4.15. If we let

$$C_m = \{(x, y) \in S \times S \mid x_i = y_i \text{ for all } i \neq m\} \cong S \times S_m,$$

we have $(X, Y) \in C_m$ with probability 1 and, assuming $X \sim \pi$,

$$P(Y \in B, X \in A)$$
$$= \int_S \pi(x) \mathbb{1}_A(x) \int_{S_m} \pi_{X_m \mid X_{\neg m}}(y_m \mid x_{\neg m})$$
$$\cdot \mathbb{1}_B(x_1, \ldots, x_{m-1}, y_m, x_{m+1}, \ldots, x_n) \, dy_m \, dx_1 \cdots dx_n.$$

Thus, the pair (X, Y) has density $\varphi(x, y_m) = \pi(x) \pi_{X_m \mid X_{\neg m}}(y_m \mid x_{\neg m})$ on C_m. From lemma 4.15 we know now that we should choose the acceptance probability $\alpha_m(x, y)$ for this move type as

$$\alpha_m(x, y) = \min\left(\frac{\gamma_m(y)\,\varphi(y, x)}{\gamma_m(x)\,\varphi(x, y)}, 1\right)$$
$$= \min\left(\frac{\gamma_m(y)\,\pi(y)\,\pi_{X_m \mid X_{\neg m}}(x_m \mid y_{\neg m})}{\gamma_m(x)\,\pi(x)\,\pi_{X_m \mid X_{\neg m}}(y_m \mid x_{\neg m})}, 1\right)$$

Since all possible transitions from x to y satisfy $x_{\neg m} = y_{\neg m}$, the acceptance probabilities can be rewritten as

$$\alpha_m(x, y) = \min\left(\frac{\gamma_m(y)\,\pi(y)\,\pi_{X_m \mid X_{\neg m}}(x_m \mid x_{\neg m})}{\gamma_m(x)\,\pi(x)\,\pi_{X_m \mid X_{\neg m}}(y_m \mid y_{\neg m})}, 1\right)$$
$$= \min\left(\frac{\pi(y)\,\pi_{X_m \mid X_{\neg m}}(x_m \mid x_{\neg m})\,\pi_{X_{\neg m}}(x_{\neg m})}{\pi(x)\,\pi_{X_m \mid X_{\neg m}}(y_m \mid y_{\neg m})\,\pi_{X_{\neg m}}(y_{\neg m})}, 1\right)$$
$$= \min\left(\frac{\pi(y)\,\pi_{X_m, X_{\neg m}}(x_m, x_{\neg m})}{\pi(x)\,\pi_{X_m, X_{\neg m}}(y_m, y_{\neg m})}, 1\right)$$
$$= \min\left(\frac{\pi(y)\,\pi(x)}{\pi(x)\,\pi(y)}, 1\right)$$
$$= 1.$$

Thus, moves of type m will always be accepted and no rejection step is necessary. The resulting algorithm is identical with the Gibbs sampler from algorithm 4.27.

4.4.2   Application to parameter estimation

In this section we illustrate the use of the Gibbs sampler in Bayesian inference, by considering the problem of sampling from the posterior distribution of the parameters in a mixture distribution, given a random sample from the mixture. The difference compared with the more generic Bayesian parameter estimation problems considered in Section 4.3 is that here we can make use of the hierarchical structure of the mixture distribution to derive a specialised MCMC algorithm for the problem under consideration.

Let $k \in \mathbb{N}$ and $\sigma > 0$ be fixed and let $\mu_1, \ldots, \mu_k \sim \varphi_\mu$ be independent random vectors in $\mathbb{R}^d$, where we assume the distribution φ_μ to have a density. Given $\mu = (\mu_1, \ldots, \mu_k)$, let $X_1, \ldots, X_n$ be an i.i.d. sample from the mixture distribution

$$\varphi_{X|\mu} = \frac{1}{k} \sum_{a=1}^{k} \mathcal{N}(\mu_a, \sigma I_d),$$

where the mixture components are normal distribution with mean μ_a and covariance matrix σI_d and where I_d denotes the d-dimensional identity matrix. In the later parts of this example we will restrict ourselves to the case $d = 2$ and we will assume that, under the prior distribution, $\mu_1, \ldots, \mu_k \sim \mathcal{U}([-10, +10] \times [-10, +10])$ i.i.d., that is we will assume that the prior density for the cluster means μ_a is given by

$$\varphi_\mu(\mu_a) = \frac{1}{20^2} \mathbb{1}_{[-10, 10] \times [-10, 10]}(\mu_a). \tag{4.25}$$

for all $\mu_a \in \mathbb{R}^2$.

Our aim is to generate samples from the posterior distribution of the μ_a, that is from the conditional distribution of $\mu_1, \ldots, \mu_k$ given observations $X_i = x_i \in \mathbb{R}^d$ for $i = 1, 2, \ldots, n$. This target distribution has density

$$p_{\mu|X}(\mu \mid x) = \frac{p_{X|\mu}(x \mid \mu) \, p_\mu(\mu)}{p_X(x)}. \tag{4.26}$$

In this section we will explain how the Gibbs sampling algorithm 4.27 can be used to generate samples from this density.

We can extend the state space of the model by introducing new variables $Y_i \in \{1, 2, \ldots, k\}$ to indicate which component the values X_i belongs to: for this we generate X_i in a two-step procedure by first generating $Y_i \sim \mathcal{U}\{1, 2, \ldots, k\}$ and then generating $X_i \sim \mathcal{N}(\mu_{Y_i}, \sigma I_d)$. Using this approach, the joint distribution of the vectors $\mu = (\mu_1, \ldots, \mu_k)$, $Y = (Y_1, \ldots, Y_n)$ and $X = (X_1, \ldots, X_n)$ is given by

$$p_{\mu,Y,X}(\mu, y, x) = \prod_{a=1}^{k} \varphi_\mu(\mu_a) \cdot \prod_{i=1}^{n} \frac{1}{k} \psi(x_i; \mu_{y_i}, \sigma I_d)$$

for all $\mu \in (\mathbb{R}^d)^k$, $y \in \{1, \ldots, k\}^n$ and $x \in (\mathbb{R}^d)^n$, where the term $1/k$ gives the probability $P(Y_i = y_i)$ and $\psi(\cdot; \mu_{y_i}, \sigma I_d)$, given by

$$\psi(\xi; \mu, \sigma I_d) = \frac{1}{(2\pi\sigma^2)^{d/2}} \exp\left(-\frac{|\xi - \mu|^2}{2\sigma^2}\right) \tag{4.27}$$

for all $\xi \in \mathbb{R}^d$, is the density of the normal distribution $\mathcal{N}(\mu, \sigma I_d)$ and $|\cdot|$ denotes the Euclidean norm in $\mathbb{R}^d$. In this extended model, both μ and Y are unknown and we want to sample from the conditional distribution of $(\mu, Y) \in (\mathbb{R}^d)^k \times \{1, \ldots, k\}^n$, given a sample $x \in (\mathbb{R}^d)^n$ of X.

To solve this extended sampling problem we can use the Gibbs sampler with state space $S = S_1 \times S_2$ where $S_1 = (\mathbb{R}^d)^k$ and $S_2 = \{1, \ldots, k\}^n$. The resulting version of algorithm 4.27 requires us to alternatingly generate samples

$$\mu^{(j)} \sim p_{\mu|Y,X}\left(\cdot \,\big|\, Y^{(j-1)}, x\right)$$

and

$$Y^{(j)} \sim p_{Y|\mu,X}\left(\cdot \,\big|\, \mu^{(j-1)}, x\right).$$

The densities $p_{\mu|Y,X}$ and $p_{Y|\mu,X}$ can be found using Bayes' rule: for $p_{\mu|Y,X}$ we have

$$
\begin{aligned}
p_{\mu|Y,X}(\mu \mid y, x) &= \frac{p_{\mu,Y,X}(\mu, y, x)}{p_{Y,X}(y, x)} \\[2mm]
&= \frac{\prod_{a=1}^{k} \varphi_\mu(\mu_a) \cdot \prod_{i=1}^{n} \frac{1}{k} \psi(x_i; \mu_{y_i}, \sigma I_d)}{p_{Y,X}(y, x)} \\[2mm]
&= \frac{1}{Z_{y,x}} \prod_{a=1}^{k} \left(\varphi_\mu(\mu_a) \cdot \prod_{\substack{i=1 \\ y_i=a}}^{n} \psi(x_i; \mu_a, \sigma I_d) \right)
\end{aligned}
\tag{4.28}
$$

for all $\mu \in (\mathbb{R}^d)^k$, where we grouped the factors $\psi(x_i; \mu_{y_i}, \sigma I_d)$ according to the values of y_i and $Z_{y,x}$ is the normalisation constant which makes the function $p_{\mu|Y,X}(\cdot \mid y, x)$ a probability density. Since the right-hand side of (4.28) is the product of individual functions of μ_a for $a = 1, 2, \ldots, k$, under the conditional distribution

$p_{\mu|Y,X}$ the components $\mu_1, \ldots, \mu_k$ are independent with densities given by the corresponding factors in the product. Using a similar calculation, we find the density $p_{Y|\mu,X}$ as

$$
\begin{aligned}
p_{Y|\mu,X}(y \mid \mu, x) &= \frac{p_{\mu,Y,X}(\mu, y, x)}{p_{\mu,X}(\mu, x)} \\
&= \frac{\prod_{a=1}^{k} \varphi_\mu(\mu_a) \cdot \prod_{i=1}^{n} \frac{1}{k} \, \psi(x_i; \mu_{y_i}, \sigma I_d)}{p_{\mu,X}(\mu, x)} \\
&= \frac{1}{Z_{\mu,x}} \prod_{i=1}^{n} \psi(x_i; \mu_{y_i}, \sigma I_d)
\end{aligned}
$$

for all $y \in \{1, 2, \ldots, k\}^n$, where $Z_{\mu,k}$ is the normalisation constant. Since the right-hand side is a product where each factor only depends on one of the components y_i, under the conditional distribution $p_{Y|\mu,X}$ the components $Y_1, \ldots, Y_n$ are independent with weights

$$
P(Y_i = a \mid \mu, X) = \frac{\psi(X_i; \mu_a, \sigma I_d)}{\sum_{\tilde{a}=1}^{k} \psi(X_i; \mu_{\tilde{a}}, \sigma I_d)} \tag{4.29}
$$

for $a = 1, 2, \ldots, k$ and for all $i = 1, 2, \ldots, n$, where the density ψ is given by (4.27).

To be more specific, we now restrict ourselves to the case $d = 2$ and we consider the prior density φ_μ given by (4.25). Then, using equation (4.28) and the definition of ψ from (4.27), we find the conditional density of μ_a given Y and X as

$$
\begin{aligned}
p_{\mu_a|Y,X}(\mu_a \mid y, x) &= \frac{1}{Z_1} \varphi_\mu(\mu_a) \prod_{\substack{i=1 \\ y_i=a}}^{n} \psi(x_i; \mu_a, \sigma I_2) \\
&= \frac{1}{Z_1} \varphi_\mu(\mu_a) \prod_{\substack{i=1 \\ y_i=a}}^{n} \frac{1}{2\pi\sigma^2} \exp\left(-\frac{|x_i - \mu_a|^2}{2\sigma^2} \right) \\
&= \frac{1}{Z_2} \varphi_\mu(\mu_a) \exp\left(-\sum_{\substack{i=1 \\ y_i=a}}^{n} \frac{|x_i - \mu_a|^2}{2\sigma^2} \right) \\
&= \frac{1}{Z_3} \varphi_\mu(\mu_a) \exp\left(-\frac{n_a}{2\sigma^2} \cdot |\mu_a|^2 + \frac{n_a}{2\sigma^2} \cdot 2\langle \mu_a, \frac{1}{n_a} \sum_{\substack{i=1 \\ y_i=a}}^{n} x_i \rangle \right),
\end{aligned}
$$

where Z_1, Z_2 and Z_3 are normalisation constants, $n_a = \sum_{i=1}^{n} \mathbb{1}_{\{a\}}(y_i)$, and $|\cdot|$ and $\langle \cdot, \cdot \rangle$ denote the Euclidean norm and inner product on $\mathbb{R}^2$, respectively. Completing the square in the exponent and substituting the definition of φ_μ from Equation (4.25) we find

$$p_{\mu_a|Y,X}(\mu_a \mid y, x)$$

$$= \frac{1}{Z_4} \mathbb{1}_{[-10,+10] \times [-10,+10]}(\mu_a) \exp \left(-\frac{1}{2\sigma^2/n_a} \left| \mu_a - \frac{1}{n_a} \sum_{\substack{i=1 \\ y_i=a}}^{n} x_i \right|^2 \right), \quad (4.30)$$

where Z_4 is the resulting normalisation constant. Thus, $p_{\mu_a|Y,X}$ is the density of the two-dimensional normal distribution $\mathcal{N} \left(\frac{1}{n_a} \sum_{i=1}^{n} \mathbb{1}_{\{a\}}(y_i)x_i, \frac{\sigma^2}{n_a} I_2 \right)$, conditioned on the value lying in the square $[-10, +10] \times [-10, +10]$. Samples from this conditional distribution can be easily generated using the rejection sampling algorithm 1.25 for conditional distributions, that is by repeatedly sampling from the unconditioned normal distribution and rejecting all values which fall outside the square. The representation from (4.30) for $p_{\mu_a|Y,X}$ is only defined for $n_a > 0$. In the case $n_a = 0$ we have no observations corresponding to the mean μ_a and consequently for this case $p_{\mu_a|Y,X}$ coincides with the prior distribution φ_μ.

As we have already seen, the components of the vector Y under the distribution $p_{Y|\mu,X}$ are independent. Their distribution is found by substituting (4.27) for $d = 2$ into equation (4.29).

Now that we have identified the conditional distributions $p_{\mu|Y,X}$ and $p_{Y|\mu,X}$, we can implement the Gibbs sampler from algorithm 4.27 to sample from the posterior distribution of μ and Y as follows:

1: **for** $a = 1, 2, \ldots, k$ **do**
2:    generate $\mu_a^{(0)} \sim \mathcal{U}([-10, 10] \times [-10, 10])$
3: **end for**
4: **for** $i = 1, 2, \ldots, n$ **do**
5:    generate $Y_i^{(0)} \sim \mathcal{U}\{1, 2, \ldots, k\}$
6: **end for**
7: **for** $j = 1, 2, 3, \ldots$ **do**
8:    **if** j is odd **then**
9:       **for** $a = 1, 2, \ldots, k$ **do**
10:         $n_a \leftarrow \sum_{i=1}^{n} \mathbb{1}_{\{a\}}(Y_i^{(j-1)})$
11:         repeatedly generate $\xi^{(j)} \sim \mathcal{N} \left(\frac{1}{n_a} \sum_{i=1}^{n} \mathbb{1}_{\{a\}}(y_i)x_i, \frac{\sigma^2}{n_a} I_2 \right)$,
            until $\xi^{(j)} \in [-10, +10] \times [-10, +10]$
12:         $\mu_a^{(j)} \leftarrow \xi^{(j)}$
13:      **end for**
14:      $Y^{(j)} \leftarrow Y^{(j-1)}$
15:    **else**
16:      $\mu^{(j)} \leftarrow \mu^{(j-1)}$
17:      **for** $i = 1, 2, \ldots, n$ **do**
18:         **for** $a = 1, 2, \ldots, k$ **do**

19:    $\qquad q_a \leftarrow \frac{1}{2\pi\sigma^2} \exp\left(-\frac{|\xi_i - \mu_a^{(j-1)}|^2}{2\sigma^2}\right)$

20:    **end for**

21:    $\qquad$ generate $Y_i^{(j)} \in \{1, 2, \ldots, k\}$ with $P(Y_i^{(j)} = a) = q_a / \sum_{b=1}^k q_b$

22:    **end for**

23: **end if**

24: **end for**

The implementation of the Gibbs sampler presented here is based on the following considerations:

- We need to decide how to generate the initial values for $\mu^{(0)}$ and $Y^{(0)}$. In the implementation given above we generate $\mu^{(0)}$ from the prior distribution. Since there is no prior distribution for the auxiliary variables $Y^{(0)}$, we use the uniform distribution on the set $\{1, 2, \ldots, k\}$ of allowed values instead.

- We need to alternatingly update μ and Y. In the implementation above we update μ when j is odd and we update Y when j is even.

- In step 11 of the algorithm some care is needed when $n_a = 0$, that is when there are no observations corresponding to cluster a. In this case the normal distribution must be replaced by the prior distribution φ_μ.

- Since the distribution of $\mu^{(j)}$ only *converges* to the posterior distribution $p_{\mu|X}$ from (4.26) as $j \to \infty$, a burn-in period should be implemented by discarding the first samples.

The result of a run of the Gibbs sampler presented above is shown in Figure 4.5.

In proposition 4.28 we have seen that the target distribution $\pi = p_{\mu|X}$ from equation (4.26) is indeed a stationary distribution of the Markov chain constructed by the algorithm described in this section, and often the distribution of the resulting Markov chain converges to a plausible posterior sample quickly. This is illustrated in Figure 4.5 where after only 100 steps of burn-in period the samples for the five cluster means μ_a are all centred well inside the five clusters. Nevertheless, it takes an extremely long time for the Markov chain to explore the complete state space. This can be seen by the fact that the posterior distribution is invariant under relabelling the cluster means and thus, in stationarity, each of the five cluster means should be found in any of the five clusters with equal probability. In the simulation depicted in Figure 4.5, no transition of cluster means between clusters is observed and indeed such transitions will not be observed in any runs of realistically achievable length.

The Markov chain generated by the algorithm converges to its stationary distribution very slowly. As a consequence, it is possible for the algorithm to get 'stuck' in states which have low probability but where all 'close by' states have even lower probability. Such states are called metastable states and, while the algorithm will eventually escape from such states, this can take an extremely long time and thus may not happen in practice. Such a problem is illustrated by Figure 4.6, where during the burn-in period two components of μ moved into the same cluster, leaving a third component to 'cover' two clusters. For typical samples from the posterior distribution

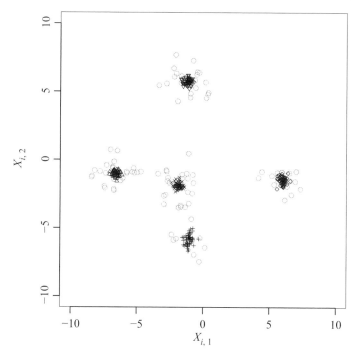

Figure 4.5   A typical sample of size 100 from the posterior distribution (4.26) of the cluster means for the parameter estimation problem described in Section 4.4.2. The grey circles give the positions of the observations $X_1, \ldots, X_{100}$, the five different kinds of smaller, black symbols give the positions of $\mu_a^{(j)}$ for $j = 1, 2, \ldots, 100$ and $a = 1, \ldots, 5$. A burn-in period of length 1000 has been used for this figure.

we would expect the components of μ to be distributed such that each cluster contains exactly one of them, but numerical experiments show that the Markov chain needs an extremely (and unachievably) long time to reach states resembling this situation. Thus, as for any MCMC method, the output of the Gibbs sampler needs to be carefully checked to verify that the distribution of the output is sufficiently close to stationarity.

4.4.3   Applications to image processing

In this section we illustrate the Gibbs sampler using an application to a simple image processing problem. Here we will represent images as a square grid of 'pixels' (short for 'picture elements'), each of which takes values in the two-element set $\{-1, +1\}$ where -1 stands for a white pixel and the value $+1$ stands for a black pixel. Thus, images in this setup are considered to be elements of the product space

$$S = \{-1, +1\}^I, \tag{4.31}$$

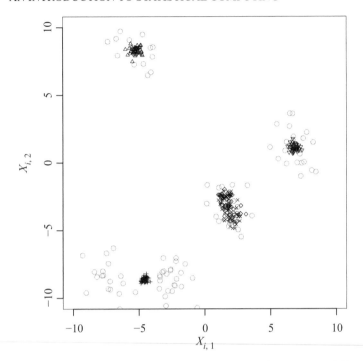

Figure 4.6   A sample of size 100, similar to Figure 4.5 but using a different seed of the random number generator. Different from Figure 4.5, the Markov chain here has not yet converged to the stationary distribution, despite the presence of a burn-in period of length 10 000: two components of μ, represented by the symbols $\times$ and $\diamond$, are concentrated in one cluster while the component $+$ is situated in the gap between two clusters. This figure was created by selecting one of a large number of runs of the algorithm.

where I is the lattice

$$I = \{1, 2, \ldots, L\} \times \{1, 2, \ldots, L\}.$$

The elements of S are vectors of the form $x = (x_i)_{i \in I}$. States $x \in S$ can be visualised as square grids of small black and white dots, where the colour at location $i \in I$ in the grid encodes the possible values -1 and $+1$ for x_i.

Definition 4.30   A probability measure on S where the weights are written in the form

$$\pi(x) = \frac{1}{Z} \exp\left(-H(x)\right) \tag{4.32}$$

for all $x \in S$ is called a *Gibbs measure*. The function $H: S \to R$ is called the *energy* and the normalisation constant

$$Z = \sum_{x \in S} \exp(-H(x))$$

is called the *partition function*.

4.4.3.1 The Ising model

The *Ising model* is a model from statistical mechanics which describes the distribution of random elements $X \in S$ where the state space S is of the form given by (4.31). Motivated by the situation of image processing, we restrict ourselves to the two-dimensional case $I \subseteq \mathbb{Z}^2$, but we note that everything in this section can be easily generalised to the case $I \subseteq \mathbb{Z}^d$ for $d \in \mathbb{N}$. For the Ising model, the distribution of X is assumed to be a Gibbs measure, as described in equation (4.32), where the energy H is given by

$$H(x) = -\beta \sum_{\substack{i,j \in I \\ i \sim j}} x_i x_j \tag{4.33}$$

for all $x \in S$. Here we write $i \sim j$ to indicate that $i, j \in I$ are nearest neighbours and the sum in the definition of H is taken over all pairs of nearest neighbours in the grid. The constant $\beta > 0$ is called the *inverse temperature*. The model can be considered for different definitions of 'neighbouring' pixels. Here we consider only nearest neighbours, that is we consider a pixel $i = (i_1, i_2)$ in the interior of the grid to have neighbours $(i_1 + 1, i_2)$, $(i_1, i_2 + 1)$, $(i_1 - 1, i_2)$ and $(i_1, i_2 - 1)$. Pixels on the edges of the grid are considered to have only three neighbours and the four corner pixels have only two neighbours. An alternative approach, often chosen to avoid complications in theoretical analysis, would be to extend the grid periodically or, equivalently, to consider the left-most column to be adjacent to the right-most column and the top-most row to be adjacent to the bottom-most row.

Samples X from the Ising model for different values of β are depicted in Figure 4.7. The figure clearly shows that with increasing β the tendency of neighbouring pixels to have the same colour increases.

To apply the Gibbs sampler from algorithm 4.27 in this situation, we need to find the conditional distributions $\pi_{X_m \mid X_{\neg m}}$, that is we need to find the distribution of the state X_m of one pixel, given the state $X_{\neg m}$ of all the other pixels. Using Bayes' rule we get

$$
\begin{aligned}
\pi_{X_m \mid X_{\neg m}}(x_m \mid x_{\neg m}) &= \frac{\pi_{X_m, X_{\neg m}}(x_m, x_{\neg m})}{\pi_{X_{\neg m}}(x_{\neg m})} \\
&= \frac{\pi_{X_m, X_{\neg m}}(x_m, x_{\neg m})}{\pi_{X_m, X_{\neg m}}(-1, x_{\neg m}) + \pi_{X_m, X_{\neg m}}(+1, x_{\neg m})}.
\end{aligned}
\tag{4.34}
$$

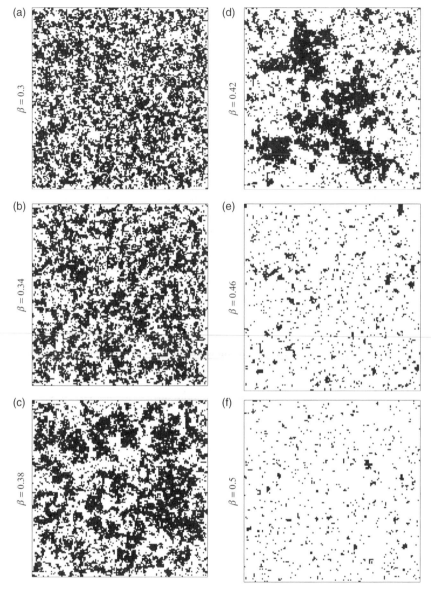

Figure 4.7   Samples from the Ising model given by equation (4.32) and equation (4.33), on the grid $I = \{1, 2, \ldots, 150\}^2$, for different values of β. The panels show the final state $X^{(N)}$ of the Gibbs sampling algorithm 4.31, for $N = 10\,000 \times 150^2$.

Here, $\pi_{X_m, X_{\neg m}}(x_m, x_{\neg m}) = \pi_X(x)$ is the probability of the state x. Using equation (4.32) and equation (4.33) we can write this probability as

$$\pi(x) = \frac{1}{Z} \exp\left(\beta \sum_{\substack{i,j \in I \\ i \sim j}} x_i x_j \right)$$

$$= \frac{1}{Z} \exp\left(\beta x_m \sum_{\substack{i \in I \\ i \sim m}} x_i \right) \exp\left(\beta \sum_{\substack{i,j \in I \\ i \sim j, i \neq m, j \neq m}} x_i x_j \right).$$

Since both Z and the final factor do not depend on the value of x_m, the corresponding terms in the numerator and denominator of equation (4.34) cancel and we get

$$\pi_{X_m | X_{\neg m}}(x_m \mid x_{\neg m}) = \frac{\exp\left(x_m \beta \sum_{i \sim m} x_i \right)}{\exp\left(\beta \sum_{i \sim m} x_i \right) + \exp\left(-\beta \sum_{i \sim m} x_i \right)}. \tag{4.35}$$

These are the required conditional probabilities to generate the new value ξ_j for location $j \in I$ in step 3 of the Gibbs sampler algorithm 4.27. The probabilities can be computed using only the values of the image at the (up to) four neighbours of j. Thus, the update step in the Gibbs algorithm for the Ising model can be performed very efficiently.

Algorithm 4.31   (Gibbs sampler for the Ising model)
    input:
        $\beta \geq 0$ (the *inverse temperature*)
        $X_0 \in \{-1, +1\}^I$ (the initial state)
    output:
        a path of a Markov chain with the Gibbs measure from (4.32) and (4.33) as its
        stationary distribution
    randomness used:
        independent samples $\xi^{(j)} \in \{-1, +1\}$ for $j \in \mathbb{N}$
    1: **for** $j = 1, 2, 3, \ldots$ **do**
    2:     $m_1 \leftarrow ((j-1) \bmod L) + 1$
    3:     $m_2 \leftarrow (\lfloor (j-1)/L \rfloor \bmod L) + 1$
    4:     $d \leftarrow X^{(j-1)}_{m_1-1, m_2} + X^{(j-1)}_{m_1+1, m_2} + X^{(j-1)}_{m_1, m_2-1} + X^{(j-1)}_{m_1, m_2+1}$
    5:     $p \leftarrow \frac{1}{1 + \exp(-2\beta d)}$
    6:     Generate $\xi^{(j)} \in \{-1, +1\}$ such that $P\left(\xi^{(j)} = +1 \right) = p$.
    7:     Define $X^{(j)} \in S$ by

$$X^{(j)}_i = \begin{cases} X^{(j-1)}_i & \text{if } i \neq (m_1, m_2) \text{ and} \\ \xi^{(j)} & \text{otherwise} \end{cases}$$

        for all $i \in I$.
    8: **end for**

In steps 2 and 3 of the algorithm, a pixel position $m = (m_1, m_2)$ is constructed such that m cycles through all possible pixel positions cyclically. When implementing step 4 of the algorithm, some care is needed when m is on the edge of the grid I: we set $X_i = 0$ for all $i \notin I$. The result of different simulations using this algorithm is shown in Figure 4.7.

As a direct consequence of proposition 4.28, the distribution of the Ising model, given by equation (4.32) and equation (4.33) is a stationary distribution of the Markov chain X constructed by algorithm 4.31. Using the results from Section 4.2.1, it is easy to check that the Markov chain constructed in algorithm 4.31 is irreducible and aperiodic and thus the distribution of $X^{(j)}$ converges to the exact distribution of the Ising model for every initial condition $X^{(0)}$.

The Ising model is well-studied in statistical mechanics and nearly everything about the behaviour of the model is known. For example, a well-known result is that for large grid sizes the behaviour of the system changes when the inverse temperature crosses the critical value

$$\beta^* = \frac{\log(1 + \sqrt{2})}{2} \approx 0.44$$

(see e.g. Pathria, 1996, Section 12.3). For $\beta < \beta^*$ typical states consist of a 'patchwork pattern' of mostly white and mostly black regions. This pattern is visible in Figure 4.7(a–d). For $\beta > \beta^*$ typical states are dominated by one colour, showing only isolated 'specks' of the opposite colour. This pattern is shown by Figure 4.7(e) and (f). By symmetry, the Markov chain will be in predominantly white states and predominantly black states for approximately equal amounts of time. Transitions between these two patterns take an extremely long time and will not be observed in practice but, for finite grid size, it is easy to show that the Markov chain is irreducible and thus, by the results from Section 4.2.1, transitions will happen on long time-scales. In contrast, for the extension of the Ising model to an infinite grid such transitions no longer happen. This change of behaviour of the system when β crosses the critical point β^* is an example of a *phase transition* and is of great interest in statistical physics. In particular, many interesting and difficult results concern the behaviour of the system at the critical point $\beta = \beta^*$.

4.4.3.2   Bayesian image analysis

We will now consider an application of Bayesian inference to the denoising of images. For this setup we assume that the original image X is described by a probability distribution on the space of all possible images, and that we have observed a noisy version Y of this image. Our aim is to reconstruct the original image X from the observation Y.

For simplicity we assume that we are in the situation of the preceding section, that is that the original image is square and consists only of black and white pixels. Thus we have $X \in S$ where S is the state space given in Equation (4.31) at the beginning of this section. We also assume that the prior distribution of the original image X is given by the Ising model from equation (4.32) and equation (4.33). Finally, for the noisy observation Y of the image X we assume that independent, $\mathcal{N}(0, \sigma^2)$-distributed random variables are added to every pixel of X, that is we have $Y \in \mathbb{R}^I$ and the conditional distribution of Y_i given the value of X is

$$Y_i \sim \mathcal{N}(X_i, \sigma^2)$$

for all $i \in I$, independently.

Using this model, the posterior distribution of X given the observations Y is found using Bayes' rule as

$$
\begin{aligned}
p_{X|Y}(x \mid y) &= \frac{p_{Y|X}(y \mid x)\, p_X(x)}{p_Y(y)} \\
&= \frac{1}{p_Y(y)} \prod_{i \in I} \frac{1}{\sqrt{2\pi\sigma^2}} \cdot \exp\left(-\frac{(y_i - x_i)^2}{2\sigma^2}\right) \cdot \frac{1}{Z} \exp\left(-H(x)\right),
\end{aligned}
\tag{4.36}
$$

where the energy function H is given in equation (4.33). Since $x_i \in \{-1, +1\}$ for all $i \in I$, we have $(y_i - x_i)^2 = -2y_i x_i + (y_i^2 + 1)$ and thus we can write the posterior distribution as the Gibbs measure

$$p_{X|Y}(x \mid y) = \frac{1}{Z_y} \exp\left(-H_y(x)\right),$$

with energy H_y given by

$$H_y(x) = -\frac{1}{\sigma^2} \sum_{i \in I} y_i x_i + H(x) = -\frac{1}{\sigma^2} \sum_{i \in I} y_i x_i + \beta \sum_{\substack{i,j \in I \\ i \sim j}} x_i x_j$$

for all $x \in I$, where β is the inverse temperature from the Ising model.

To generate samples from the posterior distribution, we can employ the Gibbs sampler from algorithm 4.27 with target distribution $\pi = p_{X|Y}$. As in equation (4.34) we have

$$\pi_{X_m | X_{\neg m}}(x_m \mid x_{\neg m}) = \frac{\pi_{X_m, X_{\neg m}}(x_m, x_{\neg m})}{\pi_{X_m, X_{\neg m}}(-1, x_{\neg m}) + \pi_{X_m, X_{\neg m}}(+1, x_{\neg m})}$$

and introducing the additional term $\sum_{i \in I} y_i x_i / \sigma^2$ into the derivation of equation (4.35) gives

$$
\begin{aligned}
&\pi_{X_m | X_{\neg m}}(x_m \mid x_{\neg m}) \\
&= \frac{\exp\left(x_m \left(\beta \sum_{i \sim m} x_i - \frac{y_m}{\sigma^2}\right)\right)}{\exp\left(\beta \sum_{i \sim m} x_i - \frac{y_m}{\sigma^2}\right) + \exp\left(-\left(\beta \sum_{i \sim m} x_i - \frac{y_m}{\sigma^2}\right)\right)}
\end{aligned}
$$

and thus

$$\pi_{X_m | X_{\neg m}}(+1 \mid x_{\neg m}) = \frac{1}{1 + \exp\left(-2\left(\beta(\sum_{i \sim m} x_i) - y_m / \sigma^2\right)\right)}.$$

Substituting these conditional probabilities into the general Gibbs sampler algorithm 4.27 leads to the following variant of algorithm 4.31.

Algorithm 4.32   (Gibbs sampler in image processing)
input:
$\beta \geq 0$ (the *inverse temperature*)
$y \in \mathbb{R}^I$ (the observed, noisy image)
$X_0 \in \{-1, +1\}^I$ (the initial state)
output:
a path of a Markov chain with the Gibbs measure from (4.32) and (4.33) as its
stationary distribution
randomness used:
independent samples $\xi^{(j)} \in \{-1, +1\}$ for $j \in \mathbb{N}$
1: **for** $j = 1, 2, 3, \ldots$ **do**
2:    $m_1 \leftarrow ((j-1) \bmod L) + 1$
3:    $m_2 \leftarrow (\lfloor (j-1)/L \rfloor \bmod L) + 1$
4:    $d \leftarrow X_{m_1-1,m_2}^{(j-1)} + X_{m_1+1,m_2}^{(j-1)} + X_{m_1,m_2-1}^{(j-1)} + X_{m_1,m_2+1}^{(j-1)}$
5:    $p \leftarrow \frac{1}{1+\exp(-2(\beta d + y_m / \sigma^2))}$
6:    Generate $\xi^{(j)} \in \{-1, +1\}$ such that $P(\xi^{(j)} = +1) = p$.
7:    Define $X^{(j)} \in S$ by

$$X_i^{(j)} = \begin{cases} X_i^{(j-1)} & \text{if } i \neq (m_1, m_2) \text{ and} \\ \xi^{(j)} & \text{otherwise,} \end{cases}$$

for all $i \in I$.
8: **end for**

In steps 2 and 3 of the algorithm, a pixel position $m = (m_1, m_2)$ is constructed such that m cycles through all possible pixel positions cyclically. When implementing step 4 of the algorithm, we set $X_i = 0$ for all $i \notin I$. The result of three simulations using this algorithm, for different values of σ, is shown in Figure 4.8.

4.5   Reversible Jump Markov Chain Monte Carlo

While MCMC methods like the Metropolis–Hastings method can be used in arbitrary spaces, so far we only have considered the case of generating samples in the Euclidean space $\mathbb{R}^d$. The Reversible Jump Markov Chain Monte Carlo (RJMCMC) method, described in this section, is concerned with a more general case, where the target

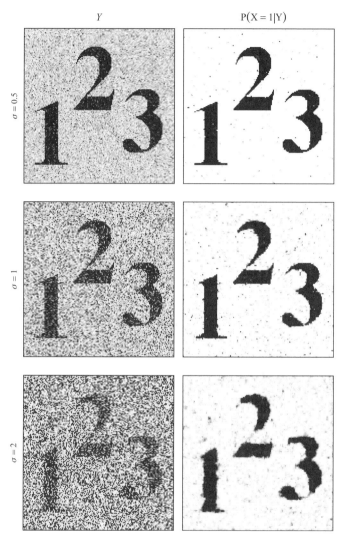

*Figure 4.8   Pairs of noisy images Y together with reconstructions generated using
algorithm 4.32 with inverse temperature β = 0.5. The grey values in the reconstructed
images give the posterior probability that the corresponding pixel in the original
image was black, ranging from black for probability 1 to white for probability 0.*

distribution lives on the disjoint union of several spaces $\mathbb{R}^{d_k}$. This setup, while
seeming artificial at first glance, is useful in various classes of practical applications:

- Bayesian inference can be used to make simultaneous inference about the
  choice of model as well as the model parameters, by assigning prior probabilities to different models. If the models under consideration have different

numbers of parameters, the basic MCMC approach becomes difficult, but RJMCMC can still be used. This approach is illustrated in Section 4.5.2.

- Many types of random geometric objects have a variable number of parameters. These include, for example random trees and random graphs. The RJMCMC meethod can be used to generate samples from the distribution of such random geometrical objects.

- Intensity functions, describing the rate of random events in time, can be modelled as piecewise constant functions, parameterised by change point positions and function values between the change points. The number of parameters in such a model depends on the number of change points. In a Bayesian setting, such an intensity function will have a random number of change points and thus the model will have a random number of parameters. Application of the RJMCMC method in this context is, for example, discussed in the article by Green (1995).

4.5.1   Description of the method

Due to the complex structure of the state space used in RJMCMC methods, more mathematical formalism is required to state the RJMCMC algorithm than was necessary in the previous sections. This section introduces the required notation and states the general RJMCMC algorithm.

We start the exposition by giving a mathematical description of the *state space*: Let I be a finite or countable set and let $d_k \in \mathbb{N}_0$ for all $k \in I$ be given. Define

$$S_k = \{k\} \times \mathbb{R}^{d_k}$$

for all $k \in I$ and

$$S = \bigcup_{k \in I} S_k.$$

Then the elements z of the space S have the form $z = (k, x)$, where $x \in \mathbb{R}^{d_k}$ and $k \in I$. Since the index k is included as the first component of all elements in S_k, the spaces S_k are disjoint and each $z \in S$ is contained in exactly one of the subspaces S_k. For a value $(k, x) \in S$, the first component, k, indicates which of the spaces $\mathbb{R}^{d_k}$ a point is in while the second component, x, gives the position in this space. The space S is the state space our target distribution will live on and the Markov chain constructed by the RJMCMC algorithm will move in.

Next, we specify the *target distribution* π on the space S. If $Z \sim \pi$, then Z can be written as $Z = (K, X)$ and we need to specify the joint distribution of $K \in I$ and $X \in \mathbb{R}^{d_K}$. To describe such a distribution we can use a density π which is split between the different subspaces S_k, that is a function $\pi(\cdot, \cdot)$ such that $\pi(k, \cdot) : \mathbb{R}^{d_k} \to [0, \infty)$ for every $k \in I$ and

$$\sum_{k \in I} \int_{\mathbb{R}^{d_k}} \pi(k, x) \, dx = 1. \tag{4.37}$$

Then we have

$$P(K = k) = \int_{\mathbb{R}^{d_k}} \pi(k, x) \, dx$$

and

$$P(K = k, X \in A) = \int_A \pi(k, x) \, dx$$

for all $A \subseteq \mathbb{R}^{d_k}$ and all $k \in I$.

Example 4.33   We can use the above formalism to construct a simple model for the number and diameter of craters on a given area of moon surface. Assume that the number K of craters is Poisson-distributed with parameter λ, and that each crater, independently, has a random diameter X given by a Pareto distribution with density

$$f(x) = \begin{cases} \frac{\alpha}{x^{\alpha+1}} & \text{if } x \geq 1 \text{ and} \\ 0 & \text{otherwise.} \end{cases}$$

(We ignore craters with size less than 1 for this example.) Then, using the notation introduced above, we have

$I = \mathbb{N}_0$	the possible numbers of craters,
$d_k = k$	one parameter per crater (the diameter),
$P(K = k) = e^{-\lambda} \frac{\lambda^k}{k!}$	the probability of having exactly k craters,
$\pi(k, x) = e^{-\lambda} \frac{\lambda^k}{k!} \prod_{i=1}^{k} f(x_i)$	the joint density of k diameters $X_1, \ldots, X_k$

for all $x \in \mathbb{R}^{d_k}$ and all $k \in I$.

Our aim is to construct a Markov chain with stationary distribution π, using the Metropolis–Hastings algorithm on the state space S. Since this Markov chain moves in S, the state at time j is described by a pair (K_j, X_j). To describe the transition probabilities of such a Markov chain, for each $(k, x) \in S$ we need to specify the distribution of (K_j, X_j) when $(K_{j-1}, X_{j-1}) = (k, x)$. It transpires that in the RJMCMC algorithm it is advantageous to first determine the value of K_j and only then, in a second step, to determine the value of X_j from the conditional distribution, conditioned on the value of K_j. Thus, the transitions of the Markov chain from $(k, x) \in S$ to $(l, y) \in S$ will be described by probability weights $b(k, x; l)$ with

$$\sum_{l \in I} b(k, x; l) = 1$$

and probability densities $p(k, x; l, \cdot)$ on $\mathbb{R}^{d_l}$ for all $(k, x) \in S$. If $(K_j, X_j)_{j \in \mathbb{N}}$ is described by b and p, then

$$P\left(K_j = l, X_j \in A \mid K_{j-1} = k, X_{j-1} = x\right) = b(k, x; l) \int_A p(k, x; l, y)\, dy$$

for all $k, l \in I$, $x \in \mathbb{R}^{d_k}$ and $A \subseteq \mathbb{R}^{d_l}$.

The next ingredient used in the RJMCMC algorithm is the idea of splitting the transition mechanism into different *move types*, as described in Section 4.1.5. We denote the set of all possible move types by M and, for a given state $(k, x) \in S_k$, the probability of choosing move $m \in M$ will be denoted by $\gamma_m(k, x)$. The probabilities $\gamma_m(k, x)$ satisfy

$$\sum_{m \in M} \gamma_m(k, x) = 1$$

for all $(k, x) \in S$. In the presence of different move types, the transition probabilities given by b and p depend on the move type m, that is instead of b and p we consider probability weights b_m and probability densities p_m for all $m \in M$.

When computing the corresponding acceptance probabilities for the Metropolis–Hastings algorithm, there are two different cases to consider: the simpler of the two cases is the case when the proposal (l, y) lies in the same space as the previous state (k, x) does, that is when we have $l = k$. In this case, occurring with with probability $b_m(k, x; k)$, the distribution of the new location X_j is given by a density $p_m(k, x; \cdot) : \mathbb{R}^{d_k} \to [0, \infty)$. We will see in proposition 4.37, that the corresponding acceptance probabilities for this case can be chosen as

$$\alpha_m(k, x; y) = \min\left(\frac{\pi(k, y)\gamma_m(k, y)b_m(k, y; k)p_m(k, y; x)}{\pi(k, x)\gamma_m(k, x)b_m(k, x; k)p_m(k, x; y)}, 1\right) \qquad (4.38)$$

for all $x, y \in \mathbb{R}^{d_k}$ and all $k \in I$. This expression is very similar to the form of the acceptance probability we found for algorithm 4.12; only the probabilities $b_m(k, \cdot; k)$ for staying in the space $k \in I$ need to be included. Many variations of this type of move are possible: for example, if the proposal can take only a discrete set of values, we can replace the densities $p_m(k, x; \cdot)$ by probability weights. Alternatively, in case the distribution of proposals is continuous but restricted to a lower dimensional subset of $\mathbb{R}^{d_k}$, we can replace $\pi(k, x)p_m(k, x; y)$ by a density $\varphi_m(k, x; y)$ on a lower dimensional subspace as described in lemma 4.15.

The case where the proposal (l, y) falls into a space $S_l \neq S_k$ is more complicated: with probability $b_m(k, x; l)$ a proposal in the space $S_l = \{l\} \times \mathbb{R}^{d_l}$ needs to be constructed. In the RJMCMC algorithm, instead of directly specifying the density of the proposal on the space $\mathbb{R}^{d_l}$, the following mechanism is used. For a transition from S_k to S_l, both $\mathbb{R}^{d_k}$ and $\mathbb{R}^{d_l}$ are temporarily extended to spaces of matching dimension, that is instead of $\mathbb{R}^{d_k}$ and $\mathbb{R}^{d_l}$ we consider $\mathbb{R}^{d_k} \times \mathbb{R}^{n_k}$ and $\mathbb{R}^{d_l} \times \mathbb{R}^{n_l}$, where

$$d_k + n_k = d_l + n_l. \tag{4.39}$$

To construct the proposal Y, we proceed as follows:

(a) Use a probability density $\psi_m(k, x, \cdot; l) : \mathbb{R}^{n_k} \to [0, \infty)$ to generate an auxiliary random variable $U \in \mathbb{R}^{n_k}$.

(b) Use a map $\varphi_m^{k \to l} : \mathbb{R}^{d_k} \times \mathbb{R}^{n_k} \to \mathbb{R}^{d_l} \times \mathbb{R}^{n_l}$ to obtain

$$(Y, V) = \varphi_m^{k \to l}(x, U).$$

While the value $V \in \mathbb{R}^{n_l}$ is not part of the proposal itself, we will see in equation (4.40), that the value V is required to compute the acceptance probability for this transition. The densities ψ_m and the maps $\varphi_m^{k \to l}$ can be chosen as part of designing the algorithm, subject to the following conditions.

Assumption 4.34   The maps $\varphi_m^{k \to l} : \mathbb{R}^{d_k} \times \mathbb{R}^{n_k} \to \mathbb{R}^{d_l} \times \mathbb{R}^{n_l}$ are bijective with continuous partial derivatives. If the move type m allows transitions from S_k to S_l, it also allows the reverse transition and the corresponding transition maps satisfy the condition

$$\varphi_m^{l \to k} = (\varphi_m^{k \to l})^{-1}.$$

We will see in proposition 4.37, that we can choose the acceptance probabilities for transitions from S_k to S_l as

$$\alpha_m(k, x, u; l, y, v)$$
$$= \min \left(\frac{\pi(l, y)\gamma_m(l, y)b_m(l, y; k)\psi_m(l, y, v; k)}{\pi(k, x)\gamma_m(k, x)b_m(k, x; l)\psi_m(k, x, u; l)} \left| \det D\varphi_m^{k \to l}(x, u) \right|, 1 \right), \tag{4.40}$$

where $D\varphi_m^{k \to l}(x, u)$ is the Jacobian matrix of $\varphi_m^{k \to l}$ as given in definition 1.35.

Example 4.35   Consider the case $d_1 = 1$ and $d_2 = 2$. To construct a move between the corresponding spaces $\mathbb{R}^{d_1}$ and $\mathbb{R}^{d_2}$, we need to first extend the dimensions of theses spaces to make them equal. For simplicity we can choose $n_1 = 1$ and $n_2 = 0$, so that $d_1 + n_1 = 1 + 1 = 2 + 0 = d_2 + n_2$. The corresponding transitions are then described by a bijective, differentiable map $\varphi_m^{1 \to 2} : \mathbb{R} \times \mathbb{R} \to \mathbb{R}^2$. If we choose

$$\varphi_m^{1 \to 2}(x, u) = \begin{pmatrix} x + u \\ x - u \end{pmatrix} \tag{4.41}$$

for all $(x, u) \in \mathbb{R} \times \mathbb{R}$ we find

$$\det D\varphi_m^{1 \to 2}(x, u) = \det \begin{pmatrix} \frac{\partial}{\partial x} \varphi_{m,1}^{1 \to 2}(x, u) & \frac{\partial}{\partial u} \varphi_{m,1}^{1 \to 2}(x, u) \\ \frac{\partial}{\partial x} \varphi_{m,2}^{1 \to 2}(x, u) & \frac{\partial}{\partial x} \varphi_{m,2}^{1 \to 2}(x, u) \end{pmatrix}$$

$$= \det \begin{pmatrix} 1 & 1 \\ 1 & -1 \end{pmatrix}$$

$$= 1 \cdot (-1) - 1 \cdot 1$$

$$= -2.$$

Thus, for a move starting from $(1, x) \in S_1$ into S_2, the proposal is constructed by first generating $U \in \mathbb{R}$ with density $\psi_m(1, x, \cdot; 2)$ and then letting $Y = (x + U, x - U)$. The move is accepted with probability $\alpha_m(1, x, U; 2, Y)$, where

$$\alpha_m(1, x, u; 2, y) = \min \left(\frac{\pi(2, y)\gamma_m(2, y)b_m(2, y; 1)}{\pi(1, x)\gamma_m(1, x)b_m(1, x; 2)\psi_m(1, x, u; 2)} \cdot 2, 1 \right).$$

The argument v and the term $\psi_m(2, y, v; 1)$ are omitted since $n_2 = 0$.

Assumption 4.34 requires us to also include a corresponding transition from $\mathbb{R}^{d_2}$ and $\mathbb{R}^{d_1}$ into the same move type. This transition is described by a function $\varphi_m^{2 \to 1} : \mathbb{R}^2 \to \mathbb{R}^1 \times \mathbb{R}^1$ and, following assumption 4.34, $\varphi_m^{2 \to 1}$ has to satisfy the condition $\varphi_m^{2 \to 1} = (\varphi_m^{1 \to 2})^{-1}$. Solving the equations $y_1 = x + u$ and $y_2 = x - u$ for x and u we find

$$\varphi_m^{2 \to 1}(y) = \begin{pmatrix} \frac{y_1 + y_2}{2} \\ \frac{y_1 - y_2}{2} \end{pmatrix}$$

for all $y \in \mathbb{R}^2$. The Jacobian determinant for this function can either be found directly, or by using the rules for the derivative of inverse functions:

$$\det D\varphi_m^{2 \to 1}(y) = \det \left(\left(D\varphi_m^{1 \to 2}(x, u) \right)^{-1} \right) = \frac{1}{\det D\varphi_m^{1 \to 2}(x, u)} = -\frac{1}{2}.$$

To describe the resulting reverse moves from S_2 to S_1 we switch notation by swapping x, u with y, v: let $(2, x) \in S_2$ be the current state and assume that a move to S_1

Table 4.1   An overview of the notation used in the RJMCMC algorithm.

I	index set for the state spaces (finite or countable)
S_k	state space with index $k \in I$: $S_k = \{k\} \times \mathbb{R}^{d_k}$
S	state space of the Markov chain: $S = \bigcup_{k \in I} S_k$
$\pi(\cdot, \cdot)$	split density of the stationary distribution, satisfying (4.37)
M	index set for the moves (finite or countable)
$\gamma_m(k, x)$	probability of choosing move $m \in M$ while in state $(k, x) \in S_k$
$b_m(k, x; l)$	probability for move m to move into space $l \in I$ when the current state is $(k, x) \in S_k$
$p_m(k, x; \cdot)$	density of the proposal when moving inside space $k \in I$ with move m
$\psi_m(k, x, \cdot; l)$	density of the auxiliary random variable $U \in \mathbb{R}^{n_k}$, when moving from space k into space l using move m
$\varphi_m^{k \to l}$	map describing moves from $\mathbb{R}^{d_k} \times \mathbb{R}^{n_k}$ to $\mathbb{R}^{d_l} \times \mathbb{R}^{n_l}$: the proposal $y \in \mathbb{R}^{d_l}$ is given by $(y, v) = \varphi_m^{k \to l}(x, u)$

was selected. Then, using $\varphi^{2 \to 1}$, the proposal $(1, Y) \in S_1$ is constructed as $Y = (x_1 + x_2)/2$ and the move is accepted with probability $\alpha_m(2, x; 1, Y, V)$ where $V = (x_1 - x_2)/2$ and

$$\alpha_m(2, x; 1, y, v) = \min \left(\frac{\pi(1, y)\gamma_m(1, y)b_m(1, y; 2)\psi_m(1, y, v; 2)}{\pi(2, x)\gamma_m(2, x)b_m(2, x; 1)} \cdot \frac{1}{2}, 1 \right).$$

For this direction of move, the auxiliary value U and and the corresponding density $\psi_m(2, x, u; 1)$ are omitted since $n_2 = 0$. To conclude this example we note that the only choices made in this example were to use two as the dimension of the extended spaces and the choice of the map $\varphi^{1 \to 2}$ in equation (4.41). All remaining expressions arise as a consequence of equation (4.40) and assumption 4.34.

Table 4.1 summarises the notation introduced for the RJMCMC method. Normally the state space S, together with the stationary distribution described by π, will be given as part of the problem whereas the set M of moves and all quantities depending on the move (indicated by a subscript m in Table 4.1) need to be chosen as part of designing the method. The resulting Metropolis–Hastings algorithm has the following form.

Algorithm 4.36   (reversible jump Markov Chain Monte Carlo)
   input:
      target distribution π
      parameters $\gamma_m, b_m, p_m \psi_m$ and $\varphi_m^{k \to l}$ as in Table 4.1
      initial values $(K_0, X_0) \in \{(k, x) \in S \mid \pi(k, x) > 0\}$

randomness used:

continuous samples with densities p_m and ψ_m

discrete samples with weights γ_m and b_m

$W_j \sim \mathcal{U}[0, 1]$ i.i.d.

output:

a sample of a Markov chain $(K_j, X_j)_{j \in \mathbb{N}}$ on S with stationary distribution given by π

1: **for** $j = 1, 2, 3, \ldots$ **do**
2:    generate $m_j \in M$ with $P(m_j = m) = \gamma_m(K_{j-1}, X_{j-1})$ for all $m \in M$
3:    generate $L_j \in I$ with $P(L_j = l) = b_{m_j}(K_{j-1}, X_{j-1}; l)$ for all $l \in I$
4:    **if** $L_j = K_{j-1}$ **then**
5:        generate Y_j with density $p_{m_j}(K_{j-1}, X_{j-1}; \cdot)$
6:        let $\alpha^{(j)} \leftarrow \alpha_{m_j}(K_{j-1}, X_{j-1}; Y_j)$, using α_m from (4.38)
7:    **else**
8:        generate U_{j-1} with density $\psi_{m_j}(K_{j-1}, X_{j-1}, \cdot; L_j)$
9:        let $(Y_j, V_j) \leftarrow \varphi_{m_j}^{K_{j-1} \to L_j}(X_{j-1}, U_{j-1})$
10:        let $\alpha^{(j)} \leftarrow \alpha_{m_j}(K_{j-1}, X_{j-1}, U_{j-1}; L_j, Y_j, V_j)$, using α_m from (4.40)
11:    **end if**
12:    generate $W_j \sim \mathcal{U}[0, 1]$
13:    **if** $W_j \leq \alpha^{(j)}$ **then**
14:        $K_j \leftarrow L_j$
15:        $X_j \leftarrow Y_j$
16:    **else**
17:        $K_j \leftarrow K_{j-1}$
18:        $X_j \leftarrow X_{j-1}$
19:    **end if**
20: **end for**

This algorithm can be used to generate samples from the target distribution π for use in a MCMC method, as described in Section 4.2. As for the basic Metropolis–Hastings method from Section 4.1.1, the resulting Markov chain is π-reversible but, to avoid technical complications, we restrict ourselves to showing that π is a stationary distribution of the process X. This is the result of the following proposition.

Proposition 4.37   Let π, γ_m, b_m, p_m ψ_m and $\varphi_m^{k \to l}$ be as described in Table 4.1 and let assumption 4.34 be satisfied. Then the Markov chain $(K_j, X_j)_{j \in \mathbb{N}}$ generated by algorithm 4.36 has a stationary distribution with

$$P(K_j = k, X_j \in A) = \int_A \pi(k, x)\, dx$$

for all $j \in \mathbb{N}$, $A \subseteq \mathbb{R}^{d_k}$ and $k \in I$.

Proof   Let $j \in \mathbb{N}$ and assume that

$$P(K_{j-1} = k, X_{j-1} \in A) = \int_A \pi(k, x)\, dx$$

for all $A \subseteq \mathbb{R}^{d_k}$ and all $k \in I$. To prove the proposition, we have to show that the probability $P(K_j = k, X_j \in A)$ equals this expression.

To determine the distribution of (K_j, X_j), we have to consider all possible values (k, x) of (K_{j-1}, X_{j-1}), all possible move types $m \in M$ and then the possible target spaces $i \in I$ for the proposed move. Systematically listing all combinations, we find

$$
P\left(K_j = l, X_j \in B\right)
= \sum_{k \in I} \int_{\mathbb{R}^{d_k}} \pi(k, x) \sum_{m \in M} \gamma_m(k, x) \sum_{i \in I} b_m(k, x; i) \, Q_{m,i}(k, x; l, B) \, dx \tag{4.42}
$$

for all $l \in I$ and $B \subseteq \mathbb{R}^{d_l}$, where $Q_{m,i}(k, x; l, B)$ is the probability of the event $X_j \in B \subseteq \mathbb{R}^{d_l}$, conditioned on the proposal moving from space S_k into space S_i using move type m. Most terms in the final sum will be zero, nonzero contributions occur only for $i = l$ if the proposal is accepted and for $l = k$ if the proposal is rejected. We discuss the two resulting cases separately.

For the first case, if $i = k$, we always stay in space S_k and the process can only reach $B \subseteq \mathbb{R}^{d_l}$ if $l = k$. In this case we have

$$
Q_{m,k}(k, x; l, B)
= \delta_{kl} \int_{\mathbb{R}^{d_k}} p_m(k, x; y) \left(\alpha_m(k, x; y) \mathbb{1}_B(y) + (1 - \alpha_m(k, x; y)) \, \mathbb{1}_B(x)\right) dy
$$

where δ_{kl} denotes the Kronecker delta. For this case, the acceptance probability α_m is given by (4.38) and using the definition of α_m we find

$$
\pi(k, x)\gamma_m(k, x)b_m(k, x; k)p_m(k, x; y) \, \alpha_m(k, x; y)
$$
$$
= \pi(k, y)\gamma_m(k, y)b_m(k, y; k)p_m(k, y; x) \, \alpha_m(k, y; x).
$$

As in the proof of proposition 4.13, we can use this symmetry to deduce

$$
\int_{\mathbb{R}^{d_k}} \pi(k, x)\gamma_m(k, x) \, b_m(k, x; k) \, Q_{m,k}(k, x; l, B) \, dx
$$
$$
= \delta_{kl} \int_B \pi(k, x)\gamma_m(k, x)b_m(k, x; k) \, dx
$$

and thus, by summing both sides over k,

$$
\sum_{k \in I} \int_{\mathbb{R}^{d_k}} \pi(k, x)\gamma_m(k, x) \, b_m(k, x; k) \, Q_{m,k}(k, x; l, B) \, dx
$$
$$
= \int_B \pi(l, x)\gamma_m(l, x) \, b_m(l, x; l) \, dx. \tag{4.43}
$$

For the second case, if $i \neq k$ in (4.42), there are two possible ways to achieve $X_j \in B$: for $l = i$ the process can reach the space S_l by accepting the proposal, whereas for $l = k$ the process can stay in the space S_l by rejecting the proposal. For all other cases, the probability $Q_{m,i}(k, x; l, B)$ in (4.42) equals 0. Thus we have

$$Q_{m,i}(k, x; l, B) = Q^{(1)}_{m,i}(k, x; l, B) + Q^{(2)}_{m,i}(k, x; l, B)$$

where, defining y and v by $(y, v) = \varphi^{k \to i}_m(x, u)$ again,

$$Q^{(1)}_{m,i}(k, x; l, B) = \delta_{li} \int_{\mathbb{R}^{n_k}} \psi_m(k, x, u; i)\alpha_m(k, x, u; i, y, v)\mathbb{1}_B(y)\, du$$

is the probability of reaching B with an accepted jump from S_k to S_l and

$$Q^{(2)}_{m,i}(k, x; l, B) = \delta_{kl} \int_{\mathbb{R}^{n_k}} \psi_m(k, x, u; i)(1 - \alpha_m(k, x, u; i, y, v))\, du \cdot \mathbb{1}_B(x)$$

is the probability of staying in space $l = k$ by rejecting a move, both conditional on the proposal using move type m and it being in space S_i.

Since $\varphi^{i \to k}_m = (\varphi^{k \to i}_m)^{-1}$ by assumption 4.34, the Jacobian of $\varphi^{i \to k}_m$ satisfies $D\varphi^{i \to k}_m(y, v) = \left(D\varphi^{k \to i}_m(x, u)\right)^{-1}$ and using the rule for the determinant of the inverse of a matrix, we have $\left|\det D\varphi^{i \to k}_m(y, v)\right| = 1/\left|\det D\varphi^{k \to i}_m(x, u)\right|$. From this relation and the definition (4.40) of α_m we find

$$\pi(k, x)\gamma_m(k, x)b_m(k, x; i)\psi_m(k, x, u; i)\, \alpha_m(k, x, u; i, y, v)$$
$$= \pi(i, y)\gamma_m(i, y)b_m(i, y; k)\psi_m(i, y, v; k)\, \alpha_m(i, y, v; k, x, u)$$
$$\cdot \left|\det D\varphi^{k \to i}_m(x, u)\right|.$$

Thus, we have

$$\int_{\mathbb{R}^{d_k}} \pi(k, x)\, \gamma_m(k, x)b_m(k, x; i)Q^{(1)}_{m,i}(k, x; l, B)\, dx$$

$$= \delta_{li} \int_{\mathbb{R}^{d_k}} \int_{\mathbb{R}^{n_k}} \pi(k, x)\gamma_m(k, x)b_m(k, x; i)\psi_m(k, x, u; i)$$
$$\cdot \alpha_m(k, x, u; i, y, v)\, \mathbb{1}_B(y)\, du\, dx$$

$$= \delta_{li} \int_{\mathbb{R}^{d_k}} \int_{\mathbb{R}^{n_k}} \pi(i, y)\gamma_m(i, y)b_m(i, y; k)\psi_m(i, y, v; k)$$
$$\cdot \alpha_m(i, y, v; k, x, u)\, \mathbb{1}_B(y)\, \left|\det D\varphi^{k \to i}_m(x, u)\right|\, du\, dx$$

$$= \delta_{li} \int_{\mathbb{R}^{d_l}} \int_{\mathbb{R}^{n_l}} \pi(i, y)\gamma_m(i, y)b_m(i, y; k)\psi_m(i, y, v; k)$$
$$\cdot \alpha_m(i, y, v; k, x, u)\, \mathbb{1}_B(y)\, dv\, dy$$

$$= \delta_{li} \int_B \pi(i, y)\, \gamma_m(i, y)b_m(i, y; k)$$

$$\cdot \int_{\mathbb{R}^{n_l}} \psi_m(i, y, v; k)\alpha_m(i, y, v; k, x, u)\, dv\, dy,$$

where the last equality uses the substitution rule for integrals (see lemma 1.36) and on the last line of the equation x and u are defined by $(x, u) = \varphi_m^{i \to k}(y, v)$. Summing this expression over $k \in I$ and $i \neq k$ we find

$$\sum_{k \in I} \int_{\mathbb{R}^{d_k}} \pi(k, x)\, \gamma_m(k, x) \sum_{i \neq k} b_m(k, x; i) Q_{m,i}^{(1)}(k, x; l, B)\, dx$$

$$= \int_B \pi(l, y)\, \gamma_m(l, y) \sum_{k \neq l} b_m(l, y; k) \tag{4.44}$$

$$\cdot \int_{\mathbb{R}^{n_l}} \psi_m(l, y, v; k) \alpha_m(l, y, v; k, x, u)\, dv\, dy.$$

For the terms involving $Q_{m,i}^{(2)}$ we find

$$\int_{\mathbb{R}^{d_k}} \pi(k, x)\, \gamma_m(k, x) b_m(k, x; i) Q_{m,i}^{(2)}(k, x; l, B)\, dx$$

$$= \delta_{kl} \int_B \pi(k, x)\, \gamma_m(k, x) b_m(k, x; i) \int_{\mathbb{R}^{n_k}} \psi_m(k, x, u; i)$$

$$\cdot (1 - \alpha_m(k, x, u; i, y, v))\, du\, dx$$

$$= \delta_{kl} \int_B \pi(k, x)\, \gamma_m(k, x) b_m(k, x; i)\, dx$$

$$- \delta_{kl} \int_B \pi(k, x)\, \gamma_m(k, x) b_m(k, x; i)$$

$$\cdot \int_{\mathbb{R}^{n_k}} \psi_m(k, x, u; i) \alpha_m(k, x, u; i, y, v)\, du\, dx$$

and summing this expression over $k \in I$ and $i \neq k$ we get

$$\sum_{k \in I} \int_{\mathbb{R}^{d_k}} \pi(k, x)\, \gamma_m(k, x) \sum_{i \neq k} b_m(k, x; i) Q_{m,i}^{(2)}(k, x; l, B)\, dx$$

$$= \int_B \pi(l, x)\, \gamma_m(l, x) \sum_{i \neq l} b_m(l, x; i)\, dx \tag{4.45}$$

$$- \int_B \pi(l, x)\, \gamma_m(l, x) \sum_{i \neq l} b_m(l, x; i)$$

$$\cdot \int_{\mathbb{R}^{n_l}} \psi_m(l, x, u; i) \alpha_m(l, x, u; i, y, v)\, du\, dx.$$

So far we have considered the terms involving $Q_{m,i}^{(1)}$ and $Q_{m,i}^{(2)}$ separately. From these results we can get the corresponding expressions for $Q_{m,i}$. Combining equation (4.44) and equation (4.45) we find

$$\sum_{k \in I} \int_{\mathbb{R}^{d_k}} \pi(k, x) \, \gamma_m(k, x) \sum_{i \neq k} b_m(k, x; i) Q_{m,i}(k, x; l, B) \, dx$$

$$= \int_B \pi(l, x) \, \gamma_m(l, x) \sum_{i \neq l} b_m(l, x; i) \, dx. \tag{4.46}$$

This is the result for the second case.

Finally, we can add the result from equation (4.43) and equation (4.46) to get

$$\sum_{k \in I} \int_{\mathbb{R}^{d_k}} \pi(k, x) \, \gamma_m(k, x) \sum_{i \in I} b_m(k, x; i) Q_{m,i}(k, x; l, B) \, dx$$

$$= \int_B \pi(l, x) \, \gamma_m(l, x) \sum_{i \in I} b_m(l, x; i) \, dx$$

$$= \int_B \pi(l, x) \, \gamma_m(l, x) \, dx$$

and summing this formula over $m \in M$ gives the right-hand side in (4.42). Thus we find

$$P\left(K_j = l, X_j \in B\right)$$

$$= \int_B \pi(l, x) \sum_{m \in M} \gamma_m(l, x) \, dx$$

$$= \int_B \pi(l, x) \, dx.$$

This shows that K_j and X_j have the correct distribution and that thus π is a stationary density of $(K_j, X_j)_{j \in \mathbb{N}}$. $\qquad \square$

As for the original Metropolis–Hastings algorithm, the target density π enters the RJMCMC algorithm only via the acceptance probabilities given by equation (4.38) and equation (4.40). Since both forms of the acceptance probabilities contain only the ratio of the two values of π, for the proposal and the current state, the algorithm can still be applied when π is only known up to a multiplicative constant. This is particularly useful for applications in Bayesian inference, where the constant probability of the observations, for example the constant Z in equation (4.21), can be omitted from the definition of π. Since the ratio of $\pi(l, y)/\pi(k, x)$ in equation (4.40) contains values of π corresponding to different subspaces $S_l \neq S_k$, the above argument only applies to global constants: if the subdensities $\pi(k, \cdot)$ contain unknown multiplicative constants which depend on the space k and where the ratio of these constants between different spaces is not known, the RJMCMC algorithm cannot be applied directly.

4.5.2 Bayesian inference for mixture distributions

In this section we illustrate the RJMCMC algorithm 4.36 with the help of an example: we consider a Bayesian inference problem for mixture distributions. For the example, we assume the following model: observations $Y_1, \ldots, Y_n$ are given from a two-dimensional mixture distribution

$$\mu = \frac{1}{k} \sum_{a=1}^{k} \mathcal{N} \left(\mu_a, r_a^2 I_2 \right), \tag{4.47}$$

where I_2 is the two-dimensional identity matrix. We assume that the number k of modes, the means μ_a and the standard deviations r_a are all random, with distributions given by

$$k \sim \text{Pois}(3) + 1 \tag{4.48}$$

as well as

$$\mu_a \sim \mathcal{U} \left([-10, +10] \times [-10, +10] \right) \tag{4.49}$$

and

$$r_a \sim \mathcal{U} \left[\frac{1}{2}, \frac{5}{2} \right] \tag{4.50}$$

for all $a \in \{1, \ldots, k\}$. Our aim is to generate samples from the posterior distribution of k, μ_a and r_a, given the data $Y_1, \ldots, Y_n$.

In this section we will use the RJMCMC algorithm to generate the required samples. In order to do so we first have to determine the state space S and the target distribution on this state space, and then we have to choose a set of moves which allows the algorithm to efficiently explore all of the state space.

4.5.2.1 State space

Since we are interested in the posterior distribution of the parameters, the state space consists of all possible parameter values. The parameters for each component of the mixture are $\mu_a \in \mathbb{R}^2$ and $r_a \in \mathbb{R}$ and thus we can choose $I = \mathbb{N}$ and

$$S_k = \left(\mathbb{R}^2 \right)^k \times \mathbb{R}^k \cong \mathbb{R}^{3k}$$

for all $k \in I$. Here, the parameter vector $(\mu, r) = (\mu_1, \ldots, \mu_k, r_1, \ldots, r_k)$ plays the role of the state vector x in the general description of the method in Section 4.5.1 and we have $d_k = 3k$ for all $k \in I$.

4.5.2.2   Target distribution

The target distribution is the posterior distribution of the parameters k, $\mu = (\mu_1, \ldots, \mu_k)$ and $r = (r_1, \ldots, r_k)$, given observations $y = (y_1, \ldots, y_n)$ for $Y = (Y_1, \ldots, Y_n)$. Using Bayes' rule and the prior distributions of k, μ_a and r_a given by equation (4.48), equation (4.49) and equation (4.50), we find the target distribution to be

$$
\begin{aligned}
\pi(k, \mu, r) \\
&= p_{k,\mu,r|Y}(k, \mu, r \mid y) \\
&= \frac{p_{Y|k,\mu,r}(y \mid k, \mu, r)\, p_{k,\mu,r}(k, \mu, r)}{p_Y(y)} \\
&= \frac{1}{Z}\, p_{Y|k,\mu,r}(y \mid k, \mu, r)\, p_{\mu|k}(\mu \mid k)\, p_{r|k}(r \mid k)\, p_k(k)
\end{aligned}
\tag{4.51}
$$

for all $(\mu, r) \in S_k$ and all $k \in I$, where $Z = p_Y(y)$ is constant, the observations have density

$$
p_{Y|k,\mu,r}(y \mid k, \mu, r) = \prod_{i=1}^{n} \frac{1}{k} \sum_{a=1}^{k} \frac{1}{2\pi r_a^2} \exp\left(-\frac{(y_{i,1} - \mu_{a,1})^2 + (y_{i,2} - \mu_{a,2})^2}{2r_a^2} \right)
$$

for all $y \in (\mathbb{R}^2)^n$ and the prior distribution is given by the densities

$$
p_{\mu|k}(\mu \mid k) = \prod_{a=1}^{k} \frac{1}{20^2} \mathbb{1}_{[-10,10] \times [-10,10]}(\mu_a)
$$

$$
p_{r|k}(r \mid k) = \prod_{a=1}^{k} \frac{1}{2} \mathbb{1}_{[1/2, 5/2]}(r_a)
$$

$$
p_k(k) = e^{-3} \frac{3^{k-1}}{(k-1)!}
$$

for all $\mu \in (\mathbb{R}^2)^k$, $r \in \mathbb{R}^k$ and $k \in \mathbb{N}$.

4.5.2.3   Move types

In order to allow the process to explore all of the state space, we need moves to change the means μ_a and variances r_a of the mixture components, and to change the number k of components. Here we use a four-element set $M = \{m_\mu, m_r, m_\pm, m_\leftrightarrow\}$ to enumerate the move types, where m_μ denotes a move which changes the means, m_r denotes a move which changes the variances and $m_\pm$ denotes a move which increases or decreases the number of mixture components by one. For simplicity we assume that $m_\pm$ only removes or adds a component at the end of the list of components. To compensate for this and to allow for arbitrary mixture components to be removed, we introduce an additional move type $m_\leftrightarrow$, which only changes the numbering of the mixture components, and otherwise leaves the state unchanged. We will now consider these four move types in detail.

We need to specify the move type probabilities $\gamma_m(k, \mu, r)$ for all move types $m \in M$, such that

$$\sum_{m \in M} \gamma_m(k, \mu, r) = 1.$$

Since k is a discrete variable, fewer moves may be necessary to explore the possible values for k than for the variables μ and r. To reflect this we choose the move probabilities as

$$\gamma_{m_\mu} = \gamma_{m_\mu}(k, \mu, r) = 4/10,$$
$$\gamma_{m_\sigma} = \gamma_{m_\sigma}(k, \mu, r) = 4/10,$$
$$\gamma_{m_\pm} = \gamma_{m_\pm}(k, \mu, r) = 1/10,$$
$$\gamma_{m_\leftrightarrow} = \gamma_{m_\pm}(k, \mu, r) = 1/10$$

for all $\mu \in \mathbb{R}^{2k}$, $r \in \mathbb{R}^k$ and $k \in \mathbb{N}$.

4.5.2.4   Move details

Moves of type m_μ only change the mixture component means μ and thus always propose values which stay inside the current subspace: we have $b_{m_\mu}(k, y; k) = 1$ and $b_{m_\mu}(k, y; l) = 0$ for all $l \neq k$. To perform a move of type m_μ, we first randomly choose an index

$$a \sim \mathcal{U}\{1, 2, \ldots, k\}$$

and then replace μ_a by

$$\tilde{\mu}_a \sim \mu_a + \mathcal{N}(0, \sigma_\mu^2 I_2);$$

all other components of μ stay unchanged. Since the increments $\tilde{\mu} - \mu$ are symmetric, we can use use equation (4.38) together with example 4.17 to determine the acceptance probability for this move type: we get

$$\alpha_{m_\mu}(k, \mu, r; \tilde{\mu}, r) = \min\left(\frac{\pi(k, \tilde{\mu}, r) \cdot \frac{4}{10} \cdot 1}{\pi(k, \mu, r) \cdot \frac{4}{10} \cdot 1}, 1\right)$$

$$= \min\left(\frac{\pi(k, \tilde{\mu}, r)}{\pi(k, \mu, r)}, 1\right),$$

$$(4.52)$$

where π is the posterior density given in equation (4.51). Since the unknown constant Z in (4.51) appears both in the numerator and denominator of (4.52), both occurrences of Z cancel and the value of Z is not required to evaluate the acceptance probability α_{m_μ}. Similar considerations apply to the acceptance probabilities for the remaining move types.

Moves of type m_r change the mixture component variances r_a^2. This move type leaves the subspaces invariant and thus we have $b_{m_r}(k, y; k) = 1$ and $b_{m_r}(k, y; l) = 0$ for all $l \neq k$. To perform a move of type m_r, we first randomly choose an index

$$a \sim \mathcal{U}\{1, 2, \ldots, k\}$$

and then replace r_a by

$$\tilde{r}_a \sim r_a + \mathcal{N}(0, \sigma_r^2);$$

all other components of r stay unchanged. Again, we can use use example 4.17 to determine the acceptance probability for this move type: we get

$$\alpha_{m_r}(k, \mu, r; \mu, \tilde{r}) = \min\left(\frac{\pi(k, \mu, \tilde{r}) \cdot \frac{4}{10} \cdot 1}{\pi(k, \mu, r) \cdot \frac{4}{10} \cdot 1}, 1\right)$$

$$= \min\left(\frac{\pi(k, \mu, \tilde{r})}{\pi(k, \mu, r)}, 1\right).$$

$$(4.53)$$

Moves of types $m_\pm$ change the number k of mixture components by exactly one. Thus, this move type always changes subspaces and we set

$$b_{m_\pm}(k, \mu, r; k + 1) = b_{m_\pm}(k, \mu, r; k - 1) = 1/2$$

for all $k > 1$. For the case $k = 1$ we cannot decrease the number of mixture components any further and so we set

$$b_{m_\pm}(1, \mu, r; 2) = 1.$$

Finally, for all other combinations of $k, l \in I$ we set $b_{m_\pm}(k, \mu, r; l) = 0$.

Since $d_k = 3k$ and $d_{k+1} = 3k + 3$, we can satisfy the dimension matching criterion (4.39) for moves from k to $l = k + 1$ by choosing $n_k = 3$ and $n_{k+1} = 0$. The transition maps $\varphi_{m_\pm}^{k \to k+1}$ and $\varphi_{m_\pm}^{k+1 \to k}$ must satisfy assumption 4.34, so we need to specify only one of the two maps and the second map is then found as the inverse of the first one. For simplicity, we choose

$$\varphi_{m_\pm}^{k \to k+1}(\mu, r, u) = (\mu_1, \ldots, \mu_k, (u_1, u_2), r_1, \ldots, r_k, u_3),$$

that is we just construct a new mean and and new standard deviation from the auxiliary sample $u \in \mathbb{R}^{n_k}$ as $\mu_{k+1} = (u_1, u_2)$ and $r_{k+1} = u_3$, and leave the remaining mixture components unchanged. To get an appropriate distribution for the newly added mixture component, we use the prior distribution to construct the density $\psi_{m_\pm}$ of the auxiliary sample U_j: we set

$$
\begin{aligned}
\psi_{m_\pm}(u) &= p_{\mu|k}\big((u_1, u_2)\,\big|\,1\big)\, p_{r|k}(u_3 \mid 1) \\
&= \frac{1}{20^2}\mathbb{1}_{[-10,10]}(u_1)\mathbb{1}_{[-10,10]}(u_2)\frac{1}{2}\mathbb{1}_{[1/2,5/2]}(u_3)
\end{aligned}
\tag{4.54}
$$

for all $u \in \mathbb{R}^{n_k}$. This density does not depend on k and x, and thus we write $\psi_{m_\pm}(u)$ instead of $\psi_{m_\pm}(k, x, u; k+1)$. Since $\varphi_{m_\pm}^{k \to k+1}$ just returns a permutation of its arguments, we have $\left|\det D\varphi_{m_\pm}^{k \to k+1}\right| = 1$. Putting everything together, we can now get the corresponding acceptance probabilities from (4.40): for $k \geq 2$ we find

$$
\begin{aligned}
&\alpha_{m_\pm}(k, \mu, r, u; k+1, \tilde{\mu}, \tilde{r}) \\
&= \min\left(\frac{\pi(k+1, \tilde{\mu}, \tilde{r}) \cdot \frac{1}{10} \cdot \frac{1}{2} \cdot 1}{\pi(k, \mu, r) \cdot \frac{1}{10} \cdot \frac{1}{2} \cdot \psi_{m_\pm}(u)}, 1\right) \\
&= \min\left(\frac{\pi(k+1, \tilde{\mu}, \tilde{r})}{\pi(k, \mu, r)\, \psi_{m_\pm}(u)}, 1\right)
\end{aligned}
\tag{4.55}
$$

and for $k = 1$ we have

$$
\begin{aligned}
&\alpha_{m_\pm}(1, \mu, r, u; 2, \tilde{\mu}, \tilde{r}) \\
&= \min\left(\frac{\pi(2, \tilde{\mu}, \tilde{r}) \cdot \frac{1}{10} \cdot \frac{1}{2} \cdot 1}{\pi(1, \mu, r) \cdot \frac{1}{10} \cdot 1 \cdot \psi_{m_\pm}(u)}, 1\right) \\
&= \min\left(\frac{\pi(2, \tilde{\mu}, \tilde{r})}{2 \cdot \pi(1, \mu, r)\, \psi_{m_\pm}(u)}, 1\right),
\end{aligned}
\tag{4.56}
$$

where π is again given by (4.51) and $\psi_{m_\pm}$ is defined in equation (4.54). Since the auxiliary sample V_j is in $\mathbb{R}^{n_{k+1}}$ with $n_{k+1} = 0$, we can omit v and its density $\psi_{m_\pm}(k+1, \mu, r, \cdot; k)$ from the formula.

The reverse transition, from k to $k-1$ mixture components, is now completely specified by assumption 4.34: we find

$$
\begin{aligned}
&\varphi_{m_\pm}^{k \to k-1}(\mu_1, \ldots, \mu_k, r_1, \ldots, r_k) \\
&= \big((\mu_1, \ldots, \mu_{k-1}, r_1, \ldots, r_{k-1}), (\mu_{k,1}, \mu_{k,2}, r_k)\big)
\end{aligned}
$$

for all $(\mu, \sigma) \in \mathbb{R}^{2k} \times \mathbb{R}^k$, that is $\varphi_{m_\pm}^{k \to k-1}$ performs a permutation and again we have $\left|\det D\varphi_{m_\pm}^{k \to k-1}\right| = 1$. This time we can omit the argument $u \in \mathbb{R}^0$ and the corresponding density $\psi_{m_\pm}(k, \mu, r, \cdot; k-1)$. The resulting form of the acceptance probability for $k \geq 3$ is

$$\alpha_{m_\pm}(k, \mu, r; k-1, \tilde{\mu}, \tilde{r}, v)$$
$$= \min\left(\frac{\pi(k-1, \tilde{\mu}, \tilde{r}) \cdot \frac{1}{10} \cdot \frac{1}{2} \cdot \psi_{m_\pm}(v)}{\pi(k, \mu, r) \cdot \frac{1}{10} \cdot \frac{1}{2} \cdot 1}, 1\right) \tag{4.57}$$
$$= \min\left(\frac{\pi(k-1, \tilde{\mu}, \tilde{r}) \, \psi_{m_\pm}(v)}{\pi(k, \mu, r)}, 1\right)$$

and for $k = 2$ we get

$$\alpha_{m_\pm}(2, \mu, r; 1, \tilde{\mu}, \tilde{r}, v)$$
$$= \min\left(\frac{\pi(1, \tilde{\mu}, \tilde{r}) \cdot \frac{1}{10} \cdot 1 \cdot \psi_{m_\pm}(v)}{\pi(2, \mu, r) \cdot \frac{1}{10} \cdot \frac{1}{2} \cdot 1}, 1\right) \tag{4.58}$$
$$= \min\left(\frac{2 \cdot \pi(1, \tilde{\mu}, \tilde{r}) \, \psi_{m_\pm}(v)}{\pi(2, \mu, r)}, 1\right).$$

Finally, moves of type $m_\leftrightarrow$ change the numbering of the components only. If $k = 1$, this move type does nothing: the proposal equals the current state, and the corresponding acceptance probability is 1. For $k > 1$, we randomly choose an index

$$a \sim \mathcal{U}\{1, 2, \ldots, k-1\}$$

and the construct the proposal as

$$\tilde{\mu} = (\mu_{k-a+1}, \ldots, \mu_k, \mu_1, \ldots, \mu_{k-a}),$$
$$\tilde{r} = (r_{k-a+1}, \ldots, r_k, r_1, \ldots, r_{k-a}).$$

Using equation (4.38) and the fact that the target distribution π does not change when the mixture components are interchanged, we find the corresponding acceptance probability as

$$\alpha_{m_\leftrightarrow}(k, \mu, r; \tilde{\mu}, \tilde{r}) = \min\left(\frac{\pi(k, \tilde{\mu}, \tilde{r}) \cdot \frac{1}{10} \cdot 1 \cdot \frac{1}{k-1}}{\pi(k, \mu, r) \cdot \frac{1}{10} \cdot 1 \cdot \frac{1}{k-1}}, 1\right) = 1.$$

This completes the description of the different move types.

4.5.2.5 Implementation

Using the state space, moves and acceptance probabilities described above, we can now implement the RJMCMC algorithm 4.36 for the Bayesian parameter estimation problem described in this section. Once the algorithm is implemented we still need to choose the standard deviations σ_μ and σ_r used in the moves of type m_μ and m_r, respectively. The parameters σ_μ and σ_r control the size of the increments in the corresponding moves and as in Section 4.1.3 we expect the acceptance rates to decrease as σ_μ and σ_r are increased. The methods from Section 4.2.2 can be used to guide tuning of these parameters. For example, the values could be chosen so that the acceptance rates are not too close to either 0 or 1. In our experiments it was also required to write the acceptance probabilities in the numerically more robust form (4.19).

Figure 4.9 shows the final state of one run of the RJMCMC algorithm for the setup considered in this section. As an example of the kind of questions which can

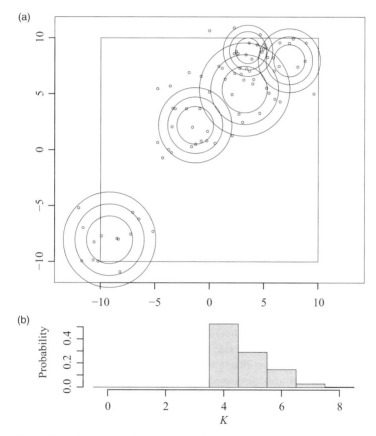

Figure 4.9   (a) The final state of a run of the RJMCMC algorithm described in Section 4.5.2. The small circles give the locations of $n = 80$ observations from the mixture distribution (4.47). The five sets of concentric circles give the locations and standard deviations of the five mixture components present in a sample from the posterior distribution (the radii are r_a, $1.5\,r_a$ and $2\,r_a$), obtained by running $N = 50\,000$ steps of the RJMCMC algorithm. (b) A histogram of the values $(K_{100\,j})_{j=1,\dots,N/100}$ observed during the run.

be answered using the RJMCMC algorithm, Figure 4.9(b) gives a histogram of the distribution of the number k of mixture components observed during the run. For the given set of observations, showing significant overlap between mixture components, all values $k \in \{4, 5, 6, 7\}$ can be observed with non-negligible probability.

4.6   Summary and further reading

In this chapter we have learned how Markov chains can be used as the basis of Monte Carlo methods. The foundation of all such methods is the Metropolis–Hastings algorithm which we have discussed in several variants. An overview over the area of MCMC methods and more pointers to the literature can be found in Gentle *et al.* (2004, Chapter II.3). Markov chains and the Metropolis–Hastings method are discussed in Chapter 7 of Robert and Casella (2004). Applications and extensions to the methods can also be found in Gilks *et al.* (1996) and Kroese *et al.* (2011). We have also covered the Gibbs sampler and the RJMCMC method; both of these methods technically still fall under the umbrella of the Metropolis–Hastings framework, but these methods are so specialised that they are often treated as separate approaches. More information about the Gibbs sampler can be found in Chapters 8–10 of Robert and Casella (2004) and the RJMCMC method, introduced in Green (1995), is for example discussed in Waagepetersen and Sorensen (2001) and Chapter 11 of Robert and Casella (2004).

In Section 4.2 we have discussed both theoretical and practical aspects related to the convergence of MCMC methods: Section 4.2.1 summarises some of the theoretical results which underpin the mathematical analysis of the resulting methods and Section 4.2.2 discusses aspects relevant to using MCMC methods in practice. Further discussion of convergence of MCMC methods can be found in Chapter 6 of Robert and Casella (2004), a survey is given in Roberts and Rosenthal (2004) and many of the technical details can be found in the monograph by Meyn and Tweedie (2009).

To illustrate the different methods introduced in this chapter we have considered a variety of Bayesian inference problems, ranging from simple Bayesian models (Section 4.3) over parameter estimation for mixture distributions (Sections 4.4.2 and 4.5.2) to applications in image processing (Section 4.4.3). More details about the use of Monte Carlo methods in Bayesian statistics can be found in Gentle *et al.* (2004, Chapter III.11) and Bayesian image analysis is, for example, discussed in Winkler (1995). In passing we also discussed the Ising model from statistical mechanics; this model is treated in many advanced texts, for example in Plischke and Bergersen (2006) and Privman (1990).

Exercises

E4.1   Implement the random walk Metropolis algorithm with target density π given by equation (4.6). Plot paths of the generated Markov chains for different values of σ.

E4.2   Implement the random walk Metropolis sampler for sampling from the target distribution $\mathcal{N}(100, 1)$ on $\mathbb{R}$, using proposals $Y_j = X_{j-1} + \varepsilon_j$ where $\varepsilon_j \sim \mathcal{N}(0, 1)$. Experiment with different starting values $X_0 \in \mathbb{R}$ and create a plot of a path $X_0, X_1, \ldots, X_N$ which illustrates the need for a burn-in period.

E4.3   Implement the Metropolis–Hastings algorithm with target density given by equation (4.6) and proposals $Y_j \sim \mathcal{N}(X_j, \sigma^2)$. For $\sigma = 1, 6, 36$, use the output of the algorithm to estimate the lag k autocorrelations $\rho_k = \text{Corr}(X_j, X_{j+k})$ for $k = 1, 2, \ldots, 100$. Create plots of the autocorrelations ρ_k as a function of k.

E4.4   For the situation of example 4.25, write a program which estimates, for a given value of σ, the average acceptance probability in the Metropolis–Hastings algorithm. Create a plot of this acceptance probability as a function of σ.

E4.5   Implement the random walk Metropolis method with $\mathcal{N}(0, \eta^2)$-distributed increments for the posterior distribution from example 4.26. The input of your program should be the variance η^2 and a list $x = (x_1, \ldots, x_n)$ of observations. The output of your program should be a path (μ_j, σ_j) of a Markov chain which has $p_{\mu,\sigma}(\cdot \mid X = x)$ as its stationary distribution. Test your program using randomly generated data.

E4.6   Implement the Gibbs sampler for sampling from the posterior distribution (4.26), using the method described in Section 4.4.2.

E4.7   Write a program to implement the Gibbs sampler for the Ising model from algorithm 4.31. Test your program by generating samples from the Ising model for $\beta = 0.30$, $\beta = 0.34$, $\beta = 0.38$, $\beta = 0.42$, $\beta = 0.46$ and $\beta = 0.50$.

E4.8   Write a program to generate samples from the posterior distribution $p_{X|Y}$ in the Bayesian image denoising problem from equation (4.36), given a noisy input image $y \in \mathbb{R}^I$. The program should follow the steps described in algorithm 4.32.

E4.9   Write a program to implement the RJMCMC method described in Section 4.5.2.

5

Beyond Monte Carlo

In this chapter we present two methods which can be used instead of Monte Carlo methods if either no statistical model is available or if the available models are too complicated to easily apply Monte Carlo methods. The first of these two methods, called Approximate Bayesian Computation (ABC), is tailored towards the case where a computer model of the studied system is available, which can be used instead of a mathematical model. The second method, Bootstrap sampling, is used in cases where only a set of observations but no model at all is available.

5.1 Approximate Bayesian Computation

ABC is a computational technique to obtain approximate parameter estimates in a Bayesian setting. The ABC approach, described in this section, is less efficient than MCMC methods are, but ABC is easier to implement than MCMC and the method requires very little theoretical knowledge about the underlying model.

Our aim in this section is to generate samples from the posterior distribution of the parameter θ in a Bayesian model, given observations $X = x$. From Section 4.3 we know that the posterior density of θ is given by

$$p_{\theta|X}(\theta \,|\, x) = \frac{1}{Z} p_{X|\theta}(x \,|\, \theta) p_{\theta}(\theta), \tag{5.1}$$

where Z is a normalising constant. The problem that ABC is aiming to solve is that in many situations the density $p_{X|\theta}(\cdot \,|\, \theta)$ can be difficult to obtain in closed form. The ABC method can be applied in situations where generating samples from this density is easier than working with the density itself. Since the resulting method avoids use of any explicit form of this density, ABC is called a *likelihood free* method.

An Introduction to Statistical Computing: A Simulation-based Approach, First Edition. Jochen Voss.
© 2014 John Wiley & Sons, Ltd. Published 2014 by John Wiley & Sons, Ltd.

The ABC method is based on the basic rejection sampling algorithm 1.19: if proposals are generated with density g and if each proposal X is accepted with probability $p(X)$, then the accepted proposals are distributed with density proportional to $p \cdot g$. In the context of Bayesian parameter estimation, we can apply this algorithm as follows:

(a) Generate samples $\theta_j \sim p_\theta$ for $j = 1, 2, 3, \ldots$.

(b) Accept each sample θ_j with probability proportional to $p_{X|\theta}(x|\theta_j)$, where x is the observed data.

By Proposition 1.20, the accepted samples are distributed with density $p(\theta|x)$ given by (5.1). Methods based on this idea are called ABC methods. Since we assume that we do not have access to an explicit formula for $p(x|\theta)$, our aim is to find a method implementing the rejection procedure described above, using only samples from the density $p(x|\theta)$ but not the density itself.

5.1.1   Basic Approximate Bayesian Computation

In this section we describe a basic version of the ABC method. We start the presentation by describing the method as an algorithm, and then give the required explanations to understand why this algorithm gives the desired result.

Algorithm 5.1   (basic Approximate Bayesian Computation)
    input:
        data $x^* \in \mathbb{R}^n$
        the prior density π for the unknown parameter $\theta \in \mathbb{R}^p$
        a summary statistic $S: \mathbb{R}^n \to \mathbb{R}^q$
        an approximation parameter $\delta > 0$
    randomness used:
        samples $\theta_j \sim p_\theta$ and $X_j \sim p_{X|\theta}(\cdot|\theta_j)$ for $j \in \mathbb{N}$
    output:
        $\theta_{j_1}, \theta_{j_2}, \ldots$ approximately distributed with density $p_{\theta|X}(\theta|x^*)$
  1: $s^* \leftarrow S(x^*)$
  2: **for** $j = 1, 2, 3, \ldots$ **do**
  3:    sample $\theta_j \sim p_\theta(\cdot)$
  4:    sample $X_j \sim p_{X|\theta}(\cdot|\theta_j)$
  5:    $S_j \leftarrow S(X_j)$
  6:    **if** $|S_j - s^*| \le \delta$ **then**
  7:        output θ_j
  8:    **end if**
  9: **end for**

In the algorithm, the summary statistic S is assumed to take values in $\mathbb{R}^q$. The dimension q is typically *much* smaller than the dimension n of the data, and often q equals the number p of parameters. The distance $|S_j - s^*|$ in line 6 of the algorithm is the Euclidean norm in $\mathbb{R}^q$. Since the algorithm considers the summary statistic $s^* = S(x^*)$ instead of the full data, the method can only be expected to work if $S(\theta)$ contains 'enough' information about θ. The optimal case for this is if S is a sufficient statistic, as described in the following definition.

Definition 5.2   A statistic $S = S(X)$ is a *sufficient statistic* for θ, if the conditional distribution of X given the value S does not depend on the parameter θ.

If S is a sufficient statistic, then the value $S(X)$ contains all information from X about θ: once the value $S(X)$ is known, the sample does not contain any additional information about θ. More information about sufficient statistics can be found in the literature, for example in Section 6.2 of the book by Casella and Berger (2001). If no summary statistic is available, the algorithm can be applied with a nonsufficient statistic S, but this will introduce an additional error. To minimise this error, S should be chosen so that it contains as much information about θ as possible.

Algorithm 5.1 involves a trade-off between speed and accuracy: if the approximation parameter $\delta > 0$ is chosen small, the distribution of the generated samples is closer to the exact posterior. On the other hand, if δ is chosen larger, generation of the samples is faster. This is described in the following proposition.

Proposition 5.3   Let $(\theta_{j_k})_{k \in \mathbb{N}}$ be the accepted output samples of Algorithm 5.1 for given data $x^* \in \mathbb{R}^n$ and $s^* = S(x^*)$. Then the following statements hold.

(a) Let $p_{S|\theta}$ be the density of $S(X)$ where $X \sim p_{X|\theta}$. Assume that for every θ the function $s \mapsto p_{S|\theta}(s|\theta)$ is continuous in a neighbourhood of s^* and that $p_{S|\theta}$ is uniformly bounded in a neighbourhood of s^*, that is

$$\sup_{\substack{s \in \mathbb{R}^q \\ |s-s^*|<\varepsilon}} \sup_{\theta \in \mathbb{R}^p} p_{S|\theta}(s|\theta) < \infty$$

for all sufficiently small $\varepsilon > 0$. Then we have

$$\lim_{\delta \downarrow 0} P\big(\theta_{j_k} \in A\big) = \int_{\mathbb{R}^p} \mathbb{1}_A(\theta)\, p_{\theta|S}^{\mathrm{ABC}}(\theta|s^*)\, d\theta \tag{5.2}$$

where $p_{\theta|S}^{\mathrm{ABC}}$ is the probability density given by

$$p_{\theta|S}^{\mathrm{ABC}}(\theta|s) \propto p_{S|\theta}(s|\theta)\, p_\theta(\theta) \tag{5.3}$$

for all $\theta \in \mathbb{R}^p$. Thus, in the limit $\delta \downarrow 0$, the samples θ_{j_k} have density $p_{\theta|S}^{\mathrm{ABC}}$.

(b) If S is a sufficient statistic, the limiting density $p_{\theta|S}^{ABC}$ from (5.3) satisfies

$$p_{\theta|S}^{ABC}(\theta|s^*) = p_{\theta|X}(\theta|x^*),$$

that is in the limit $\delta \downarrow 0$ the distribution of the output samples coincides with the posterior distribution (5.1).

(c) The average number of proposals required to generate each output sample is of order $\mathcal{O}(\delta^{-q})$.

Proof   The proposal θ_j in the algorithm has density p_θ. The probability of accepting the proposal θ_j is given by

$$P\left(|s_j - s^*| \leq \delta \,|\, \theta_j\right) = P\left(|S(X_j) - s^*| \leq \delta \,|\, \theta_j\right)$$
$$= \int_{B_\delta(s^*)} p_{S|\theta}(s_j|\theta_j)\, ds_j,$$

where $B_\delta(s^*)$ is the ball around s^* with radius δ in $\mathbb{R}^q$. Thus, by proposition 1.20, the accepted proposals θ_{j_k} have density

$$p_\delta(\theta) \propto \int_{B_\delta(s^*)} p_{S|\theta}(s_j|\theta)\, ds_j \cdot p_\theta(\theta).$$

As an abbreviation, we write

$$r_\delta(\theta) = \frac{1}{|B_\delta(s^*)|} \int_{B_\delta(s^*)} p_{S|\theta}(s_j|\theta)\, ds_j \cdot p_\theta(\theta) \tag{5.4}$$

where $|B_\delta(s^*)|$ denotes the q-dimensional volume of the ball $B_\delta(s^*)$. Using this notation we have $p_\delta(\theta) = r_\delta(\theta)/Z_\delta$. Similarly, we write

$$r(\theta) = p_{S|\theta}(s^*|\theta)\, p_\theta(\theta)$$

to get $p_{\theta|S}^{ABC}(\theta|s^*) = r(\theta)/Z$ where Z is the normalisation constant. Since $s \mapsto p_{S|\theta}(s|\theta)$ is continuous, we have

$$\frac{1}{|B_\delta(s^*)|} \int_{B_\delta(s^*)} p_{S|\theta}(s_j|\theta_j)\, ds_j \longrightarrow p_{S|\theta}(s^*|\theta_j)$$

and thus $r_\delta(\theta) \to r(\theta)$ as $\delta \downarrow 0$ for all θ. By the dominated convergence theorem from analysis (see e.g. Rudin, 1987, theorem 1.34) we then have

$$\lim_{\delta \downarrow 0} Z_\delta = \lim_{\delta \downarrow 0} \int_{\mathbb{R}^p} r_\delta(\theta)\, d\theta = \int_{\mathbb{R}^p} r(\theta)\, d\theta = Z$$

and thus

$$p_\delta(\theta) = \frac{1}{Z_\delta} r_\delta(\theta) \longrightarrow \frac{1}{Z} r(\theta) = p_{\theta|S}^{ABC}(\theta|s^*).$$

Consequently, the density of the accepted samples θ_{j_k} converges to $p_{\theta|S}^{ABC}(\theta|s^*)$ as $\delta \downarrow 0$ for every $\theta \in \mathbb{R}^p$. By Scheffé's lemma (Scheffé, 1947), this implies the convergence of probabilities in (5.2) and thus the first statement of the proposition is proved.

For the second part of the proposition, we assume that S is a sufficient statistic. By definition 5.2 of sufficiency we have $p_{X|S,\theta}(x|s, \theta) = p_{X|S}(x|s)$. Using this fact, and the fact that $S(X)$ is a deterministic function of X, we get

$$
\begin{aligned}
p_{X|\theta}(x|\theta) &= p_{X,S|\theta}(x, S(x)|\theta) \\
&= p_{X|S,\theta}(x | S(x), \theta) \cdot p_{S|\theta}(S(x)|\theta) \\
&= p_{X|S}(x | S(x)) \cdot p_{S|\theta}(S(x)|\theta).
\end{aligned}
$$

Similarly, we find

$$
p_X(x) = p_{X,S}(x, S(x)) = p_{X|S}(x | S(x)) \cdot p_S(S(x)).
$$

Using these two relations, we get

$$
\begin{aligned}
p_{\theta|X}(\theta|x) &= \frac{p_{X|\theta}(x|\theta) \cdot p_\theta(\theta)}{p_X(x)} \\
&= \frac{p_{X|S}(x | S(x)) p_{S|\theta}(S(x)|\theta) \cdot p_\theta(\theta)}{p_{X|S}(x | S(x)) p_S(S(x))} \\
&= \frac{p_{S|\theta}(S(x)|\theta) p_\theta(\theta)}{p_S(S(x))} \\
&= p_{\theta|S}^{ABC}(\theta | S(x)).
\end{aligned}
$$

This completes the proof of the second statement.

Finally, using equation (5.4) and the definition of Z_δ from the first part of the proof, we have

$$
\begin{aligned}
P(|S_j - s^*| \leq \delta) &= \int_{\mathbb{R}^p} \int_{B_\delta(s^*)} p_{S|\theta}(s_j|\theta)\, ds_j\, p_\theta(\theta)\, d\theta \\
&= |B_\delta(s^*)| \int_{\mathbb{R}^p} r_\delta(\theta)\, d\theta \\
&= |B_\delta(s^*)| Z_\delta.
\end{aligned}
$$

Thus, the number N_δ of proposals required to generate one sample is geometrically distributed with mean

$$
\mathbb{E}(N_\delta) = \frac{1}{P(|S_j - s^*| \leq \delta)} = \frac{1}{|B_\delta(s^*)| Z_\delta} = \frac{1}{\delta^q |B_1(s^*)| Z_\delta}.
$$

Since $Z_\delta \to Z$ as $\delta \downarrow 0$, we find

$$\lim_{\delta \downarrow 0} \frac{\mathbb{E}(N_\delta)}{\delta^{-q}} = \frac{1}{|B_1(s^*)| Z}$$

and consequently $\mathbb{E}(N_\delta) = \mathcal{O}(\delta^{-q})$. This completes the proof. $\qquad\square$

To illustrate use of the ABC method, we consider a very simple example. In the example, use of ABC is not really necessary, but the simplicity of the situation helps to illustrate the method.

Example 5.4   For the prior let $\mu \sim \mathcal{U}[-10, 10]$ and $\sigma \sim \text{Exp}(1)$, independently, and set $\theta = (\mu, \sigma) \in \mathbb{R}^2$. Assume that the data consist of i.i.d. values $X_1, \ldots, X_n \sim \mathcal{N}(\mu, \sigma^2)$ and that we have observed values $x = (x_1, \ldots, x_n)$ for the data. Our aim is to generate samples from the posterior distribution of θ for given observations x.

To apply the ABC method, we first have to choose a summary statistic S. From the second statement of proposition 5.3 we know that it is best to choose a sufficient statistic. In the simple case considered here, the statistic

$$S(x) = \left(\frac{1}{n} \sum_{i=1}^{n} x_i, \frac{1}{n} \sum_{i=1}^{n} x_i^2 \right).$$

is known to be sufficient for $\theta = (\mu, \sigma)$ (see, e.g. Casella and Berger, 2001, example 6.2.9). Denote the components of $S(x)$ where x is the observed data by s_1^* and s_2^*. Then we can use ABC with this summary statistic, to generate posterior samples as follows:

(a) Sample $\mu_j \sim \mathcal{U}[-10, 10]$ and $\sigma_j \sim \text{Exp}(1)$ independently.

(b) Sample $X_{j,1}, \ldots, X_{j,n} \sim \mathcal{N}(\mu_j, \sigma_j^2)$ i.i.d.

(c) Let $s_{j,1} = \frac{1}{n} \sum_{i=1}^{n} X_{j,i}$ and $s_{j,2} = \frac{1}{n} \sum_{i=1}^{n} X_{j,i}^2$.

(d) Accept $\theta_j = (\mu_j, \sigma_j^2)$ if $(s_{i,1} - s_1^*)^2 + (s_{j,2} - s_2^*)^2 \leq \delta^2$.

In this example, the posterior density can easily be computed explicitly, and thus methods like MCMC could be applied in this situation. But even here, implementation of ABC is easier than implementation of MCMC, since we do not need to perform the calculation to obtain the formula for the posterior density. The power of ABC lies in the fact that we can use the method without evaluating the posterior density (except possibly to verify the assumptions of proposition 5.3) and thus ABC can be applied in more general situations than MCMC.

The results of a simulation with $n = 20$ and $s^* = (6.989, 52.247)$ are shown in Figure 5.1. Figure 5.1(a) corresponds to $\delta = 0.05$ (generating one sample on average took 60 549 proposals), Figure 5.1(b) corresponds to $\delta = 0.15$ (6022 proposals per sample) and Figure 5.1(c) corresponds to $\delta = 0.25$ (1059 proposals per sample).

From algorithm 5.1 is is clear that in the limit $\delta \to \infty$ the distribution of the output converges to the prior distribution: in the algorithm, the proposals θ_j are generated from the prior density p_θ and if δ is large most or even all of the generated samples are output without any change. This effect is clearly visible in the right-hand column of Figure 5.1: as δ increases, the histograms get closer and closer to the Exp(1)-distribution used as the prior for σ.

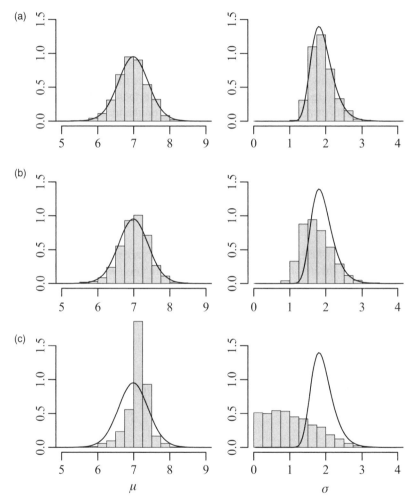

Figure 5.1   The effect of the approximation parameter δ in the ABC method from example 5.4. The histograms show the distribution of the ABC samples, the solid lines give the exact posterior density of μ and σ. The rows correspond to different values of δ: (a) $\delta = 0.05$; (b) $\delta = 0.15$; $\delta = 0.25$; The figure clearly shows that the distribution of the ABC samples gets more accurate as δ decreases (from bottom to top). On the other hand, computational cost increases with accuracy.

In practical applications it is important to scale the components of the summary statistic S so that they are all approximately of the same order of magnitude. For the generated proposals, the range of values observed for each of the components of S must have width bigger than δ, otherwise a situation can arise where for a given δ only the largest component of the summary statistic has an appreciable effect on the decision whether to accept a sample and the other components of S are effectively ignored. The required scaling causes no problems, since a rescaled version (and indeed any bijective image) of a sufficient statistic is again sufficient.

5.1.2   Approximate Bayesian Computation with regression

The basic ABC method as described in the previous section can be computationally very expensive. Many variants of ABC, aiming to reduce the computational cost, are used in application areas. In this section we describe one approach to constructing such improved variants of ABC. This approach is based on the idea of accepting a larger proportion of the samples and then to numerically compensate for the systematic error introduced by the discrepancy between the sampled values $s_j = S(X_j)$ and the observed value $s^* = S(x^*)$.

The method discussed here is based on the assumption that the samples θ_j can be written as

$$\theta_j \approx f(S_j) + \varepsilon_j$$

for all accepted j, where $f(s) = \mathbb{E}(\theta \mid S = s)$ and the ε_j are independent of each other and of the S_j. If this relation holds at least approximately, we can use the modified samples

$$
\begin{aligned}
\tilde{\theta}_j &= f(s^*) + \varepsilon_j \\
&= f(S_j) + \varepsilon_j + f(s^*) - f(S_j) \\
&= \theta_j + f(s^*) - f(S_j)
\end{aligned}
\tag{5.5}
$$

instead of θ_j in order to transform samples corresponding to $S = S_j$ into samples corresponding to the required value $S = s^*$. This idea is made more rigorous by the following result.

Lemma 5.5   Let $\varepsilon \in \mathbb{R}^p$ and $X \in \mathbb{R}^n$ be independent random variables. Furthermore, let $S \colon \mathbb{R}^n \to \mathbb{R}^q$ and $f \colon \mathbb{R}^q \to \mathbb{R}^p$ be functions and define $\theta \in \mathbb{R}^p$ by

$$\theta = f\big(S(X)\big) + \varepsilon.
\tag{5.6}$$

Then the following statements hold:

(a) S is a sufficient statistic for θ.

(b) Let $s^* \in \mathbb{R}^q$. Then the conditional distribution of θ, given $S(X) = s^*$, coincides with the distribution of $\tilde{\theta} = \theta + f(s^*) - f(S(X))$.

Proof   For the proof we restrict ourselves to the case where the distributions of X and ε have densities. In this case, since θ can be written in the form (5.6), the pair (X, θ) has joint density

$$p_{X,\theta}(x, \theta) = p_X(x) \cdot p_\varepsilon\Big(\theta - f\big(S(x)\big)\Big).$$

Consequently, the conditional density of X given θ is

$$p_{X|\theta}(x|\theta) = \frac{p_{X,\theta}(x, \theta)}{p_\theta(\theta)} = p_X(x) \cdot \frac{p_\varepsilon\Big(\theta - f\big(S(x)\big)\Big)}{p_\theta(\theta)}.$$

Thus, since the first term is independent of θ and the second term depends on x only via $S(x)$, we can use the factorisation theorem for sufficient statistics (see e.g. Casella and Berger, 2001, theorem 6.2.6) to conclude that S is sufficient for θ. This proves the first statement of the lemma.

The second statement is a consequence of the independence between X and ε: the expression $f(S(X))$, conditioned on $S(X) = s^*$ is the constant value $f(s^*)$ and, since ε is independent of X, conditioning on $S(X)$ does not change the distribution of ε. Thus, the conditional distribution of $\theta = f(S(X)) + \varepsilon$ coincides with the distribution of $\tilde{\theta} = f(s^*) + \varepsilon$. Substituting $\varepsilon = \theta - f(S(X))$ into this expression completes the proof.   □

Since we apply the correction only to samples which are accepted in the basic ABC algorithm, that is where s_j is close to s^*, we only need to consider f in a neighbourhood of s. If f is smooth and δ is chosen small enough, then f will be approximately affine. For this reason, the usual approach to modelling f is to assume that near s^* the function f is an affine function of the form

$$f(s) = \alpha + s^\top \beta$$

for all $s \in \mathbb{R}^q$, where $\alpha \in \mathbb{R}^p$ and $\beta \in \mathbb{R}^{q \times p}$. In this case, estimates $\hat{\alpha}$ and $\hat{\beta}$ for the parameters α and β can be computed from the values (s_j, θ_j), using the least squares method, that is by finding $\hat{\alpha}$ and $\hat{\beta}$ which minimise the residual sum of squares

$$r(\alpha, \beta) = \sum_{j=1}^{N} \big|\theta_j - f(s_j)\big|^2 = \sum_{j=1}^{N} \big|\theta_j - \alpha - s_j^\top \beta\big|^2.$$

To compute the solution of the multidimensional least-squares problem, we rewrite the problem in matrix notation. Using the vectors $s_j \in \mathbb{R}^q$ for $j = 1, 2, \ldots, N$, we define a matrix $S \in \mathbb{R}^{N \times (q+1)}$ by

$$S_{ji} = \begin{cases} (s_j)_i & \text{if } i \in \{1, 2, \ldots, q\} \text{ and} \\ 1 & \text{if } i = q+1 \end{cases} \tag{5.7}$$

for all $j \in \{1, 2, \ldots, N\}$ and $i \in \{1, 2, \ldots, q + 1\}$. Using the coefficients $\alpha \in \mathbb{R}^p$ and $\beta \in \mathbb{R}^{q \times p}$, we define a matrix $B \in \mathbb{R}^{(q+1) \times p}$ by

$$B_{ik} = \begin{cases} \beta_{ik} & \text{if } i \in \{1, 2, \ldots, q\} \text{ and} \\ \alpha_k & \text{if } i = q + 1 \end{cases} \tag{5.8}$$

for all $i \in \{1, 2, \ldots, q + 1\}$ and $k \in \{1, 2, \ldots, p\}$. Finally, we collect the vectors θ_j for $j = 1, 2, \ldots, N$ in a matrix $\Theta \in \mathbb{R}^{N \times p}$, given by

$$\Theta_{jk} = (\theta_j)_k \tag{5.9}$$

for all $j \in \{1, 2, \ldots, N\}$ and $k \in \{1, 2, \ldots, p\}$. Using this notation, we have $f(s_j)_k = (SB)_{jk}$ where SB is the matrix product of S and B (Figure 5.2), and the residual sum of squares can be rewritten as

$$r(\alpha, \beta) = \sum_{j=1}^{n} \sum_{k=1}^{p} (\Theta - SB)_{jk}^2.$$

The matrices Θ and S only depend on the given data while the matrix B collects the entries of the unknowns α and β as described in (5.8). Using the standard theory for multidimensional least squares, the least squares estimate for B can be found as

$$\hat{B} = (S^\top S)^{-1} S^\top \Theta,$$

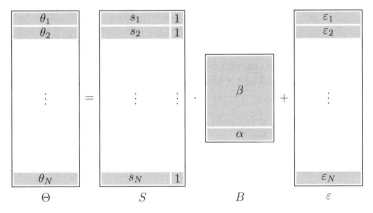

Figure 5.2   Illustration of the matrices Θ, S and B introduced in Section 5.1.2. The matrices Θ and S are given, and the aim is to estimate B (and thus α and β) using the least squares method.

where $S^\top$ is the transposed matrix and $(S^\top S)^{-1}$ is the inverse of $S^\top S$. Finally, the estimates for α and β can be extracted from $\hat{B}$ using (5.8), that is by defining $\hat{\alpha}_k = \hat{B}_{q+1,k}$ and

$$\hat{\beta}_{ik} = \hat{B}_{ik} \tag{5.10}$$

for all $j \in \{1, 2, \ldots, N\}$ and $k \in \{1, 2, \ldots, p\}$. Substituting the result of the least squares estimation into (5.5), the modified samples are then

$$\tilde{\theta}_j = \theta_j + (s^* - s_j)^\top \hat{\beta}.$$

Since $f(s) \approx \hat{\alpha} + s^\top \hat{\beta}$, we can use lemma 5.5 to conclude that the distribution of the resulting samples $\tilde{\theta}_j$ approximately coincides with the conditional distribution of θ given $S = s^*$. The resulting algorithm follows.

Algorithm 5.6 (Approximate Bayesian Computation with regression)
    input:
        data $x \in \mathbb{R}^n$
        the prior density π for the unknown parameter $\theta \in \mathbb{R}^p$
        a summary statistic $S \colon \mathbb{R}^n \to \mathbb{R}^q$
        an approximation parameter $\delta > 0$
        the output sample size N
    randomness used:
        samples $\theta_j \sim p_\theta$ and $X_j \sim p_{X|\theta}(\cdot | \theta_j)$ for $j \in \mathbb{N}$
    output:
        $\tilde{\theta}_1, \ldots, \tilde{\theta}_N$, approximately distributed with density $p_{\theta|X}(\theta | x)$
    1: $s^* \leftarrow S(x^*)$
    2: Use basic ABC to get samples $\theta_1, \ldots, \theta_N$ and $s_1, \ldots, s_N$.
    3: Compute the matrix $S \in \mathbb{R}^{N \times (q+1)}$ given by (5.7).
    4: Compute $C = S^\top \Theta \in \mathbb{R}^{(q+1) \times p}$, where Θ is given by (5.9).
    5: Compute the matrix $S^\top S \in \mathbb{R}^{(q+1) \times (q+1)}$.
    6: Solve the system of linear equations $(S^\top S)\hat{B} = C$, to find the matrix $\hat{B} \in \mathbb{R}^{(q+1) \times p}$.
    7: Construct $\hat{\beta}$ from B, using (5.10).
    8: **for** $j = 1, 2, \ldots, N$ **do**
    9:     $\tilde{\theta}_j \leftarrow \theta_j + (s^* - s_j)^\top \hat{\beta}$
    10: **end for**

The correction of samples implemented by this algorithm allows to use larger values of δ, resulting in a more efficient method, while still getting reasonably accurately distributed samples. This is illustrated by the difference between Figure 5.1 (for basic ABC) and Figure 5.3 (for ABC with regression): while the distribution of samples generated by basic ABC for $\delta = 0.25$ in Figure 5.1(c) is very different from the exact distribution (shown as solid lines), the distribution of samples

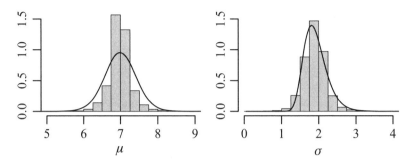

Figure 5.3   The improvements in the distribution of ABC samples which can be achieved by using algorithm 5.6 instead of algorithm 5.1 in the situation of example 5.4. The value of $\delta = 0.25$ used for this figure is the same as for Figure 5.1(c). Comparing this figure with Figure 5.1(c) shows that ABC with regression gives samples with a much improved distribution.

obtained from ABC with regression for $\delta = 0.25$ in Figure 5.3 is reasonably close to the exact distribution.

5.2   Resampling methods

Resampling methods are applied in situations where a sample of data is available for a stochastic system, but where no adequate model is available to describe the data. This can be the case, for example when the data are obtained by physical measurements of the state of a complex system. In such situations, Monte Carlo methods cannot be applied directly, because generation of samples requires a statistical model.

The methods described in this section are based on the idea of replacing the samples in Monte Carlo methods with values picked at random from the available sample of data. This procedure of reusing the data for sampling is called *resampling* and the resulting methods are called *resampling methods*. While such methods will typically be less accurate than Monte Carlo methods, their advantages are the simplicity of the resulting algorithms and the wide applicability of these methods.

5.2.1   Bootstrap estimates

The basis of all resampling methods is to replace the distribution given by a model with the 'empirical distribution' of the given data, as described in the following definition.

Definition 5.7   Given a sequence $x = (x_1, x_2, \ldots, x_M)$, the distribution of $X^* = x_K$, where the index K is random and uniformly distributed on the set $\{1, 2, \ldots, M\}$, is called the *empirical distribution* of the x_i. In this chapter we denote the empirical distribution of x by P_x^*.

In the definition, the vector x is assumed to be fixed. The randomness in X^* stems from the choice of a random element, with index $K \sim \mathcal{U}\{1, , 2, \ldots, M\}$, of this fixed sequence. Computational methods which are based on the idea of approximating an unknown 'true' distribution by an empirical distribution are called bootstrap methods.

Assume that X^* is distributed according to the empirical distribution P_x^*. Then we have

$$P(X^* = a) = \frac{1}{M} \sum_{i=1}^{M} \mathbb{1}_{\{a\}}(x_i),$$

that is under the empirical distribution, the probability that X^* equals a is given by the relative frequency of occurrences of a in the given data. Similarly, we have the relations

$$P\big(X^* \in A\big) = \frac{1}{M} \sum_{i=1}^{M} \mathbb{1}_A(x_i)$$

and

$$\mathbb{E}\big(f(X^*)\big) = \sum_{a \in \{x_1, \ldots, x_M\}} f(a) P\big(X^* = a\big)$$

$$= \sum_{a \in \{x_1, \ldots, x_M\}} f(a) \frac{1}{M} \sum_{i=1}^{M} \mathbb{1}_{\{a\}}(x_i) = \frac{1}{M} \sum_{i=1}^{M} f(x_i). \qquad (5.11)$$

Some care is needed when verifying this relation: the sums where the index a runs over the set $\{x_1, \ldots, x_M\}$ have only one term for each element of the set, even if the corresponding value occurs repeatedly in the given data.

Example 5.8 Let the data $x = (1, 2, 1, 4)$ be given and let X^* be distributed according to the empirical distribution of x. Then we have the probabilities $P(X^* = 1) = 2/4 = 1/2$ and $P(X^* = 2) = P(X^* = 4) = 1/4$. The expectation of X^* is

$$E(X^*) = 1 \cdot \frac{2}{4} + 2 \cdot \frac{1}{4} + 4 \cdot \frac{1}{4} = \frac{2 + 2 + 4}{4} = 2.$$

In the context of resampling methods, the data $X = (X_1, X_2, \ldots, X_M)$ is assumed to be an i.i.d. sample from some distribution P unknown to us. When considering $X^* \sim P_X^*$ in this situation, there are two different levels of randomness involved: first, the data X and thus the empirical distribution P_X^* is random. And secondly, for every instance of the data X, the sample X^* contains the additional randomness from the random choice of the index K in definition 5.7. In this section we write $P_X^*(\cdot)$ and $\mathbb{E}_X^*(\cdot)$ for probabilities and expectations which take X as being fixed and which only consider the randomness from generating $X^* \sim P_X^*$. Technically, P_X^* is the conditional probability and $\mathbb{E}_X^*$ is the conditional expectation, both conditioned on the value of X. The following lemma shows how these two different levels of randomness combine to create the total variance.

Lemma 5.9   Let $X = (X_1, \ldots, X_M)$ where the $X_i \sim P$ are i.i.d. Furthermore, let $X_1^*, \ldots, X_n^* \sim P_X^*$ be i.i.d. and $Y^* = f(X_1^*, \ldots, X_n^*)$ for some function f. Then

$$\mathbb{E}(Y^*) = \mathbb{E}\left(\mathbb{E}_X^*(Y^*)\right) \qquad (5.12)$$

and

$$\mathrm{Var}(Y^*) = \mathbb{E}\left(\mathrm{Var}_X^*(Y^*)\right) + \mathrm{Var}\left(\mathbb{E}_X^*(Y^*)\right). \qquad (5.13)$$

Proof   Since we have $\mathbb{E}_X^*(Y^*) = \mathbb{E}(Y^* | X)$, we can use the tower property $\mathbb{E}(\mathbb{E}(Y | X)) = \mathbb{E}(Y)$ of the conditional expectation to get (5.12). For the variance (5.13) we find

$$\begin{aligned}
\mathrm{Var}(Y^*) &= \mathbb{E}\left((Y^*)^2\right) - \mathbb{E}(Y^*)^2 \\
&= \mathbb{E}\left(\mathbb{E}((Y^*)^2 | X)\right) - \mathbb{E}\left(\mathbb{E}(Y^* | X)\right)^2 \\
&= \mathbb{E}\left(\mathbb{E}((Y^*)^2 | X) - \mathbb{E}(Y^* | X)^2\right) \\
&\quad + \mathbb{E}\left(\mathbb{E}(Y^* | X)^2\right) - \mathbb{E}\left(\mathbb{E}(Y^* | X)\right)^2 \\
&= \mathbb{E}\left(\mathrm{Var}(Y^* | X)\right) + \mathrm{Var}\left(\mathbb{E}(Y^* | X)\right) \\
&= \mathbb{E}\left(\mathrm{Var}_X^*(Y^*)\right) + \mathrm{Var}\left(\mathbb{E}_X^*(Y^*)\right).
\end{aligned}$$

This completes the proof.   □

If we consider the data $x = (x_1, \ldots, x_M)$ to be an instance of an i.i.d. sample from the distribution P, and if we let $X^* \sim P_x^*$ as well as $X \sim P$, then the law of large numbers (theorem A.8) guarantees

$$P_x^*(X^* \in A) = \frac{1}{M} \sum_{i=1}^{M} \mathbb{1}_A(x_i) \approx P(X \in A) \qquad (5.14)$$

for every set A and sufficiently large M. Similarly, we find

$$\mathbb{E}_x^*\left(f(X^*)\right) = \frac{1}{M} \sum_{i=1}^{M} f(X_i) \approx \mathbb{E}\left(f(X)\right)$$

for all functions f. These two equations show that the distribution of the random variable X^* (i.e. the empirical distribution P_x^*) can be used as an approximation to the distribution of X (i.e. the unknown distribution P).

In typical applications of the bootstrap method, the function f depends on n independent values, that is we want to compute an expectation of the form $\mathbb{E}(f(X_1, \ldots, X_n))$ where $X_1, \ldots, X_n$ are i.i.d., distributed with an unknown distribution P. In these cases, we can use the approximation

$$\mathbb{E}^*_x\big(f(X_1^*, \ldots, X_n^*)\big) \approx \mathbb{E}\big(f(X_1, \ldots, X_n)\big), \tag{5.15}$$

where $X_1^*, \ldots, X_n^*$ are independent samples from the empirical distribution P_x^*. Finally, if we are interested in any property $\theta = \theta(P)$ of the unknown distribution P we can approximate the result by considering the same property of the empirical distribution P_x^*:

$$\theta(P) \approx \theta(P_x^*). \tag{5.16}$$

For example, if $\sigma^2(P)$ is the variance of the distribution P (or equivalently, of $X \sim P$), then we can use the approximation

$$\sigma^2(P) \approx \sigma^2(P_x^*) = \mathrm{Var}_x^*(X^*)$$

where $X^* \sim P_x^*$. Since the distribution P_x^* of X^* is known, the value $\mathrm{Var}_x^*(X^*)$ can be computed from the given data x.

Equation (5.14), equation (5.15) and equation (5.16) illustrate a general method for constructing estimators from given data x. The method is based on two different (but related) substitutions:

(a) Every direct occurrence of the unknown distribution P is replaced with the empirical distribution P_x^*.

(b) All instances of random variables $X \sim P$ are replaced with bootstrap samples $X^* \sim P_x^*$.

The examples above shows only one of the two substitutions at a time, but in the next section we will see an application where both substitutions occur simultaneously. Estimates based on this methodology are called *bootstrap estimates* and samples $(X_1^*, \ldots, X_n^*)$ from the empirical distribution are called *bootstrap samples*.

Example 5.10   Let $X_1, X_2, \ldots, X_n$ be i.i.d. with $P(X_i = 1) = p$ and $P(X_i = 0) = 1 - p$. Assume that the value of p is unknown to us, but that we have a sample $x = (x_1, x_2, \ldots, x_n)$ from the distribution of the X_i. Since we do not know the value of p, we do not know the distribution P of the random variables X_i. As we have seen above, the empirical distribution P_x^* can be used to approximate the unknown distribution P: let $X^* \sim P_x^*$ and $k = \sum_{i=1}^n x_i$. Since X^* is chosen uniformly from the entries of x, we have $P(X^* = 1) = k/n$ and $P(X^* = 0) = 1 - k/n$. A bootstrap sample $X_1^*, \ldots, X_n^*$ can be constructed by generating n i.i.d. samples from the distribution P_x^*. Expressions involving X_i can be approximated using the corresponding expression for X_i^*. For example, using (5.14), we find the bootstrap estimate for $P(\sum_{i=1}^{n} X_i \leq a)$ as

$$P\left(\sum_{i=1}^{n} X_i \leq a\right) \approx P_x^*\left(\sum_{i=1}^{n} X_i^* \leq a\right)$$

$$= \sum_{j=1}^{a} P_x^*\left(\sum_{i=1}^{n} X_i^* = j\right)$$

$$= \sum_{j=1}^{a} \binom{n}{j} \left(\frac{k}{n}\right)^j \left(1 - \frac{k}{n}\right)^{n-j}$$

for all $a \in \mathbb{N}$. This result is not surprising: the bootstrap estimate equals the exact probability where p is replaced by the estimate k/n.

The difficulty in bootstrap methods lies in evaluating the expressions involving the empirical distribution P_x^* in (5.14), (5.15) and (5.16). In simple cases, such as in example 5.10, this can be done analytically but in most situations an analytical solution will not be available. Since generating samples from the empirical distribution is easy, Monte Carlo methods can be used instead. For example, in the case of approximation (5.15) we can generate independent samples $X_i^{*(j)} \sim P_x^*$ for $i = 1, 2, \ldots, n$ and $j = 1, 2, \ldots, N$, and then compute the approximation

$$\mathbb{E}_x^*\left(f(X_1^*, \ldots, X_n^*)\right) \approx \frac{1}{N} \sum_{j=1}^{N} f\left(X_1^{*(j)}, \ldots, X_n^{*(j)}\right). \tag{5.17}$$

This method leads to the following algorithm.

Algorithm 5.11 (general bootstrap estimate)
    input:
        data $x_1, x_2, \ldots, x_M \in A$ with values in some set A
        $f : A^n \to \mathbb{R}$
        $N \in \mathbb{N}$
    randomness used:
        a sequence $(K_i^{(j)})_{j \in \mathbb{N}, i=1,\ldots,n}$ with $K_i^{(j)} \sim \mathcal{U}\{1, 2, \ldots, M\}$ i.i.d.
    output:
        an estimate for $\mathbb{E}(f(X_1, \ldots, X_n))$ where $X_1, \ldots, X_n$ are i.i.d. from the distribution of the data x_i
        distribution of the data x_i
    1: $s \leftarrow 0$
    2: **for** $j = 1, 2, \ldots, N$ **do**
    3:    generate $K_1^{(j)}, \ldots, K_n^{(j)} \sim \mathcal{U}\{1, 2, \ldots, M\}$ i.i.d.

4:   let $X_i^{*(j)} \leftarrow x_{K_i^{(j)}}$ for $i = 1, 2, \ldots, n$

5:   $s \leftarrow s + f(X_1^{*(j)}, \ldots, X_n^{*(j)})$

6: **end for**

7: return s/N

As explained above, this algorithm is based on the following sequence of approximations:

$$\frac{1}{N} \sum_{j=1}^{N} f\left(X_1^{*(j)}, \ldots, X_n^{*(j)}\right) \approx \mathbb{E}_x^*\left(f(X_1^*, \ldots, X_n^*)\right)$$

$$\approx \mathbb{E}\left(f(X_1, \ldots, X_n)\right).$$

The error in the first of these two approximations goes to 0 as N increases. Since we control generation of the samples $X^{*(j)}$, we can make this error arbitrarily small at the expense of additional computation time. The error in the second approximation decreases as $M \to \infty$. Normally, the given data set $x_1, x_2, \ldots, x_M$ is fixed and cannot be easily extended; in these cases there is no way to reduce the error in the second approximation.

Proving the convergence

$$\mathbb{E}_x^*\left(f(X_1^*, \ldots, X_n^*)\right) \to \mathbb{E}\left(f(X_1, \ldots, X_n)\right)$$

for $M \to \infty$ is challenging in the general setting and results depend on regularity properties of the function f. The ususal approach is to prove convergence only for the specific forms of f arising in applications, instead of relying on general theorems.

In most applications, the number M of available samples equals the number n of arguments of the function f, that is enough data are given to compute one value $f(X_1, \ldots, X_n)$. The bootstrap estimate computed by algorithm 5.11 allows to reuse the available information to also get an estimate for the expectation of this value.

Some care is needed when using bootstrap estimates in practice: while the arguments $X_1^*, \ldots, X_n^*$ of f in Equation (5.15) have (approximately) the correct distribution, with large probability the bootstrap sample contains duplicate values and thus the arguments are not independent. On the other hand, the arguments $X_1, \ldots, X_n$ on the right-hand side of (5.15) are independent. For this reason, the bootstrap estimate (5.17) is usually biased (this is in contrast to the Monte Carlo estimate, which by proposition 3.14 is always unbiased) and sometimes it is not even clear that the total error decreases to 0 as the value of n increases.

5.2.2   Applications to statistical inference

The main application of the bootstrap method in statistical inference is to quantify the accuracy of parameter estimates.

In this section, we will consider parameters as a function of the corresponding distribution: if θ is a parameter, for example the mean or the variance, then we write $\theta(P)$ for the corresponding parameter. In statistics, there are many ways of constructing estimators for a parameter θ. One general method for constructing parameter estimators, the plug-in principle, is given in the following definition.

Definition 5.12   Consider an estimator $\hat{\theta}_n = \hat{\theta}_n(X_1, \ldots, X_n)$ for a parameter $\theta(P)$. The estimator $\hat{\theta}_n$ satisfies the *plug-in principle*, if it satisfies the relation

$$\hat{\theta}_n(x_1, \ldots, x_n) = \theta(P_x^*), \tag{5.18}$$

for all $x = (x_1, \ldots, x_n)$, where P_x^* is the empirical distribution of x. In this case, $\hat{\theta}_n$ is called the *plug-in estimator* for θ.

Since the idea of bootstrap methods is to approximate the distribution P by the empirical distribution P_x^*, plug-in estimators are particularly useful in conjunction with bootstrap methods.

Example 5.13   The plug-in estimator for the mean μ is found by taking the mean of the empirical distribution. Taking $X^* \sim P_x^*$ we get

$$\hat{\mu}(x_1, \ldots, x_n) = \mu(P_x^*) = \mathbb{E}(X^*)$$

and using equation (5.11), with $M = n$ and $f(x) = x$, we find

$$\hat{\mu}(x_1, \ldots, x_n) = \frac{1}{n} \sum_{i=1}^n x_i.$$

Thus, the plug-in estimator for the mean is just the sample average.

Example 5.14   The plug-in estimator for the variance σ^2 is the variance of the empirical distribution. Taking $X^* \sim P_x^*$ again, we get

$$\hat{\sigma}^2(x_1, \ldots, x_n) = \sigma^2(P_x^*) = \text{Var}(X^*) = \mathbb{E}((X^*)^2) - (\mathbb{E}(X^*))^2.$$

From equation (5.11) we find

$$\hat{\sigma}^2(x_1, \ldots, x_n) = \frac{1}{n} \sum_{i=1}^n x_i^2 - \left(\frac{1}{n} \sum_{i=1}^n x_i \right)^2,$$

and using the abbreviation $\bar{x} = \frac{1}{n} \sum_{i=1}^{n} x_i$ we can rewrite this as

$$\hat{\sigma}^2(x_1, \ldots, x_n) = \frac{1}{n} \sum_{i=1}^{n} x_i^2 - \bar{x}^2$$

$$= \frac{1}{n} \left(\sum_{i=1}^{n} x_i^2 - 2 \sum_{i=1}^{n} x_i \bar{x} + \sum_{i=1}^{n} \bar{x}^2 \right)$$

$$= \frac{1}{n} \sum_{i=1}^{n} (x_i - \bar{x})^2.$$

This is the plug-in estimator for the variance.

5.2.2.1   Bootstrap estimates of the bias

The bias, as given in definition 3.9, is a measure for the systematic error of an estimator. In the notation of this section, the bias of an estimator $\hat{\theta}_n$ for a parameter θ is given by

$$\text{bias}(\hat{\theta}_n) = \mathbb{E}_P(\hat{\theta}_n) - \theta(P),$$

where $\mathbb{E}_P(\cdot)$ denotes the expectation with respect to the distribution P. If the distribution P is not known, we cannot compute the bias analytically but for a given i.i.d. sample $X = (X_1, \ldots, X_n)$ from the distribution P we can find the bootstrap estimate for the bias by applying the bootstrap principle. This results in the estimator

$$\widehat{\text{bias}}^*(\hat{\theta}_n) = \mathbb{E}_X^*(\hat{\theta}_n^*) - \theta(P_X^*) \tag{5.19}$$

for the bias, where

$$\hat{\theta}_n^* = \hat{\theta}_n(X_1^*, \ldots, X_n^*) \tag{5.20}$$

and $X_1^*, \ldots, X_n^* \sim P_X^*$ are i.i.d. As before, we write $\mathbb{E}_X^*(\cdot)$ to indicate that the expectation takes $X_1, \ldots, X_n$ as fixed and only considers the additional randomness in $X_1^*, \ldots, X_n^*$.

Since, for given $X_1, \ldots, X_n$, the distribution P_X^* is known, we can evaluate the right-hand side of (5.19): If we assume that $\hat{\theta}_n$ satisfies the plug-in principle, then, by equation (5.18), we have $\theta(P_X^*) = \hat{\theta}_n(X_1, \ldots, X_n)$. For the evaluation of $\mathbb{E}^*(\hat{\theta}_n^*)$ we can use a Monte Carlo estimate: we generate samples $X_i^{*(j)} \sim P_X^*$ i.i.d. for $i = 1, \ldots, n$ and $j = 1, 2, \ldots, N$ and let

$$\hat{\theta}_n^{*(j)} = \hat{\theta}_n(X_1^{*(j)}, \ldots, X_n^{*(j)}) \tag{5.21}$$

as well as

$$\overline{\hat{\theta}_n^*} = \frac{1}{N} \sum_{j=1}^{N} \hat{\theta}_n^{*(j)}. \qquad (5.22)$$

Finally, for sufficiently large N, we can use the approximation $\overline{\hat{\theta}_n^*} \approx \mathbb{E}^*(\hat{\theta}_n^*)$. Substituting these expressions into equation (5.19), we get

$$\widehat{\text{bias}}^*(\hat{\theta}_n) \approx \overline{\hat{\theta}_n^*} - \hat{\theta}_n(X_1, \ldots, X_n).$$

This is the usual form for the bootstrap estimate of the bias of the estimator $\hat{\theta}_n$.

Some care is needed here, since there are two different levels of estimation involved: $\hat{\theta}_n$ is an estimator for $\theta(P)$ while $\widehat{\text{bias}}^*$ is an estimate for the *bias* of the estimate $\hat{\theta}_n$.

Algorithm 5.15   (bootstrap estimate of the bias)
    input:
        data $x_1, x_2, \ldots, x_n$
        the plug-in estimator $\hat{\theta}_n$ for a parameter θ
        $N \in \mathbb{N}$
    randomness used:
        $K_i^{(j)} \sim \mathcal{U}\{1, 2, \ldots, n\}$ i.i.d. for $i = 1, \ldots, n$ and $j = 1, \ldots, N$
    output:
        an estimate for the bias of $\hat{\theta}_n$
  1: **for** j=1, 2, $\ldots$, N **do**
  2:    generate $K_1^{(j)}, \ldots, K_n^{(j)} \sim \mathcal{U}\{1, 2, \ldots, n\}$ i.i.d.
  3:    let $X_i^{*(j)} \leftarrow x_{K_i^{(j)}}$ for $i = 1, 2, \ldots, n$
  4:    let $\hat{\theta}_n^{*(j)} \leftarrow \hat{\theta}_n(X_1^{*(j)}, \ldots, X_n^{*(j)})$
  5: **end for**
  6: let $\overline{\hat{\theta}_n^*} = \dfrac{1}{N} \sum_{j=1}^{N} \hat{\theta}_n^{*(j)}$
  7: return $\overline{\hat{\theta}_n^*} - \hat{\theta}_n(X_1, \ldots, X_n)$

Both, the accuracy and the computational cost of this estimate increase when the size n of the given data set and the Monte Carlo sample size N increase.

If $\hat{\theta}_n$ satisfies the plug-in principle, an elegant alternative formula for $\widehat{\text{bias}}^*$ can be derived. From (5.18) we know

$$\hat{\theta}_n(X_1^*, \ldots, X_n^*) = \theta(P_{X^*}^*)$$

and substituting this relation into Equation (5.19) we get

$$\widehat{\text{bias}}^*(\hat{\theta}_n) = \mathbb{E}_X^*\left(\theta\left(P_{X^*}^*\right)\right) - \theta\left(P_X^*\right). \tag{5.23}$$

On the other hand, substituting the definition (5.18) of a plug-in estimate into the definition of the bias gives

$$\text{bias}(\hat{\theta}_n) = \mathbb{E}_P\left(\theta(P_X^*)\right) - \theta(P). \tag{5.24}$$

Comparing the expressions in (5.23) and (5.24) we find that the bootstrap estimate of the bias makes use of the fact that, in some sense, $P_{X^*}^*$ relates to P_X^*, as P_X^* does to P.

5.2.2.2   Bootstrap estimates of the standard error

The standard error of an estimator $\hat{\theta}_n$ for a parameter $\theta(P)$, introduced in definition 3.10, is the standard deviation of $\hat{\theta}_n(X_1, \ldots, X_n)$ when the random sample $X_1, \ldots, X_n$ is distributed according to the distribution P:

$$\text{se}(\hat{\theta}_n) = \text{stdev}\left(\hat{\theta}_n(X_1, \ldots, X_n)\right). \tag{5.25}$$

Together with the bias, the standard error is a measure for the accuracy of an estimator. In this section we will see how the bootstrap method can be used to obtain an estimate of the standard error of a given estimator.

Example 5.16   Consider the estimator $\hat{\mu}(x) = \frac{1}{n}\sum_{i=1}^{n} x_i$ for the mean μ of a distribution. The standard error of this estimator is given by

$$\text{se}(\hat{\mu}_n) = \sqrt{\text{Var}\left(\frac{1}{n}\sum_{i=1}^{n} X_i\right)} = \frac{\sigma(P)}{\sqrt{n}},$$

where $X_1, \ldots, X_n \sim P$ are i.i.d. and $\sigma(P)$ denotes the standard deviation of the distribution P.

The value $\text{se}(\hat{\theta}_n)$ does not directly depend on P, but only on the values X_i. Consequently, for finding the bootstrap estimate of the standard error, it is sufficient to replace the random variables $X_1, \ldots, X_n \sim P$ in (5.25) with i.i.d. samples $X_1^*, \ldots, X_n^* \sim P_X^*$ from the empirical distribution. The resulting estimate is

$$\widehat{\text{se}}^*(\hat{\theta}_n) = \text{stdev}^*\left(\hat{\theta}_n^*\right), \tag{5.26}$$

where

$$\hat{\theta}_n^* = \hat{\theta}_n\left(X_1^*, \ldots, X_n^*\right)$$

and the symbol stdev*($\cdot$) indicates the standard deviation for fixed values of $X_1, \ldots, X_n$, taking only the randomness in the X_i^* into account. In contrast to the bootstrap estimate of the bias from the previous section, for the bootstrap estimate of the standard error we do not need to assume that the estimator $\hat{\theta}_n$ satisfies the plug-in principle.

Example 5.17   Consider the estimate $\hat{\mu}_n(x) = \sum_{i=1}^{n} x_i / n$ for the mean. From example 5.16 we know that the standard error of the mean is $\mathrm{se}(\hat{\mu}_n) = \sigma(P)/\sqrt{n}$. Instead of using the general formula (5.26), we can apply the bootstrap principle to $\mathrm{se}(\hat{\mu}_n)$ and find the bootstrap estimate for the standard error of $\hat{\mu}_n$ as

$$\widehat{\mathrm{se}}^*(\hat{\mu}_n) = \frac{\sigma(P_X^*)}{\sqrt{n}}.$$

Using the result from example 5.14 we get

$$\widehat{\mathrm{se}}^*(\hat{\mu}_n) = \frac{\sqrt{\frac{1}{n} \sum_{i=1}^{n} (X_i - \bar{X})^2}}{\sqrt{n}} = \sqrt{\frac{1}{n^2} \sum_{i=1}^{n} (X_i - \bar{X})^2}.$$

This is the bootstrap estimate for the standard error of the mean.

Only in simple situations such as the one in example 5.17 is it possible to compute the bootstrap estimate of the standard error analytically. Typically, the standard deviation on the right-hand side of equation (5.26) cannot be evaluated exactly and is approximated using a Monte Carlo estimate instead. This leads to the following expression for the bootstrap estimate of the standard error:

$$\widehat{\mathrm{se}}^*(\hat{\theta}_n) \approx \sqrt{\frac{1}{N-1} \sum_{j=1}^{N} \left(\hat{\theta}_n^{*(j)} - \overline{\hat{\theta}_n^*} \right)^2}$$

where $\overline{\hat{\theta}_n^*}$ and the $\hat{\theta}_n^{*(j)}$ are given by equation (5.21) and equation (5.22), respectively. This leads to the following algorithm.

Algorithm 5.18   (bootstrap estimate of the standard error)
    input:
        data $x_1, x_2, \ldots, x_n$
        an estimator $\hat{\theta}_n$ for a parameter θ
        $N \in \mathbb{N}$
    randomness used:
        $K_i^{(j)} \sim \mathcal{U}\{1, 2, \ldots, n\}$ i.i.d. for $i = 1, \ldots, n$ and $j = 1, \ldots, N$
    output:
        an estimate for the standard error of $\hat{\theta}_n$

1: **for** j=1, 2, ..., N **do**
2:     generate $K_1^{(j)}, \ldots, K_n^{(j)} \sim \mathcal{U}\{1, 2, \ldots, n\}$ i.i.d.
3:     let $X_i^{*(j)} \leftarrow x_{K_i^{(j)}}$ for $i = 1, 2, \ldots, n$
4:     let $\hat{\theta}_n^{*(j)} \leftarrow \hat{\theta}_n(X_1^{*(j)}, \ldots, X_n^{*(j)})$
5: **end for**
6: let $\overline{\hat{\theta}_n^*} = \dfrac{1}{N} \sum_{j=1}^{N} \hat{\theta}_n^{*(j)}$

7: return $\sqrt{\dfrac{1}{N-1} \sum_{j=1}^{N} \left(\hat{\theta}_n^{*(j)} - \overline{\hat{\theta}_n^*} \right)^2}$

Both the accuracy and the computational cost of this estimate increase when the size n of the given data set and the Monte Carlo sample size N increase.

5.2.2.3 Bootstrap confidence intervals

Confidence intervals, as introduced in Section 3.4.2, are often difficult to construct exactly. There are many different approaches to construct bootstrap approximations for confidence intervals. Here we restrict ourselves to first derive one simple bootstrap confidence interval and to then give a very short description of a more complicated bootstrap confidence interval.

A confidence interval is an interval $[U, V]$, computed from the given data $(X_1, \ldots, X_n) \sim P_\theta$, such that

$$P_\theta \big(\theta \in [U, V] \big) \geq 1 - \alpha. \tag{5.27}$$

While we only require this relation to hold for the true value of θ, this value is not known and thus $U = U(X_1, \ldots, X_n)$ and $V = V(X_1, \ldots, X_n)$ are constructed such that the relation (5.27) holds for all possible values of θ simultaneously. To construct exact confidence intervals, detailed knowledge of the family of distributions P_θ is required. In contrast, the bootstrap methods presented in this section allow to construct approximate confidence intervals, based on just the given data $X_1, \ldots, X_n$. These methods use exactly the same information as is required for computing the point estimate $\hat{\theta}_n$. Thus, using bootstrap confidence intervals allows information about accuracy to be attached to any parameter estimate.

Let $\hat{\theta}_n$ be the plug-in estimate for the parameter $\theta = \theta(P_\theta)$. To construct a confidence interval, we can try to find values a and b such that

$$P_\theta \big(\hat{\theta}_n - a \leq \theta \leq \hat{\theta}_n + b \big) = 1 - \alpha. \tag{5.28}$$

Direct application of the bootstrap principle, that is replacing P_θ with P_X^* and $X_1, \ldots, X_n \sim P_\theta$ with $X_1^*, \ldots, X_n^* \sim P_X^*$, yields the condition

$$P_X^*\left(\hat{\theta}_n^* - a \le \theta(P_X^*) \le \hat{\theta}_n^* + b\right) = 1 - \alpha, \qquad (5.29)$$

where $\hat{\theta}_n^*$ is defined in equation (5.20) and we write P_X^* to indicate that, when computing the probability, the data $X_1, \ldots, X_n$ are assumed to be fixed. We use the relation (5.29) to construct values a and b for an approximate confidence interval for θ.

Lemma 5.19 Let $\hat{\theta}_n$ be the plug-in estimator for θ. Let $\theta_{\alpha/2}^*$ and $\theta_{1-\alpha/2}^*$ be the $\alpha/2$ and $1 - \alpha/2$ quantiles of the distribution of $\hat{\theta}_n^*$ given by (5.20). Define

$$a = \theta_{1-\alpha/2}^* - \hat{\theta}_n\left(X_1, \ldots, X_n\right)$$

and

$$b = \hat{\theta}_n\left(X_1, \ldots, X_n\right) - \theta_{\alpha/2}^*.$$

Then (5.29) is satisfied.

Proof Since $\hat{\theta}_n$ is the plug-in estimator for θ, we have $\hat{\theta}_n(X_1, \ldots, X_n) = \theta(P_X^*)$. Substituting this relation into the definitions of a and b, we get

$$P^*\left(\hat{\theta}_n^* - a \le \theta(P_X^*) \le \hat{\theta}_n^* + b\right)$$
$$= P^*\left(-\theta_{1-\alpha/2}^* + \theta(P_X^*) \le \theta(P_X^*) - \hat{\theta}_n^* \le \theta(P_X^*) - \theta_{\alpha/2}^*\right)$$
$$= P^*\left(\theta_{1-\alpha/2}^* \ge \hat{\theta}_n^* \ge \theta_{\alpha/2}^*\right)$$
$$= 1 - \alpha.$$

This completes the proof. □

An approximate confidence interval $[U^*, V^*]$ can now be found by substituting the values a and b from the lemma into (5.28). The resulting boundaries are

$$U^* = \hat{\theta}_n\left(X_1, \ldots, X_n\right) - a = 2\hat{\theta}_n\left(X_1, \ldots, X_n\right) - \theta_{1-\alpha/2}^*$$

and

$$V^* = \hat{\theta}_n\left(X_1, \ldots, X_n\right) + b = 2\hat{\theta}_n\left(X_1, \ldots, X_n\right) - \theta_{\alpha/2}^*.$$

Normally, the quantiles $\theta_{\alpha/2}^*$ and $\theta_{1-\alpha/2}^*$ cannot be computed analytically and are estimated using Monte Carlo instead. In order to obtain such estimates, we generate

N samples, sort them in increasing order, and then pick out the entries with indices $N\alpha/2$ and $N(1 - \alpha/2)$, with appropriate rounding, from the sorted list. This technique is used in the following algorithm.

Algorithm 5.20   (simple bootstrap confidence interval)
    input:
        data $x_1, x_2, \ldots, x_n$
        the plug-in estimator $\hat{\theta}_n$ for a one-dimensional parameter θ
        $N \in \mathbb{N}$
        $\alpha \in (0, 1)$
    randomness used:
        $K_i^{(j)} \sim \mathcal{U}\{1, 2, \ldots, n\}$ i.i.d. for $i = 1, \ldots, n$ and $j = 1, \ldots, N$
    output:
        an approximate confidence interval $[U^*, V^*]$ for θ
  1: **for** j=1, 2, …, N **do**
  2:    generate $K_1^{(j)}, \ldots, K_n^{(j)} \sim \mathcal{U}\{1, 2, \ldots, n\}$ i.i.d.
  3:    let $X_i^{*(j)} \leftarrow x_{K_i^{(j)}}$ for $i = 1, 2, \ldots, n$
  4:    let $\hat{\theta}_n^{*(j)} \leftarrow \hat{\theta}_n(X_1^{*(j)}, \ldots, X_n^{*(j)})$
  5: **end for**
  6: let $t \leftarrow \hat{\theta}_n(X_1, \ldots, X_n)$
  7: let $l \leftarrow \lceil \frac{\alpha}{2} N \rceil$
  8: let $u \leftarrow \lceil (1 - \frac{\alpha}{2})N \rceil$
  9: let $\theta_{(1)}^*, \ldots, \theta_{(N)}^*$ be $\theta_1^*, \ldots, \theta_N^*$, sorted in increasing order
10: return $(2t - \theta_{(u)}^*, 2t - \theta_{(l)}^*)$

In the algorithm, the symbol $\lceil \cdot \rceil$ denotes the operation of rounding up a number, that is for $x \in \mathbb{R}$ the value $\lceil x \rceil$ is the smallest integer greater than or equal to x.

The parameter N in algorithm 5.20 controls the accuracy of the bootstrap estimate for the quantiles of the distribution of $\hat{\theta}_n^*$. For bigger values of N the result gets more accurate but, at the same time, the computation gets slower. The situation is slightly different from the case of Monte Carlo estimates. For the bootstrap estimates considered here, the total error does not converge to 0 even as N increases to infinity, since an additional error is introduced by the finite size n of the given data set.

One easy method of finding good values for N is to repeatedly compute a confidence interval for the same data set and to consider the standard deviation of the bounds between runs of the algorithm. This standard deviation is a measure for the error contributed by the finiteness of N. The value of N should be chosen large enough that this standard deviation is small, compared with the size of the confidence interval. Numerical experiments indicate the N should be chosen to be relatively large, for example in the situation of exercise E5.6 with $N = 4000$ the sampling error is still a noticeable percentage of the width of the estimated confidence interval.

Many variants of and improvements to the simple bootstrap confidence interval from algorithm 5.20 are possible. Here we restrict ourselves to describing one of the many variants, known as the BC_a method (BC_a stands for 'bias corrected and accelerated'), but without giving any derivation. Using a bootstrap sample $\hat{\theta}_n^{*(1)}, \ldots, \hat{\theta}_n^{*(N)}$, constructed as in (5.21), the BC_a confidence interval is defined using the following steps.

First, let $\Phi \colon \mathbb{R} \to (0, 1)$ be the distribution function of the standard normal distribution and define a value $\hat{z}$ by

$$\hat{z} = \Phi^{-1} \left(\frac{\#\{ j \mid \hat{\theta}_n^{*(j)} \leq \hat{\theta}_n \}}{N} \right).$$

Next, let $x_{(i)} = (x_1, \ldots, x_{i-1}, x_{i+1}, \ldots, x_n)$, that is the data x with x_i left out, and define $\theta_{(i)} = \hat{\theta}_{n-1}(x_{(i)})$ for $i = 1, 2, \ldots, n$, as well as

$$\overline{\theta_{(\cdot)}} = \frac{1}{n} \sum_{i=1}^{n} \theta_{(i)}.$$

From this, a value $\hat{a}$ is computed as

$$\hat{a} = \frac{1}{6} \cdot \frac{\sum_{i=1}^{n} \left(\overline{\theta_{(\cdot)}} - \hat{\theta}_{(i)} \right)^3}{\left(\sum_{i=1}^{n} \left(\overline{\theta_{(\cdot)}} - \hat{\theta}_{(i)} \right)^2 \right)^{3/2}}.$$

Finally, using $\hat{z}$ and $\hat{a}$, we define a map $q \colon (0, 1) \to (0, 1)$ by

$$q(\alpha) = \Phi \left(\hat{z} + \frac{\hat{z} + \Phi^{-1}(\alpha)}{1 - \hat{a}\left(\hat{z} + \Phi^{-1}(\alpha) \right)} \right)$$

for all $\alpha \in (0, 1)$. Then the BC_a bootstrap confidence interval $[U^*, V^*]$ is given by the boundaries

$$U^* = \theta_{q(\alpha/2)}^*$$

and

$$V^* = \theta_{q(1-\alpha/2)}^*,$$

where θ_α^* denotes the α-quantile of the distribution of the $\hat{\theta}_n^*$. As before, these quantiles can be approximated by the corresponding quantiles of the empirical distribution of $\hat{\theta}_n^{*(1)}, \ldots, \hat{\theta}_n^{*(N)}$. This leads to the following algorithm.

Algorithm 5.21   (BC$_a$ bootstrap confidence interval)

    input:

        data $x_1, x_2, \ldots, x_n$

        an estimator $\hat{\theta}_n$ for a one-dimensional parameter θ

        $N \in \mathbb{N}$

        $\alpha \in (0, 1)$

    randomness used:

        $K_i^{(j)} \sim \mathcal{U}\{1, 2, \ldots, n\}$ i.i.d. for $i = 1, \ldots, n$ and $j = 1, \ldots, N$

    output:

        an approximate confidence interval $[U^*, V^*]$ for θ

1: **for** j=1, 2, $\ldots$, N **do**

2:    generate $K_1^{(j)}, \ldots, K_n^{(j)} \sim \mathcal{U}\{1, 2, \ldots, n\}$ i.i.d.

3:    let $X_i^{*(j)} \leftarrow x_{K_i^{(j)}}$ for $i = 1, 2, \ldots, n$

4:    let $\hat{\theta}_n^{*(j)} \leftarrow \hat{\theta}_n(X_1^{*(j)}, \ldots, X_n^{*(j)})$

5: **end for**

6: let $\hat{z} = \Phi^{-1}\left(\dfrac{\#\{j \mid \hat{\theta}^{*(j)} \leq \hat{\theta}_n\}}{N}\right)$

7: **for** i=1, 2, $\ldots$, n **do**

8:    let $\theta_{(i)} \leftarrow \hat{\theta}_{n-1}(x_1, \ldots, x_{i-1}, x_{i+1}, \ldots, x_n)$

9: **end for**

10: let $\overline{\theta_{(\cdot)}} \leftarrow \frac{1}{n}\sum_{i=1}^{n} \theta_{(i)}$

11: let $\hat{a} \leftarrow \dfrac{1}{6} \cdot \dfrac{\sum_{i=1}^{n}(\overline{\theta_{(\cdot)}} - \hat{\theta}_{(i)})^3}{\left(\sum_{i=1}^{n}(\overline{\theta_{(\cdot)}} - \hat{\theta}_{(i)})^2\right)^{3/2}}$

12: let $q_l \leftarrow \Phi\left(\hat{z} + \dfrac{\hat{z} + \Phi^{-1}(\alpha/2)}{1 - \hat{a}(\hat{z} + \Phi^{-1}(\alpha/2))}\right)$

13: let $q_u \leftarrow \Phi\left(\hat{z} + \dfrac{\hat{z} + \Phi^{-1}(1 - \alpha/2)}{1 - \hat{a}(\hat{z} + \Phi^{-1}(1 - \alpha/2))}\right)$

14: let $l \leftarrow \lceil q_l N \rceil$

15: let $u \leftarrow \lceil q_u N \rceil$

16: let $\theta_{(1)}^*, \ldots, \theta_{(N)}^*$ be $\theta_1^*, \ldots, \theta_N^*$, sorted in increasing order

17: return $(\theta_{(l)}^*, \theta_{(u)}^*)$

The derivation of the specific form of bounds in the BC$_a$ method is beyond the scope of this text; the reader is referred to Efron and Tibshirani (1993, Chapter 22) and DiCiccio and Efron (1996) for discussion of the details. Instead, we study the quality of the confidence intervals from algorithm 5.20 and algorithm 5.21 with the help of a numerical experiment.

By definition, a confidence interval $[U, V]$ with confidence coefficient $1 - \alpha$ for a parameter θ satisfies

$$P\left(U(X_1, \ldots, X_n) \leq \theta \leq V(X_1, \ldots, X_n)\right) \geq 1 - \alpha.$$

Typically, the interval is constructed symmetrically in the sense that it satisfies $P(\theta < U) = \alpha/2$ and $P(\theta > V) = \alpha/2$. The remaining probability $1 - \alpha/2 - \alpha/2 = 1 - \alpha$ corresponds to the case $U \leq \theta \leq V$. For a fixed distribution P with known parameter $\theta = \theta(P)$, and for a given confidence interval, we can estimate the probabilities $P(\theta < U)$, $P(U \leq \theta \leq V)$ and $P(\theta < U)$ using the following Monte Carlo approach:

(a) Generate $X_1^{(j)}, \ldots, X_n^{(j)} \sim P$, i.i.d., for $j = 1, 2, \ldots, N$.

(b) Compute $U^{(j)} = U^{(j)}(X_1^{(j)}, \ldots, X_n^{(j)})$ and $V^{(j)} = V^{(j)}(X_1^{(j)}, \ldots, X_n^{(j)})$ for $j = 1, 2, \ldots, N$.

(c) Let

$$p_{\text{inside}} = \frac{\#\{j \mid U^{(j)} \leq \theta(P) \leq V^{(j)}\}}{N}$$

as well as

$$p_{\text{left}} = \frac{\#\{j \mid \theta(P) < U^{(j)}\}}{N}$$

and

$$p_{\text{right}} = \frac{\#\{j \mid \theta(P) > V^{(j)}\}}{N}.$$

This allows to quantify how well the confidence intervals perform for different models.

Example 5.22   We apply the above procedure to three different families of models:

- the mean μ of standard normally distributed values;

- the mean μ of Exp(1)-distributed values; and

- the variance σ^2 of standard normally distributed values.

Each of these three families is considered for three different sample sizes ($n = 10, 50, 100$), resulting in nine different models being considered in total. Approximate confidence intervals were computed using both algorithm 5.20 and algorithm 5.21. The resulting Monte Carlo estimates for the probabilities p_{left}, p_{inside} and p_{right}, using $\alpha = 0.05$, are shown in Table 5.1.

The results show, as expected, that the confidence intervals improve as the sample size increases. Both methods perform better for the symmetric problem of estimating the mean of a normal distribution than they do for the more skewed cases of the last two problems. For the last two problems, both methods result in confidence intervals which extend too far to the left and not far enough to the right, resulting in very small probabilities for the event $\theta < U$ and too large probabilities for the event $\theta > V$. For the three test problems considered here, there is no clear difference in quality of the confidence intervals constructed by the two methods.

Table 5.1 The results of a numerical experiment to measure the empirical coverage of the simple and BC_a bootstrap confidence intervals, for three different test problems and different sample sizes n. The exact setup is described in example 5.22. The experimentally determined coverage is shown in the last six columns of this table. The theoretical values are 0.025 (left), 0.95 (inside) and 0.025 (right); results close to the theoretical values are displayed in bold face.

			Simple			BC_a		
P	θ	n	Left	Inside	Right	Left	Inside	Right
		10	0.042	0.91	0.049	0.062	0.89	0.048
$\mathcal{N}(0, 1)$	μ	50	0.032	**0.94**	0.031	**0.029**	**0.94**	0.035
		100	**0.027**	**0.95**	**0.022**	**0.024**	**0.95**	**0.027**
		10	0.011	0.85	0.14	**0.021**	0.86	0.12
Exp(1)	μ	50	0.008	0.92	0.072	0.015	0.93	0.055
		100	0.014	**0.94**	0.048	**0.022**	**0.94**	0.038
		10	**0.021**	0.86	0.12	0.006	0.77	0.23
$\mathcal{N}(0, 1)$	σ^2	50	0.014	0.92	0.063	0.007	0.91	0.083
		100	0.018	**0.94**	0.038	0.011	0.92	0.069

5.3   Summary and further reading

In this chapter we have studied two different methods which can be applied when not enough mathematical structure is available to employ Monte Carlo methods. The first part of this chapter introduced the ABC method which can be used in Bayesian problems when the posterior density is not explicitly known (or is too complicated), but when a method for generating samples is still available. The second part of this chapter introduced bootstrap methods, which can be used when even generating samples is problematic.

One of the earliest papers about the ABC method, covering applications in population genetics, was by Tavaré et al. (1997). Newer developments and extensions of the method can be found, for example in Beaumont et al. (2002), Blum and François (2010) and Fearnhead and Prangle (2012).

The bootstrap method is described in various textbooks, for example in Efron and Tibshirani (1993). A wealth of practical information about bootstrap methods can be found in the monograph by Davison and Hinkley (1997).

Exercises

E5.1 Implement the ABC method described in example 5.4. Test your implementation by generating samples for $n = 20$ and $s^* = (6.989, 52.247)$.

E5.2   Implement ABC with regression from algorithm 5.6 for the problem described in example 5.4. Test your implementation by generating samples for $n = 20$ and $s^* = (6.989, 52.247)$.

E5.3   Consider the estimator

$$\hat{\sigma}^2(x_1, \ldots, x_n) = \frac{1}{n} \sum_{i=1}^{n} (x_i - \bar{x})^2$$

for the variance, where $\bar{x}$ is the average of the x_i. Write a program that computes, for given data $x = (x_1, \ldots, x_n)$, the bootstrap estimate of the bias as given in algorithm 5.15. Test your program on a data set consisting of 100 standard normally distributed values. Justify your choice of the Monte Carlo sample size N.

E5.4   Assume that we observe 2500 tosses of a biased coin and that we get the following sequence of heads and tails:

H H T H H H H T H T T T H T T H T H T H T T T H H H H T T H T . . . T H T H,

containing 1600 heads and 900 tails in total. From these observations, we estimate the probability p of head as $\hat{p} = 1600/2500 = 16/25$. Compute the bootstrap estimate of the standard error of $\hat{p}$.

E5.5   Implement algorithm 5.20 for computing bootstrap confidence intervals. Test your program by computing bootstrap confidence intervals for the mean of a normal distribution and by comparing the output of your program with the theoretically exact confidence interval for this case.

E5.6   (a)   For $n = 25$, create a sample of n independent standard normally distributed random variables $X_1, \ldots, X_n$.

(b)   For fixed N, repeatedly apply algorithm 5.20 to this data set to get different estimates for a confidence interval for the variance. Numerically determine the standard deviation of both bounds as well as the average width of the interval.

(c)   Repeat the previous step for $N = 1000$, $N = 2000$ and $N = 4000$ and comment on your results.

E5.7   Implement the BC_a method from algorithm 5.21.

E5.8   Write a program which numerically determines the coverage probabilities for the simple bootstrap confidence interval (algorithm 5.20) and for the BC_a bootstrap confidence interval (algorithm 5.21), as described in example 5.22. Use this program to estimate the probability $P(\theta \in [U, V])$ for the following three problems:

(a)   Estimating confidence intervals for the mean μ of standard normally distributed values.

(b)   Estimating confidence intervals for the mean μ of Exp(1)-distributed values.

(c)   Estimating confidence intervals for the variance σ^2 of standard normally distributed values.

Compare the quality of the two different kinds of confidence intervals. Determine how the quality of the confidence intervals depends on the sample size n.

6

Continuous-time models

In this chapter we will pick up the thread we left in Chapter 2 and will return to the topic of simulating statistical models. The models we are interested in here are described by *continuous-time processes*, that is by stochastic processes where time is represented by a bounded or unbounded interval of real numbers. There are two main motivations for using continuous-time processes. First, physical time is continuous: physical quantities like the temperature at a point or the location of a particle in space are in principle defined for all times. Thus, continuous-time models are appropriate for describing physical processes. Secondly, in mathematical models continuous-time processes can arise as the limit of discrete-time processes when the distance between the time steps converges to zero. In such cases, analysis of the limiting continuous-time process is typically much easier than analysis of the underlying discrete-time process.

In this chapter we will introduce the most important classes of continuous-time stochastic processes and we will study how these processes can be simulated on a computer. In the second part of this chapter we will revisit the Monte Carlo techniques introduced in Chapter 3 and we will study how statistical continuous-time models can be studied by simulation.

6.1   Time discretisation

Compared with the situation of discrete-time processes, for example the Markov chains considered in Section 2.3, simulation in continuous-time introduces new challenges. Consider a stochastic process $(X_t)_{t\in I}$ where $I \subseteq \mathbb{R}$ is a time interval, for example $I = [0, \infty)$ or $I = [0, T]$ for some time horizon $T > 0$. Even if the time interval I is bounded, the trajectory $(X_t)_{t\in I}$ consists of uncountably many values. Since computers only have finite storage capacity, it is impossible to store the whole trajectory of a continuous-time process on a computer. Even computing values for all X_t would take an infinite amount of time. For these reasons, we restrict ourselves

An Introduction to Statistical Computing: A Simulation-based Approach, First Edition. Jochen Voss.
© 2014 John Wiley & Sons, Ltd. Published 2014 by John Wiley & Sons, Ltd.

to simulate X only for times $t \in I_n$ where $I_n = \{t_1, t_2, \ldots, t_n\} \subset I$ is finite. This procedure is called *time discretisation*.

In many cases we can simulate the process X by iterating through the times $t_1, t_2, \ldots, t_n \in I_n$: we first simulate X_{t_1}, next we use the value of X_{t_1} to simulate X_{t_2}, then we use the values X_{t_1} and X_{t_2} to simulate X_{t_3} and so on. The final step in this procedure is to use the values $X_{t_1}, \ldots, X_{t_{n-1}}$ to simulate X_{t_n}. One problem with this approach is that often the distribution of X_{t_k} does not only depend on X_{t_i} for $i = 1, 2, \ldots, k - 1$, but also on (unknown to us) values X_t where $t \notin I_n$. For this reason, most continuous-time processes cannot be simulated exactly on a computer and we have to resort to approximate solutions instead. The error introduced by these approximations is called *discretisation error*.

6.2   Brownian motion

Brownian motion forms a basic building block for many kinds of continuous-time stochastic processes.

Definition 6.1   An $\mathbb{R}^d$-valued stochastic process $(B_t)_{t \geq 0}$ is a *Brownian motion* (also called a *Wiener process*), if it satisfies the following three conditions:

(a) $B_0 = 0$;

(b) for every $t \geq 0$ and $h > 0$, the increment $B_{t+h} - B_t$ is $\mathcal{N}(0, hI_d)$-distributed and independent of $(B_s)_{0 \leq s \leq t}$; and

(c) the map $t \mapsto B_t$ is continuous.

In the definition, $\mathcal{N}(0, hI_d)$ denotes the d-dimensional normal distribution with covariance matrix hI_d (see Section 2.1) and I_d is the d-dimensional identity matrix. For a one-dimensional Brownian motion, the distribution of the increments of B simplifies to $B_{t+h} - B_t \sim \mathcal{N}(0, h)$.

In this book we denote Brownian motion by B. Some authors prefer the name 'Wiener process' over 'Brownian motion', and then use W to denote this process. Both conventions are widely used in the literature.

From the definition of a Brownian motion B we see that $t \mapsto B_t$ is a random continuous function of time with values in $\mathbb{R}^d$. One path of a one-dimensional Brownian motion is shown in Figure 6.1.

In many cases, models based on Brownian motion are a good fit for processes which are, on a microscopic scale, driven by small, random, independent, additive contributions. For such models, there are two possible categories:

- If the variance of the individual contributions is small (compared with the inverse of their number), then the macroscopic behaviour of the system will be deterministic.

- If the variance of the individual contributions is larger (comparable with the inverse of the number of contributions), the resulting system will show random

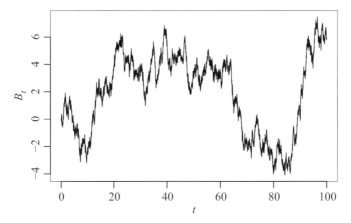

Figure 6.1   One instance of a one-dimensional Brownian motion, simulated until time T = 100.

behaviour on a macroscopic scale. In these cases often the central limit theorem (theorem A.9) applies and macroscopic increments are approximately normally distributed.

Models based on Brownian motion are used for situations falling into the second category.

Example 6.2   A tiny particle, like a grain of dust, suspended in a liquid is subject to 'impulses' by the individual water molecules hitting the particle. These impacts happen with very high frequency and due to the complicated dynamics of the water molecules can be considered to be random. This effect causes a visible motion of the suspended particle which can be described by a three-dimensional Brownian motion.

Example 6.3   The price of a share of stock of a company is determined by individual trades of this stock. For stocks with a high enough volume of trading, the stock price can be described as a (transformed) Brownian motion.

For both of these examples it is important to keep in mind that models based on Brownian motion only form approximations. Use of these approximations is justified by the fact that the resulting models describe the systems well on a wide range of timescales. Nevertheless, since normally distributed increments are only observed when sufficiently many of the microscopic contributions are combined, the models will break down on very short timescales. Similarly, on very long timescales new effects can occur; for example the physical particle considered in example 6.2 will eventually hit the boundary of the enclosing container, whereas the Brownian motion used as a model here does not include information about the container boundaries.

6.2.1 Properties

In this section we briefly state some of the most important properties of Brownian motion. Using definition 6.1, we can immediately derive the following results:

- We have $B_t = B_t - B_0 \sim \mathcal{N}(0, t I_d)$ and in particular the expectation of B_t is $\mathbb{E}(B_t) = 0$.

- For the one-dimensional case we find $B_t \sim \mathcal{N}(0, t)$ and thus the standard deviation of B_t is $\sqrt{t}$.

- Similarly, for any dimension, we find that B_t has the same distribution as $\sqrt{t} B_1$. Thus, the magnitude of $|B_t|$ only grows like $\sqrt{t}$ as t increases.

Another basic result, relating the one-dimensional case to the d-dimensional case, is given in the following lemma.

Lemma 6.4   The d-dimensional process $B = (B_t^{(1)}, \ldots, B_t^{(d)})_{t \geq 0}$ is a Brownian motion if and only if the components $B^{(i)}$ for $i = 1, 2, \ldots, d$ are independent, one-dimensional Brownian motions.

Proof   This follows directly from the definition of a Brownian motion. We check the three conditions from definition 6.1 one by one. First, the vector B_0 is zero if and only if all d of its components are zero. Secondly, $B_{t+h} - B_t \sim \mathcal{N}(0, h I_d)$ if and only if $B_{t+h}^{(i)} - B_t^{(i)} \sim \mathcal{N}(0, h)$ for all $i = 1, 2, \ldots, d$. And, finally, the map $t \mapsto B_t$ is continuous if and only if all of its components are continuous.   □

To conclude this section, in the following lemma we collect three invariance properties of Brownian motion.

Lemma 6.5   Let B be a Brownian motion. Then the following statements hold.

(a) The process $(-B_t)_{t \geq 0}$ is a Brownian motion.

(b) For every $s \geq 0$, the process $(B_{s+t} - B_s)_{t \geq 0}$ is a Brownian motion, independent of $(B_r)_{0 \leq r \leq s}$. (This is called the *Markov property* of Brownian motion.)

(c) For every $c > 0$, the process $(\sqrt{c} B_{t/c})_{t \geq 0}$ is a Brownian motion. (This is called the *scaling property* of Brownian motion.)

Proof   All three statements follow directly from the definition of a Brownian motion. For the first statement, let $X_t = -B_t$ for all $t \geq 0$. Then we have $X_0 = -B_0 = -0 = 0$, the increments $X_{t+h} - X_t = -(B_{t+h} - B_t)$ are $\mathcal{N}(0, h I_d)$ distributed since $B_{t+h} - B_t \sim \mathcal{N}(0, h I_d)$, and $t \mapsto X_t = -B_t$ is continuous since $t \mapsto B_t$ is continuous. Thus $X = -B$ satisfies the conditions of definition 6.1 and consequently is a Brownian motion.

For the second statement, let $s \geq 0$ and define $Y_t = B_{s+t} - B_s$ for all $t \geq 0$. Then $Y_0 = B_{s+0} - B_s = 0$ and

$$
\begin{aligned}
Y_{t+h} - Y_t &= (B_{s+t+h} - B_s) - (B_{s+t} - B_s) \\
&= B_{s+t+h} - B_{s+t} \\
&\sim \mathcal{N}(0, h I_d),
\end{aligned}
$$

since B is a Brownian motion. Furthermore, $t \mapsto Y_t = B_{s+t} - B_s$ is continuous and thus Y is a Brownian motion. Again, since B is a Brownian motion, the random variables $Y_t = B_{s+t} - B_s$ are independent of $(B_r)_{0 \leq r \leq s}$.

Finally, for the third statement, let $Z_t = \sqrt{c} B_{t/c}$ for all $t \geq 0$. Clearly $Z_0 = \sqrt{c} B_{0/c} = 0$. For the increments we have

$$
Z_{t+h} - Z_t = \sqrt{c} B_{(t+h)/c} - \sqrt{c} B_{t/c} = \sqrt{c}\left(B_{t/c+h/c} - B_{t/c}\right)
$$

and, since $B_{t/c+h/c} - B_{t/c} \sim \mathcal{N}(0, h/c I_d)$, we find $Z_{t+h} - Z_t \sim \mathcal{N}(0, h I_d)$. Again, the continuity of B implies continuity of Z and thus the process Z is a Brownian motion. This completes the proof of the lemma.   □

6.2.2   Direct simulation

As we have seen at the start of this chapter, we cannot hope to simulate all of the infinitely many values $(B_t)_{t \geq 0}$ on a computer simultaneously. Instead, we restrict ourselves to simulating the values of B for times $0 = t_0 < t_1 < \cdots < t_n$. For a Brownian motion, this can be easily done by using the first two conditions from definition 6.1: we have $B_0 = 0$ and B_{t_i} can be computed from $B_{t_{i-1}}$ by adding an $\mathcal{N}(0, t_i - t_{i-1})$-distributed random value, independent of all values computed so far. This method is described in the following algorithm.

Algorithm 6.6   (Brownian motion)
    input:
        sample times $0 = t_0 < t_1 < \cdots < t_n$
    randomness used:
        an i.i.d. sequence $(\varepsilon_i)_{i=1,2,\ldots,n}$ with distribution $\mathcal{N}(0, 1)$
    output:
        a sample of $B_{t_0}, \ldots, B_{t_n}$, that is a discretised path of a Brownian motion
    1: $B_0 \leftarrow 0$
    2: **for** $i = 1, 2, \ldots, n$ **do**
    3:    generate $\varepsilon_i \sim \mathcal{N}(0, 1)$
    4:    $\Delta B_i \leftarrow \sqrt{t_i - t_{i-1}}\, \varepsilon_i$
    5:    $B_{t_i} \leftarrow B_{t_{i-1}} + \Delta B_i$
    6: **end for**
    7: return $(B_{t_i})_{i=0,1,\ldots,n}$

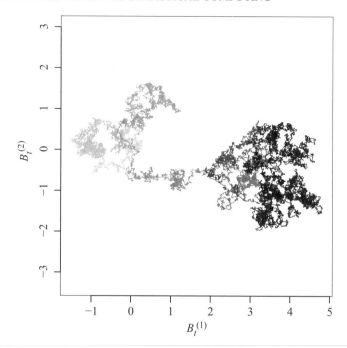

Figure 6.2   Path of a two-dimensional Brownian motion, simulated until time $T =$ 10. The shades indicate time, ranging from $t = 0$ (light grey) to $t = T$ (black).

While algorithm 6.6 only covers the one-dimensional case, it can also be used to simulate a d-dimensional Brownian motion $B = (B^{(1)}, \ldots, B^{(d)})$ for $d > 1$: by lemma 6.4 it suffices to simulate the individual components $B^{(i)}$ for $i = 1, 2, \ldots, d$, independently.

Since the paths of B are continuous, for large n we can assume that the values of B in between grid points are close to the values at the neighbouring grid points. By using small grid spacing and connecting the values B_{t_i} using straight lines, we can obtain a good approximation to a path of a Brownian motion. Results of such simulations are shown in Figure 6.1 and Figure 6.2.

6.2.3   Interpolation and Brownian bridges

If we have already simulated values of a Brownian motion B for a set of times, it is possible to refine the simulated path afterwards by simulating values of B for additional times. This method is called *interpolation* of the Brownian path. Since these additional simulations need to be compatible with the already sampled values, some care is needed when implementing this method.

Assume that we know the values of B at times $0 = t_0 < t_1 < \cdots < t_n$ and that we want to simulate an additional value for B at time s with $t_{i-1} < s < t_i$ for some $i \in \{2, 3, \ldots, n\}$. As an abbreviation we write $r = t_{i-1}$ and $t = t_i$. By the

Markov property of Brownian motion (lemma 6.5, part (b)), the increment $B_s - B_r$ is independent of $(B_u)_{0 \le u \le r}$. Similarly, $(B_u - B_t)_{u \ge t}$ is independent of $(B_u)_{0 \le u \le t}$ and thus of $B_s - B_r$. Consequently, we only need to take the value $B_t = B_{t_i}$ into account when sampling the increment $B_s - B_r$; by independence the remaining B_{t_j} with $j \ne i$ do not affect the distribution of $B_s - B_r$.

Assume that B is one-dimensional with $B_r = a$. Since $(B_{u+r} - B_r)_{u \ge 0}$ is a Brownian motion independent of B_r, we have $B_s \sim \mathcal{N}(a, s - r)$ and $B_t - B_s \sim \mathcal{N}(0, t - s)$, independently of each other. Thus, the joint density of (B_s, B_t) is

$$f_{B_s, B_t}(x, y) = \frac{1}{\sqrt{2\pi(s - r)}} \exp\left(-\frac{(x - a)^2}{2(s - r)}\right) \frac{1}{\sqrt{2\pi(t - s)}} \exp\left(-\frac{(y - x)^2}{2(t - s)}\right)$$

for all $x, y \in \mathbb{R}$. We need to condition this distribution on the, already sampled, value of B_t. By definition of the conditional density $f_{B_s | B_t}$ from (A.5) we have

$$f_{B_s | B_t}(x | b) = c_1 f_{B_s, B_t}(x, b)$$

$$= c_1 \frac{1}{\sqrt{2\pi(s - r)}} \exp\left(-\frac{(x - a)^2}{2(s - r)}\right)$$

$$\frac{1}{\sqrt{2\pi(t - s)}} \exp\left(-\frac{(b - x)^2}{2(t - s)}\right)$$

$$= c_2 \exp\left(-\frac{(x - a)^2}{2(s - r)} - \frac{(b - x)^2}{2(t - s)}\right),$$

where c_1 and c_2 are constants. By expanding the expressions inside the exponential and then completing the square we find

$$f_{B_s | B_t}(x | b) = c_3 \exp\left(-\frac{(x - \mu)^2}{2\sigma^2}\right),$$

where c_3 is a constant,

$$\mu = \frac{t - s}{t - r}a + \frac{s - r}{t - r}b \quad \text{and} \quad \sigma^2 = \frac{(t - s)(s - r)}{t - r}.$$

Since $f_{B_s | B_t}$ is a probability density, we know $c_3 = 1/\sqrt{2\pi\sigma^2}$ and thus $f_{B_s | B_t}$ is the density of a $\mathcal{N}(\mu, \sigma^2)$-distribution.

The argument presented above shows that, given $B_r = a$ and $B_t = b$, the value B_s for $r < s < t$ satisfies

$$B_s \sim \mathcal{N}\left(\frac{t - s}{t - r}a + \frac{s - r}{t - r}b, \frac{(t - s)(s - r)}{t - r}\right). \qquad (6.1)$$

The expectation of B_s for $r < s < t$ is found by linear interpolation between B_r and B_t (corresponding to the dashed line in Figure 6.3). The variance of B_s is biggest at

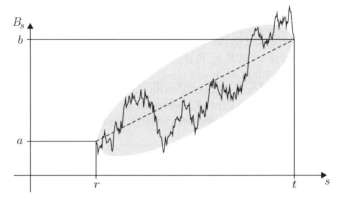

Figure 6.3 Path of a Brownian bridge $(B_s)_{s\in[r,t]}$, *interpolated between the points* (r, a) *and* (t, b). *The dashed line gives the mean of the Brownian bridge, the shaded region has a width of one standard deviation around the mean.*

the centre of the interval $[r, t]$ and converges to 0 when s approaches either of the boundary points; the corresponding standard deviation is represented by the shaded region in Figure 6.3.

Using equation (6.1), the procedure for interpolating a Brownian path is as follows: let $0 = t_0 < t_1 < \cdots < t_n$ be the times where the values B_{t_i} are already known and let $t > 0$.

(a) If $t = t_i$ for an index $i \in \{0, 1, \ldots, n\}$, we return the already sampled value $B_t = B_{t_i}$.

(b) If $t > t_n$, return $B_t \sim \mathcal{N}(B_{t_n}, t - t_n)$.

(c) If $t_{i-1} < t < t_i$, return

$$B_t \sim \mathcal{N}\left(\frac{t_i - t}{t_i - t_{i-1}} B_{t_{i-1}} + \frac{t - t_{i-1}}{t_i - t_{i-1}} B_{t_i}, \frac{(t_i - t)(t - t_{i-1})}{t_i - t_{i-1}}\right).$$

For the last two cases, if the procedure is repeated, we need to add the newly sampled value to the list of already known values for B. The result of a simulation using this method is shown in Figure 6.4.

The method we used to interpolate a Brownian path between already sampled points describes a continuous-time stochastic process on the time interval $[r, t]$.

Definition 6.7 A *Brownian bridge* between (r, a) and (t, b), where $r < t$ are times and $a, b \in \mathbb{R}^d$, is the continuous-time stochastic process on the time interval $[r, t]$, obtained by conditioning a Brownian motion B on the events $B_r = a$ and $B_t = b$.

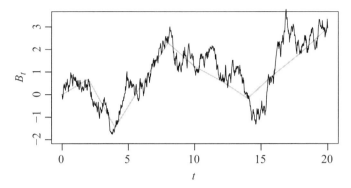

Figure 6.4   One instance of a one-dimensional Brownian motion, simulated using the interpolation method described in Section 6.2.3. The grey 'skeleton' of the path is simulated first. The refined path, shown in black, is sampled later using interpolation.

A Brownian bridge can be simulated by interpolating a Brownian path between $B_r = a$ and $B_t = b$, as described above. The distribution of a Brownian bridge at a fixed time $s \in [r, t]$ is described by equation (6.1).

A second method for sampling a path of a Brownian bridge is given by the following procedure: we can first sample a path B of an ordinary Brownian motion, and then define a process X by

$$X_s = B_{s-r} + a - \frac{s - r}{t - r}(B_{t-r} - b + a). \tag{6.2}$$

One can prove that the resulting process $(X_s)_{s \in [r,t]}$ is again a Brownian bridge. This method is often more efficient than the interpolation method, when many samples from the path of a Brownian bridge are required.

6.3   Geometric Brownian motion

Geometric Brownian motion is a continuous-time stochastic process which is derived from Brownian motion and can, for example, be used as a simple model for stock prices.

Definition 6.8   A continuous-time stochastic process $X = (X_t)_{t \geq 0}$ is a *geometric Brownian motion*, if

$$X_t = X_0 \exp\big(\alpha B_t + \beta t\big) \tag{6.3}$$

for all $t \geq 0$, where $\alpha > 0$ and $\beta \in \mathbb{R}$ are constants and B is a one-dimensional Brownian motion, independent of X_0.

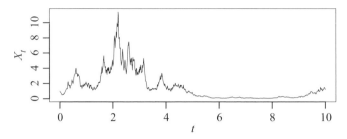

Figure 6.5   One path of a geometrical Brownian motion with $X_0 = 1$, $\alpha = 1$ and $\beta = -0.1$, simulated until time $T = 10$.

Since we know how to generate samples of a Brownian motion from section 6.2.2, generating samples of a geometric Brownian motion is easy: we can simply simulate B_t, for example using algorithm 6.6, and then separately compute $X_0 \exp(\alpha B_t + \beta t)$ for every t-value used in the simulation (the result of one such simulation is shown in Figure 6.5). Nevertheless, we will see below that some care is needed when using the simulated values for Monte Carlo estimates.

Lemma 6.9   Let X be a geometric Brownian motion with parameters α and β. Then $\mathbb{E}(X_t) = \mathbb{E}(X_0) \exp\big((\alpha^2/2 + \beta)t\big)$ for all $t \geq 0$.

Proof   Since we know $B_t \sim \mathcal{N}(0, t)$, we can compute the expectation of X_t using integration:

$$
\begin{aligned}
\mathbb{E}(X_t) &= \mathbb{E}(X_0)\, \mathbb{E}\left(\exp\big(\alpha B_t + \beta t\big)\right) \\
&= \mathbb{E}(X_0) \int_{\mathbb{R}} \exp(\alpha y + \beta t)\, \frac{1}{\sqrt{2\pi t}} \exp\left(-\frac{y^2}{2t}\right) dy \\
&= \mathbb{E}(X_0) \frac{e^{\beta t}}{\sqrt{2\pi t}} \int_{\mathbb{R}} \exp\left(-\frac{y^2}{2t} + \alpha y\right) dy.
\end{aligned}
$$

By completing the square we find

$$
\begin{aligned}
\mathbb{E}(X_t) &= \mathbb{E}(X_0) \frac{\exp\big((\alpha^2/2 + \beta)t\big)}{\sqrt{2\pi t}} \int_{\mathbb{R}} \exp\left(-\frac{1}{2t}\big(y^2 - 2y\alpha t + \alpha^2 t^2\big)\right) dy \\
&= \mathbb{E}(X_0) \exp\big((\alpha^2/2 + \beta)t\big) \frac{1}{\sqrt{2\pi t}} \int_{\mathbb{R}} \exp\left(-\frac{(y - \alpha t)^2}{2t}\right) dy \\
&= \mathbb{E}(X_0) \exp\big((\alpha^2/2 + \beta)t\big).
\end{aligned}
\tag{6.4}
$$

This completes the proof.   □

As a consequence of lemma 6.9 we see that the conditions for X_t to have constant expectation on the one hand and for the exponent $\alpha B_t + \beta t$ in (6.3) to have constant expectation on the other hand are different. For $\beta = -\alpha^2/2$ we find that the process X_t satisfies

$$\mathbb{E}(X_t) = \mathbb{E}(X_0)\exp\left(\left(\frac{\alpha^2}{2} - \frac{\alpha^2}{2}\right)t\right) = \mathbb{E}(X_0),$$

that is for $\beta = -\alpha^2/2$ the geometric Brownian motion has constant expectation. On the other hand, the exponent $\alpha B_t + \beta t$ satisfies for this case

$$\mathbb{E}(\alpha B_t + \beta t) = \mathbb{E}\left(\alpha B_t - \frac{\alpha^2}{2}t\right) = -\frac{\alpha^2}{2}t$$

and thus $\mathbb{E}(\alpha B_t + \beta t)$ is not constant but converges to $-\infty$ as $t \to \infty$.

The difference in behaviour of $\mathbb{E}(X_t)$ and $\mathbb{E}(\alpha B_t + \beta t)$ is caused by the fact that positive fluctuations of $\alpha B_t + \beta t$ are greatly amplified by the exponential in the definition (6.3) whereas negative fluctuations are damped down. A more quantitative result can be obtained by studying the integral in equation (6.4): the integration variable y runs over all possible values of B_t and the integrand $\exp\left(-(y - \alpha t)^2/2t\right)$ in the last integral combined the map which transforms B into X with the density for B_t. It is easy to check that the main contribution of this integral comes from the region $y \approx \alpha t \pm \sqrt{t}$. Thus, the main contribution to $\mathbb{E}(X_t)$ comes from values of $B_t \approx \alpha t \pm \sqrt{t}$. This corresponds to very unlikely events which, when they occur, make huge a contribution to the expectation: The probability of $B_t \approx \alpha t \pm \sqrt{t}$ can be estimated as

$$\begin{aligned} P\left(\alpha t - \sqrt{t} \leq B_t \leq \alpha t + \sqrt{t}\right) &= P\left(\alpha\sqrt{t} - 1 \leq \frac{B_t}{\sqrt{t}} \leq \alpha\sqrt{t} + 1\right) \\ &= \frac{1}{\sqrt{2\pi}}\int_{-1}^{1}\exp\left(-\frac{1}{2}(x + \alpha\sqrt{t})^2\right)dx \quad (6.5) \\ &\approx \frac{2}{\sqrt{2\pi}}\exp\left(-\frac{\alpha^2 t}{2}\right). \end{aligned}$$

As a consequence, Monte Carlo methods cannot be used to estimate $\mathbb{E}(X_t)$ for large values of t. In order to get a reasonable estimate for $\mathbb{E}(X_t)$, at least a few Monte Carlo samples need to fall into the region $B_t \approx \alpha t \pm \sqrt{t}$. From equation (6.5) we see that this requires sample size $N \gg \exp(\alpha^2 t/2)$ and for large values of $\alpha^2 t$ such sample sizes will no longer be practical. This effect is illustrated in Figure 6.6. A solution to this problem would be to use importance sampling and to replace the i.i.d. copies $B_t \sim \mathcal{N}(0, t)$ used in the basic Monte Carlo method with $\mathcal{N}(\alpha t, t)$-distributed proposals.

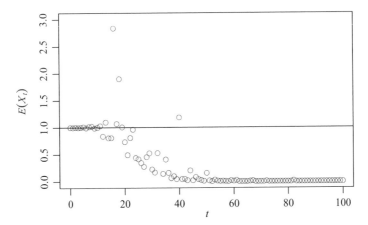

Figure 6.6   Monte Carlo estimates for the expectation of the geometric Brownian motion $X_t = \exp(B_t - t/2)$, where B is a Brownian motion, using $N = 10^6$ Monte Carlo samples for each value of t. The estimates for different t are displayed as circles whereas the exact value of the expectation, $\mathbb{E}(X_t) = 1$ by lemma 6.9, is given by the horizontal line. The figure shows that good estimates are only obtained for $t < 15$.

6.4   Stochastic differential equations

A wide class of continuous-time stochastic processes $X = (X_t)_{t \geq 0}$ can be described as solutions to stochastic differential equations (SDEs) of the form

$$dX_t = \mu(t, X_t)\, dt + \sigma(t, X_t)\, dB_t \quad \text{for all } t > 0,$$
$$X_0 = x_0,$$

(6.6)

where $B = (B_t)_{t \geq 0}$ is a Brownian motion with values in $\mathbb{R}^m$ and $\mu\colon [0, \infty) \times \mathbb{R}^d \to \mathbb{R}^d$ as well as $\sigma\colon [0, \infty) \times \mathbb{R}^d \to \mathbb{R}^{d \times m}$ are functions, and the initial condition $x_0 \in \mathbb{R}^d$ can be random or deterministic.

6.4.1   Introduction

In this section we will give a very short introduction to SDEs, mainly by giving an intuitive idea about the properties of processes described by equations such as (6.6). We start by giving an informal explanation of different aspects of equation (6.6).

- The stochastic process $X = (X_t)_{t \geq 0}$ is the 'unknown' in equation (6.6). *Solving the SDE* means to find a stochastic process X such that (6.6) is satisfied. (We will discuss below what this means.) Since the Brownian motion B on the right-hand side of (6.6) is random, the solution X is random, too.

- $x_0 \in \mathbb{R}^d$ is called the *initial value* of the SDE.

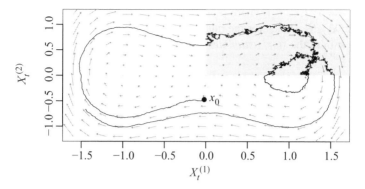

Figure 6.7   One path of a two-dimensional SDE. The drift μ is indicated by the grey arrows; the diffusion coefficient is large inside the grey rectangle and small outside it.

- The function $\mu \colon [0, \infty) \times \mathbb{R}^d \to \mathbb{R}^d$ is called the *drift* of the SDE. Given the current time t and the current value X_t, it determines the direction of mean change of the process just after time t:

$$\mathbb{E}\big(X_{t+h} \,\big|\, X_t\big) \approx X_t + \mu(t, X_t)h$$

  as $h \downarrow 0$. The effect of the drift is illustrated in Figure 6.7.

- The matrix-valued function $\sigma \colon [0, \infty) \times \mathbb{R}^d \to \mathbb{R}^{d \times m}$ is called the *diffusion coefficient* of the SDE. It determines the amount of random fluctuations X is subject to at any given time and place. Conditioned on the value of X_t, the covariance matrix of X_{t+h} satisfies

$$\mathrm{Cov}\big(X_{t+h} \,\big|\, X_t\big) \approx \sigma(t, X_t)\sigma(t, X_t)^\top h$$

  as $h \downarrow 0$.

- The *stochastic differentials* dX_t, dt and dB_t describe the increments of the processes X, t and B. Formally, we have $dX_t = X_t - X_0$, $dt = t - 0 = t$ and $dB_t = B_t - B_0 = B_t$.

- The 'products' $\mu(t, X_t)\, dt$ and $\sigma(t, X_t)\, dB_t$ are shorthand notations for integrals. The first of these terms, $\mu(t, X_t)\, dt$, stands for the vector with components

$$\big(\mu(t, X_t)\, dt\big)_i = \int_0^t \mu_i(s, X_s)\, ds$$

for $i = 1, 2, \ldots, d$ where μ_i is the ith component of the vector μ. Similarly, $\sigma(t, X_t) \, dB_t$ stands for the random vector where the components are given by the sum of the *stochastic integrals*

$$\left(\sigma(t, X_t) \, dB_t\right)_i = \sum_{j=1}^{m} \int_0^t \sigma_{ij}(s, X_s) \, dB_s^{(j)} \tag{6.7}$$

for $i = 1, 2, \ldots, d$. Here, σ_{ij} stands for the element in row i, column j of the matrix σ, the values $B^{(1)}, \ldots, B^{(m)}$ are the components of the Brownian motion B, and the sum over j is part of a matrix-vector product between the matrix-valued function σ and the vector-valued process B. We will defer explanation of the stochastic integrals in (6.7) until Section 6.4.2.

Using the notation introduced, a continuous-time stochastic process X is defined to be a solution of the SDE (6.6), if $X = (X^{(1)}, \ldots, X^{(d)})$ satisfies

$$X_t^{(i)} = x_0^{(i)} + \int_0^t \mu_i(s, X_s) \, ds + \sum_{j=1}^{m} \int_0^t \sigma_{ij}(s, X_s) \, dB_s^{(j)} \tag{6.8}$$

for all $t \geq 0$ and all $i = 1, 2, \ldots, d$.

In order for the SDE (6.6) to have a solution, that is for a process X which satisfies (6.8) to exist, assumptions on the drift μ and the diffusion coefficient σ are required. We will not consider these conditions here and instead refer to the literature given at the end of this chapter, for example the books by Mao (2007, Section 5.2) and Kloeden and Platen (1999, Section 4.5).

6.4.2   Stochastic analysis

Some technical detail is required to give a mathematically rigorous definition of the stochastic integrals in equation (6.7). We omit the rigorous definition here and refer to the references given at the end of this chapter for details. Instead, we restrict ourselves to heuristic explanations of the most important aspects.

6.4.2.1   Ito integrals

The stochastic integral, also called the *Ito integral*, of the integrand Y with a integrator X is given by the limit

$$\int_0^T Y_t \, dX_t = \lim_{n \to \infty} \sum_{i=0}^{n-1} Y_{t_i^{(n)}} \left(X_{t_{i+1}^{(n)}} - X_{t_i^{(n)}}\right) \tag{6.9}$$

where $t_i^{(n)} = iT/n$ for $i = 0, 1, \ldots, n$. Here, X and Y are stochastic processes. In (6.7) this relation is used with the integrand $\sigma_{ij}(s, X_s)$ instead of Y and with the integrator B instead of X. Since X and Y are random, the value of the stochastic integral (6.9) is a random variable. Equation (6.9) is in analogy to the approximation of the ordinary Riemann integral by Riemann sums:

$$\int_0^T f(t)\,dt = \lim_{n\to\infty} \sum_{i=0}^{n-1} f(t_i)(t_{i+1} - t_i).$$

(6.10)

While the characterisation of a stochastic integral given in Equation (6.9) is not enough to be able to give mathematically rigorous proofs of statements involving stochastic integrals, it suffices to motivate the numerical methods introduced in this chapter. Also, equation (6.9) can be used as the basis of numerical methods to compute stochastic integrals.

An important special case of equation (6.9) is when Y is constant, that is $Y_t = c$ for all $t \in [0, T]$. In this case, we have

$$\int_0^T c\,dX_t = c \lim_{n\to\infty} \sum_{i=0}^{n-1} \left(X_{t_{i+1}^{(n)}} - X_{t_i^{(n)}}\right) = c\left(X_{t_n} - X_{t_0}\right) = c\left(X_T - X_0\right).$$

6.4.2.2   Ito's formula

An important tool from stochastic analysis is *Ito's formula*: This formula allows to evaluate the stochastic differentials of functions of a stochastic process. If X is a stochastic process with values in $\mathbb{R}^d$ and $f: [0, \infty) \times \mathbb{R}^d \to \mathbb{R}$ is differentiable with respect to the first argument and differentiable twice with respect to the second argument, then the stochastic differentials of the process $f(t, X_t)$ satisfy

$$d\big(f(t, X_t)\big) = \frac{\partial}{\partial t} f(t, X_t)\,dt + \sum_{i=1}^{d} \frac{\partial}{\partial x_i} f(t, X_t)\,dX_t^{(i)}$$
$$+ \frac{1}{2} \sum_{i,j=1}^{d} \frac{\partial^2}{\partial x_i \partial x_j} f(t, X_t)\,dX_t^{(i)}\,dX_t^{(j)},$$

(6.11)

where the product $dX_t^{(i)} dX_t^{(j)}$ is determined by the rules

$$dB_t^{(i)}\,dB_t^{(j)} = \begin{cases} dt & \text{if } i = j \text{ and} \\ 0 & \text{otherwise,} \end{cases}$$

$$dB_t^{(i)}\,dt = 0.$$

In particular, if X solves the SDE (6.6), we have

$$dX_t^{(i)} = \mu_i(t, X_t)\,dt + \sum_{k=1}^{d} \sigma_{ik}(t, X_t)\,dB_t^{(k)}$$

and thus

$$dX_t^{(i)} dX_t^{(j)} = \mu_i(t, X_t)\mu_j(t, X_t)\,dt\,dt + \mu_i(t, X_t)\sum_{l=1}^{d}\sigma_{jl}(t, X_t)\,dB_t^{(l)}\,dt$$

$$+ \sum_{k=1}^{d}\sigma_{ik}(t, X_t)\mu_j(t, X_t)\,dB_t^{(k)}\,dt$$

$$+ \sum_{k,l=1}^{d}\sigma_{ik}(t, X_t)\sigma_{jl}(t, X_t)\,dB_t^{(k)}\,dB_t^{(l)}$$

$$= \sum_{k=1}^{d}\sigma_{ik}(t, X_t)\sigma_{jk}(t, X_t)\,dt$$

$$= (\sigma\sigma^\top)_{ij}(t, X_t)\,dt.$$

These expressions can be substituted into (6.11).

One important special case is the case of a one-dimensional stochastic process X. In this case, equation (6.11) simplifies to

$$d\big(f(t, X_t)\big) = \frac{\partial}{\partial t}f(t, X_t)\,dt + \frac{\partial}{\partial x}f(t, X_t)\,dX_t$$

$$+ \frac{1}{2}\frac{\partial^2}{\partial x^2}f(t, X_t)\,dX_t\,dX_t. \tag{6.12}$$

Furthermore, if f does not depend on t, we get

$$d\big(f(X_t)\big) = f'(X_t)\,dX_t + \frac{1}{2}f''(X_t)\,dX_t\,dX_t. \tag{6.13}$$

Finally, if X solves a one-dimensional SDE of the form (6.6), the product $dX_t\,dX_t$ can be written as

$$dX_t\,dX_t = \sigma(t, X_t)^2\,dt$$

and we get

$$d\big(f(X_t)\big) = f'(X_t)\,dX_t + \frac{1}{2}f''(X_t)\sigma(t, X_t)^2\,dt$$

$$= f'(X_t)\big(\mu(t, X_t)\,dt + \sigma(t, X_t)\,dB_t\big) + \frac{1}{2}f''(X_t)\sigma(t, X_t)^2\,dt \tag{6.14}$$

$$= \left(f'(X_t)\mu(t, X_t) + \frac{1}{2}f''(X_t)\sigma(t, X_t)^2\right)dt + f'(X_t)\sigma(t, X_t)\,dB_t.$$

Example 6.10 Let $X = B$ be a one-dimensional Brownian motion and $f(x) = x^2$. In this case, since $f'(x) = 2x$ and $f''(x) = 2$, we can use Ito's formula in the form of equation (6.13) to find

$$d\left(B_t^2\right) = 2B_t\, dB_t + \frac{1}{2}2\, dB\, dB = 2B_t\, dB_t + dt.$$

This equation can be written in integral notation as

$$B_t^2 - B_0^2 = 2\int_0^t B_s\, dB_s + (t - 0)$$

and, since $B_0 = 0$, we can rewrite this as

$$\int_0^t B_s\, dB_s = \frac{1}{2}\left(B_t^2 - t\right).$$

Thus, with the help of Ito's formula we have found the exact value of the stochastic integral $\int_0^t B_s\, dB_s$.

Example 6.11 Let $X_t = x_0 \exp\left(\alpha B_t + \beta t\right)$ be a geometric Brownian motion with $x_0 \in \mathbb{R}$. Then we can write $X_t = f(t, B_t)$ for all $t \geq 0$, where

$$f(t, x) = x_0 \exp\left(\alpha x + \beta t\right).$$

In order to find the stochastic differential dX, we need to compute the derivatives of f: we get

$$\frac{\partial}{\partial t} f(t, x) = x_0 \beta \exp\left(\alpha x + \beta t\right) = \beta f(t, x),$$

$$\frac{\partial}{\partial x} f(t, x) = x_0 \alpha \exp\left(\alpha x + \beta t\right) = \alpha f(t, x),$$

$$\frac{\partial^2}{\partial x^2} f(t, x) = x_0 \alpha^2 \exp\left(\alpha x + \beta t\right) = \alpha^2 f(t, x).$$

Now we can apply Ito's formula in the form of equation (6.12) to get

$$dX_t = d\left(f(t, B_t)\right)$$
$$= \frac{\partial}{\partial t} f(t, B_t)\, dt + \frac{\partial}{\partial x} f(t, B_t)\, dB_t + \frac{1}{2}\frac{\partial^2}{\partial x^2} f(t, B_t)\, dB_t\, dB_t$$
$$= \beta f(t, B_t)\, dt + \alpha f(t, B_t)\, dB_t + \frac{1}{2}\alpha^2 f(t, B_t)\, dt$$

$$= \left(\beta + \frac{\alpha^2}{2} \right) f(t, B_t) \, dt + \alpha f(t, B_t) \, dB_t$$

$$= \left(\beta + \frac{\alpha^2}{2} \right) X_t \, dt + \alpha X_t \, dB_t.$$

Thus, we have found that the geometric Brownian motion X is a solution of the SDE

$$dX_t = \left(\frac{\alpha^2}{2} + \beta \right) X_t \, dt + \alpha X_t \, dB_t$$

$$X_0 = x_0.$$

6.4.2.3 Stratonovich integrals

Looking back at the definition of the Ito integral in equation (6.9) we can see that we had a choice when we chose the analogy to the Riemann integral from equation (6.10): The approximation in (6.10) still works if $f(t_i)$ is replaced with $f(t_{i+1})$, or with any $f(\tilde{t}_i)$ where $\tilde{t}_i \in [t_i, t_{i+1}]$. It transpires that the stochastic integral in (6.9) is much less robust: the integrand Y must be evaluated at the left-most point t_i of each discretisation interval $[t_i, t_{i+1}]$. If Y_{t_i} in (6.9) is, for example, replaced with $Y_{t_{i+1}}$, one can show that for many integrators X the approximation converges to a different limit as $n \to \infty$. This is caused by the irregular paths featured by typical integrators, for example, for $X = B$ as before.

By choosing different times to evaluate the integrand of a stochastic integral, different kinds of stochastic integrals can be obtained. The most common choice, after the Ito integral, is called the *Stratonovich integral*. This integral is obtained by evaluating the integrand at the centre of each discretisation interval, and it is denoted by an additional $\circ$ in front of the integrand dX:

$$\int_0^T f(t) \circ dX_t = \lim_{n \to \infty} \sum_{i=0}^{n-1} f \left(\frac{t_i^{(n)} + t_{i+1}^{(n)}}{2} \right) \left(X_{t_{i+1}^{(n)}} - X_{t_i^{(n)}} \right). \tag{6.15}$$

Correspondingly, by replacing the Ito integral in (6.8) with a Stratonovich integral, *Stratonovich SDEs* of the form

$$dX_t = \mu(t, X_t) \, dt + \sigma(t, X_t) \circ dB_t$$
$$X_0 = x_0 \tag{6.16}$$

can be defined.

Studying the difference between the Ito integral (6.9) and the Stratonovich integral (6.15) more closely, one can show that a process X solves the Stratonovich

SDE (6.16) with differentiable diffusion coefficient σ, if and only if it solves the Ito SDE

$$dX_t = \tilde{\mu}(t, X_t)\, dt + \sigma(t, X_t)\, dB_t$$
$$X_0 = x_0,$$

where the drift vector $\tilde{\mu} \colon [0, \infty) \times \mathbb{R}^d \to \mathbb{R}^d$ has components

$$\tilde{\mu}_i(t, x) = \mu_i(t, x) + \frac{1}{2} \sum_{j=1}^{m} \sum_{k=1}^{d} \sigma_{kj}(t, x) \frac{\partial}{\partial x_k} \sigma_{ij}(t, x)$$

for all $t \geq 0$ and $x \in \mathbb{R}^d$ and for all $i = 1, 2, \ldots, d$. This allows to convert any Stratonovich SDE into an Ito SDE with the same solutions. For this reason, in the rest of this chapter we will only discuss Ito SDEs.

6.4.3   Discretisation schemes

In this section we will discuss methods to simulate solutions of an SDE using a computer. We consider the SDE

$$dX_t = \mu(t, X_t)\, dt + \sigma(t, X_t)\, dB_t \tag{6.17}$$
$$X_0 = x_0$$

where $B = (B_t)_{t \geq 0}$ is an m-dimensional Brownian motion, $\mu \colon [0, \infty) \times \mathbb{R}^d \to \mathbb{R}^d$ is the drift, and the diffusion coefficient is given by $\sigma \colon [0, \infty) \times \mathbb{R}^d \to \mathbb{R}^{d \times m}$. Our aim is to simulate values of X at times $0 = t_0 < t_1 < \cdots < t_n$. We will proceed, starting with $X_0 = x_0$, by successively computing $X_{t_1}, X_{t_2}, \ldots$ until time t_n is reached.

The amount of change of X over the time interval $[t_i, t_{i+1}]$ can be found from the definition (6.8) of a solution: subtracting the expressions for X_{t_i} from the expression for $X_{t_{i+1}}$, we find

$$X_{t_{i+1}}^{(j)} - X_{t_i}^{(j)} = \int_{t_i}^{t_{i+1}} \mu_j(t, X_t)\, dt + \sum_{k=1}^{m} \int_{t_i}^{t_{i+1}} \sigma_{jk}(t, X_t)\, dB_t^{(k)} \tag{6.18}$$

for $j = 1, 2, \ldots, d$. Assume that we have already computed $X_0, X_{t_1}, \ldots, X_{t_i}$ and we want to compute $X_{t_{i+1}}$. Then we have to solve the following two problems:

- The integrals on the right-hand side of (6.18) depend on the values X_t for $t \in [t_i, t_{i+1}]$. This is a problem, since we only know the (approximate) value of X_{t_i} and the values of X_t for $t > t_i$ are unknown.

- Even if X was known, it is still not clear how the integrals can be solved for nontrivial functions μ and σ.

There are different approaches to solving these problems, leading to different numerical schemes.

6.4.3.1   The Euler–Maruyama scheme

The idea behind the *Euler–Maruyama scheme* (sometimes called the *stochastic Euler scheme* or just the *Euler scheme*) is to evaluate the drift and diffusion coefficient in (6.18) at time t_i instead of time t for all $t \in [t_i, t_{i+1}]$. Then the integrals can be approximated by

$$\int_{t_i}^{t_{i+1}} \mu_j(t, X_t)\, dt \approx \int_{t_i}^{t_{i+1}} \mu_j(t_i, X_{t_i})\, dt = \mu_j(t_i, X_{t_i})(t_{i+1} - t_i)$$

and

$$\sum_{k=1}^{m} \int_{t_i}^{t_{i+1}} \sigma_{jk}(t, X_t)\, dB_t^{(k)} \approx \sum_{k=1}^{m} \int_{t_i}^{t_{i+1}} \sigma_{jk}(t_i, X_{t_i})\, dB_t^{(k)}$$

$$= \sum_{k=1}^{m} \sigma_{jk}(t_i, X_{t_i})\left(B_{t_{i+1}}^{(k)} - B_{t_i}^{(k)}\right)$$

$$= \left(\sigma(t_i, X_{t_i})\left(B_{t_{i+1}} - B_{t_i}\right)\right)_j.$$

where the last expression uses matrix-vector multiplication to simplify notation. Substituting these approximations into equation (6.18) allows us to compute approximate values

$$X_{t_{i+1}} - X_{t_i} \approx \mu_j(t_i, X_{t_i})(t_{i+1} - t_i) + \sigma(t_i, X_{t_i})\left(B_{t_{i+1}} - B_{t_i}\right) \qquad (6.19)$$

for $i = 0, 1, \ldots, n - 1$. This leads to the following algorithm for computing approximate solutions to SDEs.

Algorithm 6.12   (Euler–Maruyama scheme)
     input:
         the drift $\mu\colon [0, \infty) \times \mathbb{R}^d \to \mathbb{R}^d$
         the diffusion coefficient $\sigma\colon [0, \infty) \times \mathbb{R}^d \to \mathbb{R}^{m \times d}$
         the initial value $x_0 \in \mathbb{R}^d$
         the time horizon $T \geq 0$
         the discretisation parameter $n \in \mathbb{N}$
     randomness used:
         samples $B_0, B_h, B_{2h}, \ldots, B_{nh}$ from a d-dimensional Brownian motion,
         where $h = T/n$
     output:
         an approximation $(X_{ih}^{\mathrm{EM}})_{i=0,\ldots,n}$ to a solution of (6.17)
     1: $h \leftarrow T/n$
     2: $X_0^{\mathrm{EM}} \leftarrow x_0$

3: **for** $i = 0, 1, \ldots, n - 1$ **do**
4:     $\Delta B_i \leftarrow B_{(i+1)h} - B_{ih}$
5:     $X^{EM}_{(i+1)h} \leftarrow X^{EM}_{ih} + \mu(ih, X^{EM}_{ih})\, h + \sigma(ih, X^{EM}_{ih})\, \Delta B_i$
6: **end for**
7: return $(X^{EM}_{ih})_{i=0,1,\ldots,n}$

The algorithm can either use a given path of a Brownian motion, or alternatively the increments ΔB_i can be directly sampled in the algorithm by generating $\Delta B_i \sim \mathcal{N}(0, hI_d)$ independently in each iteration of the loop.

Lemma 6.13   The computational cost for computing a path using the Euler–Maruyama scheme from algorithm 6.12 is of order $C(n) = \mathcal{O}(n)$.

Proof   Each of the iterations of the loop has the same cost, so the total cost is $C(n) = a + bn$, where a is the cost of the assignments at the start of the algorithm and b is the cost per iteration of the loop.   □

Since the approximations underlying the Euler method get more accurate when the size of the time step decreases, we expect the 'error' of the Euler–Maruyama approximation to go to 0 as the discretisation parameter n increases. This convergence is discussed in Section 6.4.4.

Example 6.14   For $\alpha, \beta \in \mathbb{R}$, consider the one-dimensional SDE

$$dX_t = \left(\frac{\alpha^2}{2} + \beta\right) X_t\, dt + \alpha X_t\, dB_t. \tag{6.20}$$

The Euler–Maruyama scheme for this SDE is

$$X^{EM}_{(i+1)h} = X^{EM}_{ih} + \left(\frac{\alpha^2}{2} + \beta\right) X^{EM}_{ih}\, h + \alpha\, X^{EM}_{ih}\, \Delta B_i$$

for $i = 0, 1, \ldots, n - 1$, where $X^{EM}_0 = x_0$ and $\Delta B_i = B_{(i+1)h} - B_{ih}$. We can obtain numerical solutions of SDE (6.20) by iteratively computing X^{EM}_{ih} for $i = 0, 1, \ldots, n - 1$.

When applying algorithm 6.12, there is a trade-off between speed and accuracy to be made: larger values of the discretisation parameter n lead to more accurate results, smaller values of n lead to faster execution.

6.4.3.2   The Milstein scheme

The Milstein scheme is an improved version of the Euler–Maruyama scheme. It uses a slightly more complicated discretisation mechanism to achieve more accurate results. In this section we only discuss the one-dimensional case. More specifically, we consider

$$
\begin{aligned}
dX_t &= \mu(X_t)\,dt + \sigma(X_t)\,dB_t \quad \text{for all } t > 0, \\
X_0 &= x_0,
\end{aligned}
\tag{6.21}
$$

where $B = (B_t)_{t \geq 0}$ is a one-dimensional Brownian motion, $\mu\colon \mathbb{R} \to \mathbb{R}$ is the drift and $\sigma\colon \mathbb{R} \to \mathbb{R}$ is the diffusion coefficient.

Like the Euler–Maruyama scheme, the *Milstein scheme* is based on equation (6.18), but instead of the approximation $\sigma(X_t) \approx \sigma(X_{t_i})$ for all $t \in [t_i, t_{i+1}]$, the Milstein scheme uses an improved approximation: since the paths of the Brownian motion B are very rough, for $t \approx t_i$ the Brownian increment $|B_t - B_{t_i}|$ is much bigger than $|t - t_i|$. Consequently, the second term in the approximation (6.19) dominates and we have

$$
X_t - X_{t_i} \approx \sigma(X_{t_i}) \cdot \left(B_t - B_{t_i} \right).
$$

Using this approximation, together with first-order Taylor approximation for σ, we find

$$
\begin{aligned}
\sigma(X_t) &= \sigma(X_{t_i} + X_t - X_{t_i}) \\
&\approx \sigma(X_{t_i}) + \sigma'(X_{t_i})(X_t - X_{t_i}) \\
&\approx \sigma(X_{t_i}) + \sigma'(X_{t_i})\sigma(X_{t_i}) \cdot \left(B_t - B_{t_i} \right).
\end{aligned}
$$

Using this expression, the second integral in (6.18) can be approximated as

$$
\begin{aligned}
\int_{t_i}^{t_{i+1}} \sigma(X_t)\,dB_t &\approx \int_{t_i}^{t_{i+1}} \sigma(X_{t_i}) + \sigma'(X_{t_i})\sigma(X_{t_i}) \cdot \left(B_t - B_{t_i} \right) dB_t \\
&= \sigma(X_{t_i})\left(B_{t_{i+1}} - B_{t_i} \right) + \sigma'(X_{t_i})\sigma(X_{t_i}) \int_{t_i}^{t_{i+1}} \left(B_t - B_{t_i} \right) dB_t \\
&= \sigma(X_{t_i})\left(B_{t_{i+1}} - B_{t_i} \right) + \sigma'(X_{t_i})\sigma(X_{t_i}) \frac{1}{2} \left(\left(B_{t_{i+1}} - B_{t_i} \right)^2 - (t_{i+1} - t_i) \right),
\end{aligned}
$$

where the value of the last stochastic integral is found similar to the one in example 6.10. Based on this argument, the Milstein scheme replaces (6.19) by

$$
\begin{aligned}
X_{t_{i+1}} - X_{t_i} &\approx \mu_j(X_{t_i})\,h_i + \sigma(X_{t_i})\,\Delta B_i \\
&\quad + \frac{1}{2}\sigma(X_{ih})\sigma'(X_{ih})\left(\Delta B_i^2 - h_i \right),
\end{aligned}
\tag{6.22}
$$

where σ' is the derivative of σ, $\Delta B_i = B_{t_{i+1}} - B_{t_i}$, and $h_i = t_{i+1} - t_i$ for $i = 0, 1, \ldots, n - 1$. Comparing this expression with equation (6.19) shows that the only difference between the Euler–Maruyama and Milstein methods is the term in the second line of equation (6.22). In the special case where σ is constant, we have $\sigma' = 0$ and both methods coincide.

Algorithm 6.15   (Milstein scheme)
    input:
        the drift $\mu \colon \mathbb{R} \to \mathbb{R}$
        the diffusion coefficient $\sigma \colon \mathbb{R} \to \mathbb{R}$ and its derivative σ'
        the initial value $X_0 \in \mathbb{R}$
        the time horizon $T \geq 0$
        the discretisation parameter $n \in \mathbb{N}$
    randomness used:
        samples $B_0, B_h, B_{2h}, \ldots, B_{nh}$ from a one-dimensional Brownian motion,
        where $h = T/n$
    output:
        an approximation $(X_{ih}^{\mathrm{MIL}})_{i=0,\ldots,n}$ to a solution of (6.21)
    1: $h \leftarrow T/n$
    2: **for** $i = 0, 1, \ldots, n-1$ **do**
    3:   generate $\Delta B_i = B_{(i+1)h} - B_{ih}$
    4:   $X_{(i+1)h}^{\mathrm{MIL}} = X_{ih}^{\mathrm{MIL}} + \mu(X_{ih}^{\mathrm{MIL}}) h + \sigma(X_{ih}^{\mathrm{MIL}}) \Delta B_i$
$$+ \tfrac{1}{2}\sigma(X_{ih}^{\mathrm{MIL}})\sigma'(X_{ih}^{\mathrm{MIL}})\left(\Delta B_i^2 - h\right)$$
    5: **end for**
    6: return $(X_{ih}^{\mathrm{MIL}})_{i=0,1,\ldots,n}$

As before, there is a trade-off between speed of the algorithm and accuracy of the obtained approximation: larger values of the discretisation parameter n lead to more accurate results, smaller values of n lead to faster execution.

Lemma 6.16   The computational cost for computing a path from the Milstein scheme in algorithm 6.1 is of order $C(n) = \mathcal{O}(n)$.

Proof   Each of the iterations of the loop has the same cost, so the total cost is $C(n) = a + bn$ where a is the cost of the assignments at the start of the algorithm and b is the cost per iteration of the loop.   □

We defer discussion of the error of the Milstein method until Section 6.4.4.

Example 6.17   Continuing from example 6.14, we consider again the SDE

$$dX_t = \left(\frac{\alpha^2}{2} + \beta\right) X_t \, dt + \alpha X_t \, dB_t.$$

Since the diffusion coefficient is $\sigma(x) = \alpha x$, we have $\sigma'(x) = \alpha$ and $\sigma(x)\sigma'(x) = \alpha^2 x$. Thus the Milstein scheme for this SDE is

$$X_{(i+1)h}^{\mathrm{MIL}} = X_{ih}^{\mathrm{MIL}} + \left(\frac{\alpha^2}{2} + \beta \right) X_{ih}^{\mathrm{MIL}} h + \alpha \, X_{ih}^{\mathrm{MIL}} \, \Delta B_i$$
$$+ \frac{\alpha^2}{2} X_{ih}^{\mathrm{MIL}} \left(\Delta B_i^2 - h \right)$$

for $i = 0, 1, \ldots, n - 1$, where $X_0^{\mathrm{MIL}} = x_0$ and $\Delta B_i = B_{(i+1)h} - B_{ih}$.

6.4.4   Discretisation error

Since numerical methods for SDEs replace the increments from equation (6.18) by approximations, a numerically obtained solution will not exactly coincide with the exact solution. In this section we will discuss the resulting discretisation error.

Since both the exact solution and the numerical approximation are random, different ways of quantifying the error can be considered.

6.4.4.1   Strong error

Let X be the exact solution of the SDE (6.17) and let $\hat{X}$ be a numerical approximation to X, using the same Brownian motion as X does. Then the *strong error* of the approximation $\hat{X}$ at time T is given by

$$e_{\mathrm{strong}} = \mathbb{E} \left| \hat{X}_T - X_T \right| \qquad (6.23)$$

for all sufficiently large n. The quantity e_{strong} measures the average distance between the exact and the approximate solution. The strong error is small, if the values of the numerical solution are close to the values of the exact solution, that is if the paths of the SDE are approximated well.

Different numerical methods can be compared by studying how the strong error decreases when the discretisation parameter increases. This needs to be balanced to the corresponding increase in computational cost.

Let X^{EM} be the Euler–Maruyama approximation computed by algorithm 6.12 with discretisation parameter n. Then, under appropriate assumptions on the drift μ and the diffusion coefficient σ, there is a constant $c > 0$ such that

$$e_{\mathrm{strong}}^{\mathrm{EM}} = \mathbb{E} \left| X_T^{\mathrm{EM}} - X_T \right| \leq \frac{c}{\sqrt{n}}.$$

Let X^{MIL} be the Milstein approximation, computed by algorithm 6.15 with discretisation parameter n. Then, under appropriate assumptions on the drift μ and the diffusion coefficient σ, there is a constant $c > 0$ such that the strong error of the Milstein scheme satisfies

$$e_{\text{strong}}^{\text{MIL}} = \mathbb{E} \left| X_T^{\text{MIL}} - X_T \right| \leq \frac{c}{n}$$

for all sufficiently large n. As n gets large, this bound decays faster than the bound $c/\sqrt{n}$ for the Euler–Maruyama method. Thus, for large n, the Milstein scheme has a significantly smaller strong error than the Euler–Maruyama scheme.

Example 6.18   SDE (6.20) from example 6.14 is simple enough that it can be solved explicitly: from example 6.11 we know that the exact solution of (6.20) is the geometric Brownian motion

$$X_t = X_0 \exp \left(\alpha B_t + \beta t \right).$$

This knowledge of the exact solution allows us to 'measure' the error of different discretisation schemes: comparing a numerical solution $\hat{X}_T$ to the exact value X_T allows to determine $|\hat{X}_T - X_T|$. By repeating this experiment for different paths of the Brownian motion B and taking averages, we can obtain a Monte Carlo estimate for the error $e_{\text{strong}} = \mathbb{E} |\hat{X}_T - X_T|$. Repeating this estimation procedure for different values of n allows then to determine the dependence of $e_{\text{strong}}^{\text{EM}}$ on the discretisation parameter n.

Figure 6.8 shows the result of such simulations, for $\hat{X} = X^{\text{EM}}$ and $\hat{X} = X^{\text{MIL}}$, respectively. The figure confirms that we have indeed $e_{\text{strong}}^{\text{EM}} \approx c/\sqrt{n}$ and $e_{\text{strong}}^{\text{MIL}} \approx c/n$.

6.4.4.2   Weak error

In cases where we solve the SDE for use in a Monte Carlo estimate, it is only important that the *distribution* of X_T^{EM} is accurate, whereas it is less important that the *values* of X_T^{EM} (as a function of B) are accurate. For this reason, a second error criterion is considered in this section.

As before, let X be the exact solution of an SDE and let $\hat{X}$ be a numerical approximation to X. Furthermore, let $\mathcal{A}$ be a class of functions from $\mathbb{R}^d$ to $\mathbb{R}$ (e.g. all bounded, twice differentiable functions). Then the error in the distribution of $\hat{X}_T$ can be quantified by considering

$$e_{\text{weak}}(f) = \left| \mathbb{E} \left(f(\hat{X}_T) \right) - \mathbb{E} \left(f(X_T) \right) \right| \tag{6.24}$$

for all $f \in \mathcal{A}$ and all sufficiently large n. The quantity

$$e_{\text{weak}} = \sup_{f \in \mathcal{A}} e_{\text{weak}}(f)$$

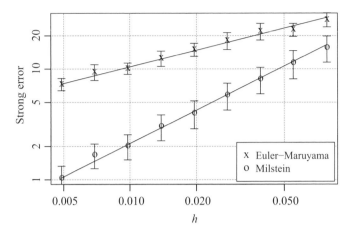

Figure 6.8   Strong error of the Euler–Maruyama and Milstein schemes, for the SDE (6.20) with $\alpha = 1$, $\beta = 0.4$ and $T = 5$. The plot shows the estimated strong error for both methods as a function of the step size $h = T/n$, for different values of n. The points marked with 'x' are Monte Carlo estimates for $e_{\text{strong}}^{\text{EM}}$, the points marked with 'o' are Monte Carlo estimates for $e_{\text{strong}}^{\text{MIL}}$, and the bars give 95% confidence intervals for the Monte Carlo estimates. Finally, the solid lines are fitted curves of the forms $c/\sqrt{n}$ (for the Euler–Maruyama method) and c/n (for the Milstein method), respectively.

is called the *weak error* of $\hat{X}$ at time T. In this context, the functions $f \in \mathcal{A}$ are called *test functions*. The weak error is small, if the distribution of the numerical solution is close to the distribution of the exact solution.

Let X^{EM} be the Euler–Maruyama approximation with discretisation parameter n. Then, under appropriate assumptions on the drift μ, the diffusion coefficient σ and the class $\mathcal{A}$, there is a constant $c > 0$ such that

$$e_{\text{weak}}^{\text{EM}} = \sup_{f \in \mathcal{A}} \left| \mathbb{E}\left(f(X_T^{\text{EM}}) \right) - \mathbb{E}\left(f(X_T) \right) \right| \leq \frac{c}{n} \qquad (6.25)$$

for all $f \in \mathcal{A}$ and all sufficiently large n. Similarly, let X^{MIL} be the Milstein approximation to X. Then, under appropriate assumptions on the drift μ, the diffusion coefficient σ and the class $\mathcal{A}$, there is a constant $c > 0$ such that

$$e_{\text{weak}}^{\text{MIL}}(f) = \sup_{f \in \mathcal{A}} \left| \mathbb{E}\left(f(X_T^{\text{MIL}}) \right) - \mathbb{E}\left(f(X_T) \right) \right| \leq \frac{c}{n} \qquad (6.26)$$

for all $f \in \mathcal{A}$ and all sufficiently large n. This is the same rate of convergence as for the Euler–Maruyama scheme. Thus we would expect the weak errors for the two schemes to stay comparable as n increases.

Example 6.19   Continuing from example 6.18, it is possible to numerically estimate $e_{\text{weak}}^{\text{EM}}(f)$ for solutions of (6.20). The exact expectation is given by

$$\mathbb{E}\left(f(X_T)\right) = \mathbb{E}\left(f\left(X_0 \exp(\alpha B_T + \beta T)\right)\right).$$

Since $B_T \sim \mathcal{N}(0, T)$, for simple functions f and deterministic X_0 this expectation can be calculated analytically. On the other hand, the expectation $\mathbb{E}\left(f(X_T^{\text{EM}})\right)$ can be estimated using Monte Carlo integration.

For this example we consider the case of $X_0 = 1$ and $f(x) = \mathbb{1}_{(-\infty, a]}(x)$ for some constant a. Then we have

$$\mathbb{E}\left(f(X_T)\right) = P\left(\exp\left(\alpha B_T + \beta t\right) \leq a\right) = P\left(B_T \leq \frac{\log(a) - \beta t}{\alpha}\right). \quad (6.27)$$

Figure 6.9 shows a simulation where a is chosen such that the expectation in (6.27) equals 0.6.

Some care is needed when performing numerical experiments of this kind: since $e_{\text{weak}}(f)$ is typically much smaller than either of the values $\mathbb{E}\left(f(\hat{X}_T)\right)$ and $\mathbb{E}\left(f(X_T)\right)$, the Monte Carlo estimates for the expectation need to be performed with high accuracy and thus require huge sample sizes. In addition, for unbounded test functions f, the

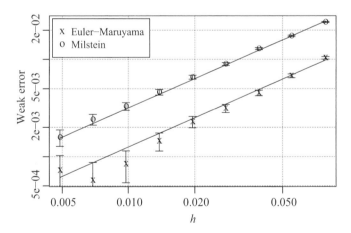

Figure 6.9   Weak error for the Euler–Maruyama and Milstein schemes, for the SDE (6.20) with $\alpha = 1$, $\beta = 0.4$ and $T = 5$. The plot shows the estimated weak error for both methods as a function of the step size $h = T/n$, for different values of n. The points marked 'x' are Monte Carlo estimates for $e_{\text{weak}}^{\text{EM}}(f)$, the points marked 'o' are Monte Carlo estimates for $e_{\text{weak}}^{\text{MIL}}(f)$, and the bars give 95% confidence intervals for the Monte Carlo estimates. The test function used is $f(x) = \mathbb{1}_{(-\infty, a]}(x)$ for some constant a. The solid lines correspond to $0.6275/n$ (for the Euler–Maruyama method) and $1.584/n$ (for the Milstein method), respectively.

SDE (6.20) is susceptible to problems such as the one illustrated in Figure 6.6, potentially making Monte Carlo estimation very difficult.

6.4.4.3   Qualitative behaviour of solutions

The strong and weak errors discussed so far describe convergence of numerical approximations to the exact solutions as the discretisation parameter increases. In this section we will discuss two effects which, for fixed discretisation parameter, can affect the qualitative behaviour of numerical solutions.

One case where numerical solutions can be qualitatively different from the exact solution is when the exact solution is known to be positive whereas the numerical solution may become negative. This problem could, for example, occur when the SDE in question models a stock price or an interest rate. We illustrate this problem here for the case of the one-dimensional Euler–Maruyama scheme: assume that the current value of the simulation is $X_{ih}^{\mathrm{EM}} = x$. Then the next value is given by

$$X_{(i+1)h}^{\mathrm{EM}} = x + \mu(x)\,h + \sigma(x)\,\Delta B_i$$
$$\sim \mathcal{N}\left(x + \mu(x)h, \sigma(x)^2 h\right).$$

Since $X_{(i+1)h}^{\mathrm{EM}}$ is normally distributed, it takes negative values with positive probability. Usually, the probability of this happening is negligible, but in cases where the current state x is close to 0 or when μ or σ is large, the problem can appear with high enough probability to be seen in practice. This effect is illustrated in Figure 6.10.

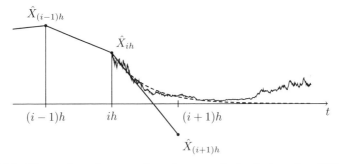

Figure 6.10   Illustration of a case where the numerical solution of an SDE becomes negative while the exact solution stays positive. The dashed line shows the solution of the equation with the noise term removed, starting at $(ih, \hat{X}_{ih})$. A method such as the Euler–Maruyama scheme will choose the next approximation point $((i + 1)h, \hat{X}_{(i+1)h})$ by following the tangent of this line and then adding the random term coming from the Brownian increment. If h is large enough, this can lead to $\hat{X}_{(i+1)h} < 0$. The thin, rough line gives a path of the exact SDE, starting $(ih, \hat{X}_{ih})$, for comparison.

If this effect causes problems, for example in cases when μ or σ are only defined for $x > 0$, there are several ad hoc ways to force the numerical solution to be positive:

- One can replace negative values of $\hat{X}_{ih}$ with $|\hat{X}_{ih}|$ throughout the algorithm. For example, for the Euler method, one could replace the update step by

$$\hat{X}_{(i+1)h} = \left| \hat{X}_{ih} + \mu(ih, \hat{X}_{ih}) h + \sigma(ih, \hat{X}_{ih}) \Delta B_i \right|,$$

with the modulus added to keep the solution positive. This method may introduce a bias, because solutions are only ever modified to take larger values.

- One can consider $Y_t = \log(X_t)$ instead of X. Since the function $f(x) = \log(x)$ has derivatives $f'(x) = 1/x$ and $f''(x) = -1/x^2$, Ito's formula in the form of equation (6.14) can be used to derive and SDE for Y:

$$dY_t = d(\log(X_t))$$
$$= \left(\frac{\mu(t, X_t)}{X_t} - \frac{\sigma(t, X_t)^2}{2X_t^2} \right) dt + \frac{\sigma(t, X_t)}{X_t} dB_t$$
$$= \left(\frac{\mu(t, e^{Y_t})}{e^{Y_t}} - \frac{\sigma(t, e^{Y_t})^2}{2e^{2Y_t}} \right) dt + \frac{\sigma(t, e^{Y_t})}{e^{Y_t}} dB_t$$
$$= \tilde{\mu}(t, Y_t) dt + \tilde{\sigma}(t, Y_t) dB_t,$$

where

$$\tilde{\mu}(t, y) = \frac{\mu(t, e^y)}{e^y} - \frac{\sigma(t, e^y)^2}{2e^{2y}} \quad \text{and} \quad \tilde{\sigma}(t, y) = \frac{\sigma(t, e^y)}{e^y}.$$

Another class of problems of numerical solutions of SDEs comes from the lack of stability of some numerical methods: it can happen that the numerical solution 'explodes', while the exact solution of the SDE stays bounded. This effect is illustrated in Figure 6.11.

We illustrate the problem of numerical instability here with the help of an example. Consider the one-dimensional SDE

$$dX_t = -X^3 dt + dB_t.$$

The Euler–Maruyama scheme with step size $h > 0$ for this SDE is given by

$$\hat{X}_{(i+1)h} = \hat{X}_{ih} - \hat{X}_{ih}^3 h + \Delta B_i = \hat{X}_{ih} \left(1 - \hat{X}_{ih}^2 h \right) + \Delta B_i.$$

Assume now that the discretised solution reaches by chance a state $\hat{X}_{ih} = x$ with

$$|x| \geq \sqrt{\frac{2 + \varepsilon}{h}}$$

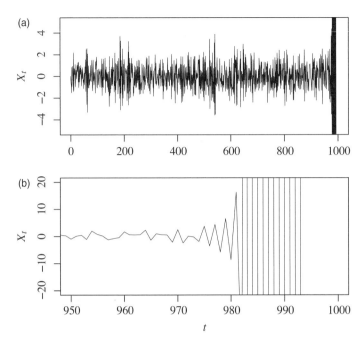

Figure 6.11   Numerical solution of an SDE where the Euler scheme is unstable. (a) A simulation of the SDE until the method becomes unstable. (b) A 'zoomed in' view of the last few steps before the numerical solution explodes.

for some $\varepsilon > 0$. Then we have $\hat{X}_{ih}^2 h \geq 2 + \varepsilon$. If h is small, for this x the drift $-x^3$ will be much bigger than the diffusion term ΔB_i and thus we find

$$\left|\hat{X}_{(i+1)h}\right| \approx \left|\hat{X}_{ih}\right|\left|1 - \hat{X}_{ih}^2 h\right| \geq (1 + \varepsilon)\left|\hat{X}_{ih}\right|,$$

where $\hat{X}_{(i+1)h}$ and $\hat{X}_{ih}$ have opposite signs. Since this implies $\hat{X}_{(i+1)h} > \hat{X}_{ih}$, the same argument applies again for the next step of the discretisation, and we find

$$\left|\hat{X}_{(i+k)h}\right| \gtrsim (1 + \varepsilon)^k\left|\hat{X}_{ih}\right|$$

and consequently $\hat{X}_{(i+k)h}$ diverges exponentially fast. While in theory the values $\hat{X}_{(i+k)h}$ stay finite (but very big) numbers, in a numerical simulation the resulting numbers will quickly leave the range of values which can be represented on a computer.

In cases where numerical instability is a problem, the best solution is often to decrease the step size h of the discretisation scheme. With decreasing h, the probability that the approximation hits an unstable state decays normally very quickly, so that the instability, even when present theoretically, is not seen in practice for small enough h.

6.5   Monte Carlo estimates

One situation where numerical solutions of SDEs are employed is the computation of Monte Carlo estimates.

6.5.1   Basic Monte Carlo

Assume that $X = (X_t)_{t \in [0,T]}$ is given as the solution of an SDE and that we want to compute $\mathbb{E}(f(X))$. In order to estimate this quantity using Monte Carlo integration, we generate independent samples $X^{(n,1)}, X^{(n,2)}, \ldots, X^{(n,N)}$, using numerical approximations to X, obtained by repeatedly solving a discretised SDE with discretisation parameter n. Then we can use the approximation

$$\mathbb{E}(f(X)) \approx \frac{1}{N} \sum_{j=1}^{N} f(X^{(n,j)}) = Z_{n,N}. \tag{6.28}$$

Here we allow for the function f to depend on the whole path of X until time T. We can choose, for example,

$$f(X) = \sup_{t \in [0,T]} X_t$$

to get the maximum of a one-dimensional path, or $f(X) = |X_T|^2$ to get the second moment of the final point of the path.

As for all Monte Carlo methods, there is a trade-off between accuracy of the result and computational cost. One notable feature of Monte Carlo estimation for SDEs is that the result is not only affected by the Monte Carlo error, but also by discretisation error.

Lemma 6.20   The computational cost of computing the Monte Carlo estimate $Z_{n,N}$ from (6.28) is of order $C = \mathcal{O}(nN)$. The mean squared error of the estimate is

$$\mathrm{MSE}(Z_{n,N}) = \frac{\sigma^2}{N} + e_{\mathrm{weak}}(f)^2,$$

where $\sigma^2 = \mathrm{Var}\left(f(X_T^{(n)})\right)$ is the sample variance corresponding to the endpoint of the numerical solution of the SDE with discretisation parameter n and $e_{\mathrm{weak}}(f)$ is the weak error of the approximation scheme used.

Proof   The cost of computing a single solution $X^{(n,j)}$ with discretisation parameter n is of order $\mathcal{O}(n)$. For $Z_{n,N}$ we have to compute N such solutions and add them up, leading to a total cost of order $\mathcal{O}(nN)$.

By lemma 3.12, the mean squared error of the estimator $Z_{n,N}$ is

$$\text{MSE}(Z_{n,N}) = \mathbb{E}\left((Z_{n,N} - \mathbb{E}(f(X_T)))^2\right) = \text{Var}(Z_{n,N}) + \text{bias}(Z_{n,N})^2.$$

The variance of $Z_{n,N}$ is given by

$$\text{Var}(Z_{n,N}) = \frac{\text{Var}\left(f(X_T^{(n,1)})\right)}{N} = \frac{\sigma^2}{N}.$$

The bias of $Z_{n,N}$ can be found as

$$\begin{aligned}
\left|\text{bias}(Z_{n,N})\right| &= \left|\mathbb{E}(Z_{n,N}) - \mathbb{E}(f(X_T))\right| \\
&= \left|\mathbb{E}\left(f(X_T^{(n,1)})\right) - \mathbb{E}(f(X_T))\right| \\
&= e_{\text{weak}}(f).
\end{aligned}$$

Substituting these two expressions into the formula for the mean squared error completes the proof.   □

The variance of $Z_{n,N}$, given by the term σ^2/N, corresponds to Monte Carlo error. It decreases to 0 as the Monte Carlo sample size N increases. The bias of $Z_{n,N}$, corresponding to $e_{\text{weak}}(f)^2$ in the lemma, decreases to 0 as the grid parameter n increases.

Example 6.21   Consider the process

$$\begin{aligned}
dX_t &= -X_t\, dt + dB_t \\
X_0 &= 0.
\end{aligned} \tag{6.29}$$

Assume that we want to estimate the probability that the process X exceeds the level $c > 0$ before time T, that is we want to estimate

$$p = P\left(\sup_{t \in [0,T]} X_t > c\right).$$

A basic Monte Carlo estimate for this probability is obtained as follows:

(a) Choose a discretisation parameter n and let $h = T/n$.

(b) Simulate solutions $(X_{ih}^{(j)})_{i=0,1,\dots,n}$ of (6.29) for $j = 1, 2, \dots, N$. The Euler–Maruyama scheme (algorithm 6.12) can be used for this purpose.

(c) For each path $X^{(j)}$, determine

$$\mathbb{1}_{(c,\infty)}\left(\sup_i X_{ih}^{(j)}\right) = \begin{cases} 1 & \text{if } X_{ih}^{(j)} > c \text{ for at least one } i \in \{0, 1, 2, \ldots, n\} \\ 0 & \text{otherwise.} \end{cases}$$

(d) Compute the estimate p^{MC} for p as

$$p^{\text{MC}} = \frac{1}{N} \sum_{j=1}^{N} \mathbb{1}_{(c,\infty)}(\sup_i X_{ih}^{(j)}).$$

A numerical experiment, using 200000 simulated paths of SDE (6.29), results in the estimate

$$P\left(\sup_{t \in [0,5]} X_t > 3.2\right) \approx 1.45 \cdot 10^{-4}.$$

An estimated confidence interval for this probability, obtained ignoring the bias and only considering the standard deviation of the Monte Carlo samples, is $[0.92 \cdot 10^{-4}, 1.98 \cdot 10^{-4}]$.

We can optimise the parameters N and n to minimise the computational cost required to achieve a given error. For the Euler–Maruyama scheme and the Milstein scheme the rate of decay of the weak error is given in equation (6.25) and equation (6.26), respectively. The weak error typically decays as a/n for some constant $a > 0$. Thus, the mean squared error of the estimator $Z_{n,N}$ satisfies

$$\text{MSE}(Z_{n,N}) \approx \frac{\sigma^2}{N} + \frac{a^2}{n^2} \tag{6.30}$$

where n is the grid parameter used for discretising the SDE and N is the sample size for the Monte Carlo estimate. This error needs to be balanced against the cost of computing the estimate $Z_{n,N}$, given by

$$C(n, N) = bnN \tag{6.31}$$

for some constant $b > 0$.

Our aim is now to tune the parameters N and n to minimise the computational cost while, at the same time, keeping the mean squared error below a specified level. For this, we will use the following general result.

Theorem 6.22 (optimisation under constraints). Let $f, g_1, \ldots, g_n \colon \mathbb{R}^d \to \mathbb{R}$ be continuously differentiable functions. Define $C \subseteq \mathbb{R}^d$ by

$$C = \left\{x \in \mathbb{R}^d \mid g_1(x) = g_2(x) = \cdots = g_n(x) = 0\right\}$$

and let $x^* \in C$ be a global maximum or minimum of the restriction of f to the set C, that is $f(x^*) \geq f(x)$ for all $x \in C$. Then x^* satisfies

$$\nabla f(x^*) = \lambda_1 \nabla g_1(x^*) + \lambda_2 \nabla g_2(x^*) + \cdots + \lambda_n \nabla g_n(x^*) \tag{6.32}$$

for some $\lambda_1, \ldots, \lambda_n \in \mathbb{R}$. The numbers λ_i are called *Lagrange multipliers*.

While the criterion given in theorem 6.22 is only necessary but not sufficient, it can be used to identify candidates for maxima or minima. The relation (6.32) is a system of d equations and the constraints $g_i(x^*) = 0$ provide another n equations, giving $d + n$ equations in total. On the other hand, the unknown vector x^* has d components and we also have to identify the Lagrange multipliers $\lambda_1, \ldots, \lambda_n$. Thus the total number of unknowns, $d + n$, matches the number of equations. As a consequence, in many cases the resulting set of equations has only finitely many solutions and by a systematic search we can determine which of these solutions corresponds to the maximum or minimum of f.

Taking n and N to be continuous variables for simplicity, we can apply theorem 6.22 to find the minimum of the cost function (6.31) under the constraint $\text{MSE}(Z_{n,N}) = \varepsilon^2$, where the mean squared error is given by equation (6.30). For the partial derivatives, comprising the gradients in (6.32) we find

$$\frac{\partial}{\partial n} \text{MSE}(Z_{n,N}) = -\frac{2a^2}{n^3}, \qquad \frac{\partial}{\partial n} C(n, N) = bN$$

and

$$\frac{\partial}{\partial N} \text{MSE}(Z_{n,N}) = -\frac{\sigma^2}{N^2}, \qquad \frac{\partial}{\partial N} C(n, N) = bn.$$

Thus, the minimum of C under the constraint $\text{MSE}(Z_{n,N}) = \varepsilon^2$ satisfies the equations

$$-\frac{2a^2}{(n^*)^3} = \lambda \cdot bN^*, \qquad -\frac{\sigma^2}{(N^*)^2} = \lambda \cdot bn^*,$$

where λ is the Lagrange multiplier, as well as the constraint

$$\frac{\sigma^2}{N^*} + \frac{a^2}{(n^*)^2} = \varepsilon^2.$$

The unique solution of this system of equations is given by

$$n^* = \sqrt{3}\, a \cdot \frac{1}{\varepsilon}, \qquad N^* = \frac{3\sigma^2}{2} \cdot \frac{1}{\varepsilon^2}, \tag{6.33}$$

where $\lambda = -4\,\varepsilon^5/\sqrt{243}\,ab\sigma^2$, and substituting these values into (6.31) gives

$$C(n^*, N^*) = \frac{\sqrt{27}\,ab\sigma^2}{\varepsilon^3} = \mathcal{O}\left(1/\varepsilon^3\right). \qquad (6.34)$$

This is the optimal cost for Monte Carlo estimates of $\mathbb{E}\left(f(X_T)\right)$ with mean squared error ε^2, where X_T is the end-point of the solution of an SDE.

From equation (6.33), assuming that σ^2 and a are both of order 1, we find that the optimal way to balance the parameters N and n is to choose n approximately equal to $\sqrt{N}$. While the variance, controlled by N, is easy to determine numerically, the bias, controlled by n, is difficult to measure. For this reason, in practice it might make sense to choose n bigger than $\sqrt{N}$ in order to avoid the risk of an unnoticed bias affecting the Monte Carlo estimates for SDEs.

6.5.2   Variance reduction methods

The error in Monte Carlo estimates such as (6.28) can be significantly reduced by employing the variance reduction techniques from Section 3.3.

6.5.2.1   Antithetic paths

Pairs of antithetic paths can be easily generated for SDEs by using the fact that, if B is a Brownian motion, $-B$ is also a Brownian motion (see lemma 6.5): thus, solving the SDE using the Brownian motions B and $B' = -B$ gives rise to two paths X and X' of the SDE. Since B_t and B_t' are negatively correlated, typically X_t and X_t' will also be negatively correlated for all $t \in [0, T]$.

Example 6.23   For illustration, Figure 6.12 (a) shows two paths of the SDE

$$\begin{aligned} dX_t &= (0.8 - X_t)\,dt + X_t(1 - X_t)\,dB_t \\ X_0 &= 0, \end{aligned} \qquad (6.35)$$

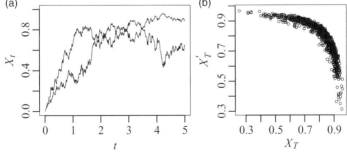

Figure 6.12   Illustration of the antithetic variables method for SDEs. (a) Two paths X and X' of SDE (6.35), computed using Brownian paths B and $-B$. (b) A scatter plot of 1000 pairs (X_T, X_T'), illustrating that the values are negatively correlated.

one path X computed from a Brownian path B and one path X' computed from $-B$. One can clearly see that often, if one path moves up, the other path moves down. Figure 6.12(b) shows a scatter plot of 1000 pairs (X_T, X'_T), clearly illustrating that the values are negatively correlated. In this example, the numerical value for the correlation coefficient is -0.787. From proposition 3.14 and proposition 3.27 we see that for fixed sample size N, the root-mean squared error of this antithetic variables method satisfies

$$\text{RMSE}^{\text{AV}} = \sqrt{1 - 0.787}\,\text{RMSE}^{\text{MC}} = 0.461\,\text{RMSE}^{\text{MC}}.$$

Thus, for the same number of SDEs solved, the antithetic variables method in this example has less than half the error than the corresponding basic Monte Carlo estimate.

6.5.2.2 Importance sampling

In this section we will study how to apply the importance sampling method from Section 3.3.1 to Monte Carlo estimates for SDEs. At first this seems an impossible task, since the solution of an SDE is a *random function* and not just a random number, but it transpires that the method still works in many cases.

Consider solutions to the two SDEs

$$dX_t = \mu(t, X_t)\,dt + \sigma(t, X_t)\,dB_t \tag{6.36}$$

and

$$dY_t = \tilde{\mu}(t, Y_t)\,dt + \sigma(t, Y_t)\,dB_t, \tag{6.37}$$

on the time interval $t \in [0, T]$, where $B = (B_t)_{t \geq 0}$ is a d-dimensional Brownian motion, $\mu, \tilde{\mu} \colon [0, \infty) \times \mathbb{R}^d \to \mathbb{R}^d$ are different drift functions and the common diffusion coefficient $\sigma \colon [0, \infty) \times \mathbb{R}^d \to \mathbb{R}^{d \times d}$ is invertible. We assume that both processes start at the same initial value $X_0 = Y_0$.

Importance sampling for SDEs is based on the following consequence of Girsanov's theorem (see e.g. Mao, 2007, theorem 2.2). Here we state the result without proof.

Theorem 6.24   Assume that the SDEs (6.36) and (6.37) have solutions X and Y, respectively, up to time $T > 0$. Let f be a function which maps paths $X \colon [0, T] \to \mathbb{R}^d$ to real numbers. Then, under additional technical assumptions, we have

$$\mathbb{E}(f(X)) = \mathbb{E}\left(f(Y)\frac{\varphi(Y)}{\psi(Y)}\right), \tag{6.38}$$

where

$$\varphi(Y) = \exp\left(\int_0^T a(t, Y_t)^{-1} \mu(t, Y_t) \, dY_t - \frac{1}{2} \int_0^T \mu(t, Y_t)^\top a(t, Y_t)^{-1} \mu(t, Y_t) \, dt \right)$$

and

$$\psi(Y) = \exp\left(\int_0^T a(t, Y_t)^{-1} \tilde{\mu}(t, Y_t) \, dY_t - \frac{1}{2} \int_0^T \tilde{\mu}(t, Y_t)^\top a(t, Y_t)^{-1} \tilde{\mu}(t, Y_t) \, dt \right)$$

with

$$a(t, x) = \sigma(t, x)\sigma(t, x)^\top \in \mathbb{R}^{d \times d}.$$

The relation (6.38) from the theorem replaces (3.13) in the basic importance sampling method. When the theorem is used for importance sampling, the paths Y will be generated using an approximation scheme for the SDE (6.37), and the integrals in the expressions for φ and ψ are evaluated using the approximations (6.9) and (6.10).

An important special case is the situation where the SDEs for X and Y are one-dimensional, μ and $\tilde{\mu}$ do not depend on time, and σ is constant. In this case, the expression for φ simplifies to

$$\varphi(Y) = \exp\left(\frac{1}{\sigma^2} \int_0^T \mu(Y_t) \, dY_t - \frac{1}{2\sigma^2} \int_0^T \mu(Y_t)^2 \, dt \right)$$

and ψ is the corresponding expression obtained by replacing μ with $\tilde{\mu}$. Thus, the factor $\varphi(Y)/\psi(Y)$ from (6.38) can be found as

$$\frac{\varphi(Y)}{\psi(Y)} = \exp\left(\frac{1}{\sigma^2} \int_0^T (\mu(Y_t) - \tilde{\mu}(Y_t)) \, dY_t - \frac{1}{2\sigma^2} \int_0^T \left(\mu(Y_t)^2 - \tilde{\mu}(Y_t)^2 \right) \, dt \right).$$
$$(6.39)$$

Example 6.25 In example 6.21, we considered the probability that the solution of the SDE (6.29) exceeds level c before time T. Since the drift $\mu(x) = -x$ of this SDE drives the process towards 0, large values of X_t are very unlikely. Thus, when c is big, a large number of Monte Carlo samples is required to see a sufficient number of cases where $\mathbb{1}_{(c,\infty)}(\sup_i X_{ih}^{(j)}) = 1$.

To reduce the required number of samples, we can use importance sampling with samples from a modified process Y. The process Y should be chosen so that Y is 'similar' to X but that the event $\sup_{t \in [0,1]} Y_t > c$ happens with probability higher than p. For this example we will use solutions of the SDE

$$dY_t = -qY_t \, dt + dB_t$$
$$Y_0 = 0$$
$$(6.40)$$

with constant $q \in [0, 1)$. Since the drift is weaker, this process will stay less close to 0 and thus is more likely to exceed level c. The drift of Y is $\tilde{\mu}(x) = -qx$ and consequently we find $\mu(x) - \tilde{\mu}(x) = -x - (-qx) = -(1 - q)x$ as well as $\mu(x)^2 - \tilde{\mu}(x)^2 = x^2 - q^2 x^2 = (1 - q^2)x^2$. Substituting these expressions into (6.39), we find

$$\frac{\varphi(Y)}{\psi(Y)} = \exp\left(-(1 - q)\int_0^T Y_t \, dY_t - \frac{1 - q^2}{2}\int_0^T Y_t^2 \, dt\right).$$

A numerical experiment, using 200000 simulated paths of SDE (6.40), results in the estimate

$$P\left(\sup_{t \in [0,5]} X_t > 3.2\right) \approx 1.55 \cdot 10^{-4}.$$

An estimated confidence interval, obtained by considering the standard deviation of the Monte Carlo samples, for this probability is $[1.41 \cdot 10^{-4}, 1.70 \cdot 10^{-4}]$. Comparing these estimates with the corresponding results from example 6.21 shows that the importance sampling estimate, for the same number of samples, has significantly smaller error.

Another important variance reduction technique for Monte Carlo estimates for solutions of SDEs is described in the next section.

6.5.3 Multilevel Monte Carlo estimates

The variance reduction methods we have discussed so far are applications of the general methods from Section 3.3 to the problem of estimating expectations for paths of SDEs. In contrast, the multilevel Monte Carlo approach, discussed in the rest of this section, is specific to situations where discretisation error is involved.

Multilevel Monte Carlo methods, by cleverly balancing the effects of discretisation error and Monte Carlo error, allow us to reduce the computational cost required to compute an estimate with a given level of error. Let X be the solution of an SDE and let $\varepsilon > 0$. In equation (6.34) we have seen that the basic Monte Carlo estimate for $\mathbb{E}(f(X_T))$ requires computational cost of order $\mathcal{O}(1/\varepsilon^3)$ in order to bring the root-mean squared error down to $\varepsilon > 0$. We will see that multilevel Monte Carlo methods can reduce this cost to nearly $\mathcal{O}(1/\varepsilon^2)$.

Let

$$Y_i = f\left(X_T^{(n_i)}\right)$$

where $X_T^{(n_i)}$ is the approximation for X_T with discretisation parameter n_i. We assume that $n_1 < n_2 < \ldots < n_k$ so that the approximations Y_i get more accurate as i increases. Also, except for the smallest values of i we expect $Y_i \approx Y_{i-1}$, when computed from the same path of the underlying Brownian motion. Thus the two values will be strongly correlated. We will exploit this correlation by using a variant of the control variates method (see Section 3.3.3): we can expand Y_k in a telescopic sum as

$$Y_k = \sum_{i=1}^{k} (Y_i - Y_{i-1})$$

where $Y_0 = 0$. For large enough k we have

$$\mathbb{E}\left(f(X_T)\right) \approx \mathbb{E}\left(Y_k\right) = \sum_{i=1}^{k} \mathbb{E}\left(Y_i - Y_{i-1}\right).$$

Since we can obtain values of a Brownian path on the coarse and on the fine grid simultaneously, each of the terms on the right-hand side can be estimated using Monte Carlo integration: we get

$$\mathbb{E}\left(Y_i - Y_{i-1}\right) \approx \frac{1}{N_i} \sum_{j=1}^{N_i} \left(Y_i^{(i,j)} - Y_{i-1}^{(i,j)}\right).$$

The resulting multilevel Monte Carlo estimate for $\mathbb{E}\left(f(X_T)\right)$ is then

$$Z_{N_1,\dots,N_k} = \sum_{i=1}^{k} \frac{1}{N_i} \sum_{j=1}^{N_i} \left(Y_i^{(i,j)} - Y_{i-1}^{(i,j)}\right). \tag{6.41}$$

Here, $Y_i^{(i,j)} = f\left(X_T^{(n_i,i,j)}\right)$ and $Y_{i-1}^{(i,j)} = f\left(X_T^{(n_{i-1},i,j)}\right)$ where $X^{(n_i,i,j)}$ and $X^{(n_{i-1},i,j)}$ are numerical solutions of the SDE with discretisation parameters n_i and n_{i-1}, respectively. The two solutions $X^{(n_i,i,j)}$ and $X^{(n_{i-1},i,j)}$ are both computed using the same Brownian motion $B^{(i,j)}$. The Brownian motions $B^{(i,j)}$ for different (i,j) are independent.

Algorithm 6.26   (multilevel Monte Carlo estimate)
   input:
     a function f
     an SDE on the time interval $t \in [0, T]$
     $k \in \mathbb{N}, n_1 < \cdots < n_k$ and $N_1, \dots, N_k \in \mathbb{N}$
   randomness used:
     independent Brownian paths $B^{(i,j)}$ for $i = 1, \dots, k, j = 1, \dots, N_i$
   output:
     an estimate for $\mathbb{E}\left(f(X_T)\right)$ where X solves the given SDE
   1: $s \leftarrow 0$
   2: **for** $i = 1, 2, \dots, k$ **do**
   3:   $s_i \leftarrow 0$

4:    **for** $j = 1, 2, \ldots, N_i$ **do**
5:        Generate a Brownian path $B^{(i,j)}$.
6:        Compute a solution $(X_t^{(n_i,i,j)})_{t \in [0,T]}$ of the SDE, using discretisation
         parameter n_i and the Brownian path $B^{(i,j)}$.
7:        $Y_i^{(i,j)} \leftarrow f\left(X_T^{(n_i,i,j)}\right)$
8:        **if** $i > 1$ **then**
9:            Compute a solution $(X_t^{(n_{i-1},i,j)})_{t \in [0,T]}$ of the SDE, using discretisation
             parameter n_{i-1} and the Brownian path $B^{(i,j)}$.
10:            $Y_{i-1}^{(i,j)} \leftarrow f\left(X_T^{(n_{i-1},i,j)}\right)$
11:        **else**
12:            $Y_{i-1}^{(i,j)} \leftarrow 0$
13:        **end if**
14:        $s_i \leftarrow s_i + Y_i^{(i,j)} - Y_{i-1}^{(i,j)}$
15:    **end for**
16:    $s \leftarrow s + s_i/N_i$
17: **end for**
18: return s

A convenient choice for the discretisation parameters n_i in the algorithm is $n_i = m^i$. Then $X^{(n_i,i,j)}$ and $X^{(n_{i-1},i,j)}$ can be simulated jointly, where $X^{(n_i,i,j)}$ uses time steps of length T/n_i and m of these time steps together form one time step for the simulation of $X^{(n_{i-1},i,j)}$.

Lemma 6.27   The computational cost for obtaining the multilevel Monte Carlo estimate $Z_{N_1,\ldots,N_k}$ from (6.41) is

$$C(N_1, \ldots, N_k) \propto \sum_{i=1}^{k} n_i N_i.$$

The mean squared error of the estimate satisfies

$$\text{MSE}(Z_{N_1,\ldots,N_k}) = \sum_{i=1}^{k} \frac{\sigma_i^2}{N_i} + \left(e_{\text{weak}}^{(n_k)}(f)\right)^2,$$

where $\sigma_i^2 = \text{Var}(Y_i - Y_{i-1})$ and $e_{\text{weak}}^{(n_k)}(f)$ is the weak error from (6.24) for discretisation parameter n_k.

Proof   The statement about the computational cost follows from the fact that the cost of simulating one path at level i is proportional to the discretisation parameter n_i.

For the statement about the mean squared error we first note that, by lemma 3.12, the mean squared error of the estimate is

$$\mathrm{MSE}(Z_{N_1,\ldots,N_k}) = \mathrm{Var}(Z_{N_1,\ldots,N_k}) + \mathrm{bias}(Z_{N_1,\ldots,N_k})^2.$$

For the variance we find

$$\mathrm{Var}(Z_{N_1,\ldots,N_k}) = \sum_{i=1}^{k} \frac{1}{N_i^2} \sum_{j=1}^{N_i} \mathrm{Var}\left(Y_i^{(i,j)} - Y_{i-1}^{(i,j)}\right) = \sum_{i=1}^{k} \frac{\sigma_i^2}{N_i}$$

and the bias is given by

$$\mathrm{bias}(Z_{N_1,\ldots,N_k}) = \mathbb{E}\left(Z_{N_1,\ldots,N_k}\right) - \mathbb{E}\left(f(X_T)\right)$$

$$= \sum_{i=1}^{k} \frac{1}{N_i} \sum_{j=1}^{N_i} \left(\mathbb{E}\left(Y_i^{(i,j)}\right) - \mathbb{E}\left(Y_{i-1}^{(i,j)}\right)\right) - \mathbb{E}\left(f(X_T)\right)$$

$$= \sum_{i=1}^{k} \left(\mathbb{E}\left(Y_i\right) - \mathbb{E}\left(Y_{i-1}\right)\right) - \mathbb{E}\left(f(X_T)\right)$$

$$= \mathbb{E}\left(f(X_T^{(n_k)})\right) - \mathbb{E}\left(f(X_T)\right).$$

Substituting the definition of $e_{\mathrm{weak}}^{(n_k)}(f)$ completes the proof. $\qquad\square$

The parameters $k \in \mathbb{N}$ and $N_1, \ldots, N_k$ can be chosen to minimise the numerical error for a given computational cost C. Since the weak error $e_{\mathrm{weak}}^{(n_i)}(f)$ is independent of the values $N_1, \ldots, N_k$, it suffices to minimise

$$V(N_1, \ldots, N_k) = \sum_{i=1}^{k} \frac{\sigma_i^2}{N_i}$$

under the constraint of fixed cost $C(N_1, \ldots, N_k) = C$. For simplicity, we take $N_1, \ldots, N_k$ to be continuous variables. Then, using minimisation under constraints (theorem 6.22), we find that the minimum satisfies the equations

$$0 = \frac{\partial V}{\partial N_i}(N_1, \ldots, N_k) - \lambda \frac{\partial C}{\partial N_i}(N_1, \ldots, N_k) = -\frac{\sigma_i^2}{N_i^2} - \lambda c n_i$$

for $i = 1, 2, \ldots, k$, where λ is the Lagrange multiplier. The solution of these equations is

$$N_i \sim \sqrt{\sigma_i^2 / n_i} \qquad (6.42)$$

for all $i = 1, 2, \ldots, k$, where the common constant is chosen so that the condition $C(N_1, \ldots, N_k) = C$ is satisfied.

Since

$$Y_i - Y_{i-1} = \left(f(X_T^{(n_i)}) - f(X_T) \right) - \left(f(X_T^{(n_{i-1})}) - f(X_T) \right),$$

the variances σ_i^2 depend on the path-wise accuracy of the approximations $X_T^{(n_i)}$ and $X_T^{(n_{i-1})}$ for X_T (i.e. on the strong error of the numerical scheme for solving the SDE), as well as on the regularity of the function f. For the Euler–Maruyama scheme and Lipschitz continuous f, one can show $\sigma_i^2 = \mathcal{O}(1/n_i)$.

Example 6.28   Let $n_i = m^i$ for some $m \in \mathbb{N}$ and assume $\sigma_i^2 \sim 1/n_i$ as well as $e_{\text{weak}}^{(n)} \sim 1/n$. The condition on the weak error is, for example, satisfied for the Euler–Maruyama scheme and the Milstein scheme. In this case, the optimality condition (6.42) turns into $N_i \sim 1/n_i$.

For $\varepsilon > 0$, set

$$k \approx \frac{\log(\varepsilon^{-1})}{\log(m)} \quad \text{and} \quad N_i \sim \frac{k}{n_i \varepsilon^2}.$$

Here, k is chosen in order to get a bias of order $\mathcal{O}(\varepsilon^2)$ and the form of the sample sizes N_i is dictated by (6.42). Then we have

$$\sum_{i=1}^{k} \frac{\sigma_i^2}{N_i} \sim \sum_{i=1}^{k} \frac{1}{n_i} \cdot \frac{n_i \varepsilon^2}{k} = \varepsilon^2$$

and

$$\left(e_{\text{weak}}^{(n_k)}(f) \right)^2 \sim \frac{1}{n_k^2} = m^{-2k} \approx \exp\left(-2\frac{\log(\varepsilon^{-1})}{\log(m)} \log(m) \right) = \varepsilon^2$$

and thus, by lemma 6.27, the mean squared error satisfies

$$\text{MSE}(Z_{N_1,\ldots,N_k}) \sim \varepsilon^2,$$

On the other hand, the computational cost of the resulting estimate is

$$C(N_1, \ldots, N_k) \sim \sum_{i=1}^{k} n_i N_i \sim \sum_{i=1}^{k} n_i \frac{k}{n_i \varepsilon^2} = \frac{k^2}{\varepsilon^2} \sim \frac{\log(\varepsilon)^2}{\varepsilon^2}.$$

Since $\log(\varepsilon)^2$ grows only very slowly as $\varepsilon \downarrow 0$, the cost of the method is nearly $\mathcal{O}(1/\varepsilon^2)$. For small values of ε, this computational cost will be much lower than the cost $\mathcal{O}(1/\varepsilon^3)$, from (6.34), for the basic Monte Carlo estimate with the same error.

The results from this section suggest the following procedure for choosing the parameters of a multilevel Monte Carlo method: first, choose n_k large enough that the weak error of solutions with discretisation parameter n_k is expected to be smaller than the level of error we are willing to tolerate. Next, choose discretisation parameters $n_1 < n_2 < \cdots < n_k$ for the intermediate levels. Normally, these value are chosen so that $n_i \approx m n_{i-1}$ for all i and some constant m. Experiments (e.g. example 6.30) indicate that it is best not to use too many intermediate levels. Finally, let $N_i \approx L/n_i$ where the constant $L \geq n_k$ is chosen large enough to keep the Monte Carlo variance of the samples small. The execution time of the program will be proportional to L, the Monte Carlo variance will decay as $1/L$.

6.6   Application to option pricing

In this section we illustrate the techniques from Chapter 6 with the help of an example from financial mathematics. In the Heston model (Heston, 1993), the joint evolution of a stock price S_t and of the corresponding volatility $\sqrt{V_t}$ at time t is described by a system of two stochastic differential equations:

$$
\begin{aligned}
dS_t &= r S_t \, dt + S_t \sqrt{V_t} \, dB_t^{(1)}, \\
dV_t &= \lambda(\sigma^2 - V_t) \, dt + \xi \sqrt{V_t} \left(\rho \, dB_t^{(1)} + \sqrt{1 - \rho^2} \, dB_t^{(2)} \right).
\end{aligned}
\tag{6.43}
$$

Here, $r \geq 1$ (the interest rate), $\lambda > 0$, $\sigma^2 > 0$, $\xi > 0$ and the correlation $\rho \in (-1, 1)$ are fixed parameters and $B^{(1)}$ and $B^{(2)}$ are two independent Brownian motions. The SDEs start at time $t = 0$ (assumed to be the current time) with given values S_0 and V_0. Our aim is to numerically estimate the price of a call option with expiry time $T > 0$ and strike price $K > 0$. This price is given by the expectation

$$
C = \mathbb{E}\left(e^{-rT} \max(S_T - K, 0) \right),
\tag{6.44}
$$

where S_T is the solution of (6.43) at time T.

We start our analysis by rewriting (6.43) as a two-dimensional SDE. By lemma 6.4, the process $B_t = (B_t^{(1)}, B_t^{(2)})$ is a two-dimensional Brownian motion. Defining $X_t = (S_t, V_t)$, we have

$$
dX = \mu(X_t) \, dt + \sigma(X_t) \, dB_t
$$

with

$$
\mu(x) = \begin{pmatrix} r x_1 \\ \lambda(\sigma^2 - x_2) \end{pmatrix} \quad \text{and} \quad \sigma(x) = \begin{pmatrix} x_1 \sqrt{x_2} & 0 \\ \rho \xi \sqrt{x_2} & (1 - \rho^2)\xi \sqrt{x_2} \end{pmatrix}
$$

for all $x \in \mathbb{R}^2_+$. Thus, the system (6.43) of SDEs fits into the framework discussed in this chapter. To simulate solutions of this SDE numerically, we can use the Euler–Maruyama scheme described in algorithm 6.12. Specialised to (6.43), the algorithm has the following form.

Algorithm 6.29 (Euler–Maruyama scheme for the Heston model)
    input:
        interest rate $r \geq 1$
        parameters $\lambda > 0$, $\sigma^2 > 0$, $\xi > 0$ and $\rho \in (-1, 1)$
        initial values S_0, $V_0 > 0$
        time horizon $T \geq 0$
        discretisation parameter $n \in \mathbb{N}$
    randomness used:
        $\Delta B_i^{(j)} \sim \mathcal{N}(0, h)$ i.i.d. for $i = 0, \ldots, n - 1$ and $j \in \{1, 2\}$, where
        $h = T/n$
    output:
        approximate solutions $(S_{ih}^{(n)})_{i=0,\ldots,n}$ and $(V_{ih}^{(n)})_{i=0,\ldots,n}$ to (6.43)
1: $h \leftarrow T/n$
2: $S_0^{(n)} \leftarrow S_0$
3: $V_0^{(n)} \leftarrow V_0$
4: **for** $i = 0, 1, \ldots, n - 1$ **do**
5:    $S_{(i+1)h}^{(n)} \leftarrow S_{ih}^{(n)} + r S_{ih}^{(n)} h + S_{ih}^{(n)} \sqrt{V_{ih}^{(n)}}\, \Delta B_i^{(1)}$
6:    $V_{(i+1)h}^{(n)} \leftarrow V_{ih}^{(n)} + \lambda \left(\sigma^2 - V_{ih}^{(n)} \right) h + \xi \sqrt{V_{ih}^{(n)}} \left(\rho \Delta B_i^{(1)} + \sqrt{1 - \rho^2}\, \Delta B_i^{(2)} \right)$
6: **end for**
7: return $(S_{ih}^{(n)})_{i=0,1,\ldots,n}$ and $(V_{ih}^{(n)})_{i=0,1,\ldots,n}$

This algorithm can easily be implemented on a computer. One problem with this approach is that there is the possibility that the numerical approximations for S and V may take negative values. Details of this problem are discussed near the end of Section 6.4.4. Once negative values occur for V_t, the algorithm cannot continue since the next iteration will involve computing the value $\sqrt{V_t}$, which is not defined for $V_t < 0$. One solution to this problem is to replace $\sqrt{V_{ih}^{(n)}}$ with $\sqrt{|V_{ih}^{(n)}|}$ throughout the algorithm. Then the solution for V can still take negative values, but the algorithm can continue and due to the drift term the process V will return to positive values soon. A pair (S, V) of paths of a simulation using this algorithm is shown in Figure 6.13.

Now that we are able to simulate paths from the Heston model, our next aim is to use these paths to estimate the expectation C from (6.44). For this task, we employ Monte Carlo estimation as described in Section 6.5. Given a discretisation parameter n and a Monte Carlo sample size N, the basic Monte Carlo estimate can be computed as follows:

(a) Simulate solutions $(S_{ih}^{(j)})_{i=0,1,\ldots,n}$ to (6.43) for $j = 1, 2, \ldots, N$, independently, using algorithm 6.29 with step size $h = T/n$.

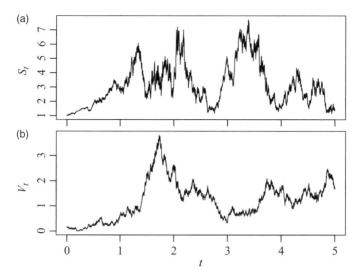

Figure 6.13   A pair (S_t, V_t) of paths from the Heston model (6.43). The evolution of (a) the stock price S_t and (b) the instantaneous variance V_t is given by the SDEs (6.43). The solutions were simulated using the Euler–Maruyama scheme from algorithm 6.29.

(b) Compute the estimate C^{MC} for C as

$$C^{MC} = \frac{1}{N} \sum_{j=1}^{N} f\left(S_T^{(j)}\right),$$

where $f(s) = e^{-rT} \max(s - K, 0)$ for all $s \in \mathbb{R}$.

As an example, the histogram in Figure 6.14 shows the distribution of 10000 samples of $S_T^{(j)}$. A Monte Carlo estimate for C can be obtained by applying f to each of these samples and then averaging the results.

From Section 6.5.1 we know that the computational cost of computing C^{MC} is proportional to nN and the mean squared error satisfies

$$\text{MSE}(C^{MC}) \approx \frac{\sigma^2}{N^2} + \left(e_{\text{weak}}^{(n)}(f)\right)^2, \tag{6.45}$$

where $\sigma^2 = \text{Var}\left(f(S_T^{(n)})\right)$ and $e_{\text{weak}}^{(n)}(f)$ is the weak error for the Euler–Maruyama discretisation with discretisation parameter n. From (6.33) we know that the optimal balance of N and n is approximately to choose $N \approx n^2$. While the weak error is difficult to determine, the mean squared Monte Carlo error σ^2/N^2 can be easily estimated together with C, by taking the sample variance of the Monte Carlo sample.

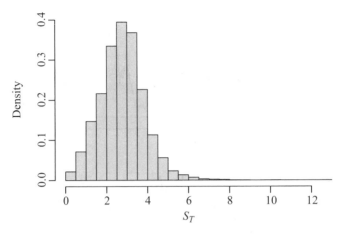

Figure 6.14   A histogram showing the distribution of the stock price S_T at time $T = 1$, obtained by generating $10\,000$ samples with initial values $S_0 = 1$, $V_0 = 0.16$ and parameters $r = 1.02$, $\lambda = 1$, $\sigma = 0.5$, $\xi = 1$ and $\rho = -0.5$.

In order to compute estimates faster or more accurately, we can use the variance reduction methods from Section 6.5.2. As an example, here we consider the multilevel Monte Carlo method (see Section 6.5.3). From lemma 6.27 we know that the computational cost of this method is proportional to $\sum_{i=1}^{k} n_i N_i$ and the mean squared error of this method satisfies

$$\text{MSE}(Z^{\text{MLMC}}) = \sum_{i=1}^{k} \frac{\sigma_i^2}{N_i} + \left(e_{\text{weak}}^{(n_k)}(f)\right)^2, \tag{6.46}$$

where n_i is the discretisation parameter on level i, N_i is the number of Monte Carlo samples on level i and σ_i^2 is the variance of these samples. From example 6.28 we know that for the Euler method the optimal balance between N_i and n_i is to choose N_i proportional to $1/n_i$.

As for the mean squared error of the basic Monte Carlo estimate, the bias, controlled by the weak error $e_{\text{weak}}^{(n_k)}(f)$, is difficult to estimate while the Monte Carlo variance $\sum_{i=1}^{k} \sigma_i^2/N_i$ can be estimated together with C at nearly no extra cost. If the discretisation parameter n for the Monte Carlo estimate is the same as the discretisation parameter n_k for the finest grid in the multilevel estimate, the two bias terms in (6.45) and (6.46) are identical. Thus, for $n = n_k$ the *difference* in mean squared error can easily be estimated numerically.

Example 6.30   We compare the Monte Carlo estimate with $n = 256$ and $N = 256^2$ to different multilevel Monte Carlo estimates, with $n_i = m^i$. The multilevel estimates we consider correspond to $m = 2$, $m = 4$ and $m = 16$. The corresponding numbers of levels are chosen so that $n_k = n$, that is we consider $k = 8$, $k = 4$ and $k = 2$.

The Monte Carlo sample sizes N_i are taken to be $N_i = L/n_i$, where L is chosen so that the computation time for the corresponding multilevel estimate is close to the computation time for the basic Monte Carlo estimate. The results are summarised as:

Method	m	Estimate	Error	Time
MC	—	0.1138	$0.7965 \cdot 10^{-6}$	1.000
MLMC	2	0.1137	$1.1570 \cdot 10^{-6}$	1.011
MLMC	4	0.1145	$0.5315 \cdot 10^{-6}$	1.004
MLMC	16	0.1133	$0.4532 \cdot 10^{-6}$	1.013

The column 'Time' gives the computation time (measured CPU time), relative to the time for the Monte Carlo estimate and 'Error' lists the Monte Carlo variance of the estimate.

The table clearly shows that, in this example, the mean squared error for $m = 4$ and $m = 16$ is smaller than the mean squared error for the basic Monte Carlo estimate, indicating that the estimate from these two methods will be more accurate than the basic Monte Carlo estimate, while taking the same amount of time to compute. The theory in Section 6.5.2 shows that the advantage of multilevel methods will become more pronounced as n increases; for larger n, the multilevel method with $m = 2$ will also become more accurate than the corresponding Monte Carlo method.

6.7   Summary and further reading

In this chapter we have discussed continuous-time stochastic processes, including Brownian motion and the solutions of SDEs. Following the topics of interest for this book, we have focused on computational aspects, leaving out most of the underlying theory. Many texts about continuous-time stochastic processes are available to provide more theoretical detail. Brownian motion is, for example, covered in Rogers and Williams (2000), Karatzas and Shreve (1991) and, in great detail, in Mörters and Peres (2010) and Borodin and Salminen (1996). SDEs are covered in textbooks such as Mao (2007), Karatzas and Shreve (1991) and Øksendal (2003), numerical methods are discussed in Kloeden and Platen (1999) and in the very accessible review by Higham (2001).

In the second part of the chapter, simulation of continuous-time processes gave the opportunity to review the Monte Carlo methods from Chapter 3 in a different context: we considered both basic Monte Carlo estimates and the variance reduction techiques from Section 3.3 for continuous-time models. Finally, we have studied the multilevel Monte Carlo method which is well-adapted to the problem of computing Monte Carlo estimates in the presence of discretisation errors. More details about multilevel Monte Carlo estimates can be found in the publications by Giles (2008a,b).

Exercises

E6.1 Write a program to simulate the values of a one-dimensional Brownian motion at times $t_1 < t_2 < \ldots < t_n$. Test your program by plotting the graph of one path of a Brownian motion.

E6.2 Write a program to simulate a path of a two-dimensional Brownian motion. Test your program by generating a plot of one such path, similar to the one in Figure 6.2.

E6.3 Write a function which takes time t as an argument and computes the value B_t of a Brownian path. Repeated calls to the function should return samples for *the same* Brownian path.

E6.4 Write a program to sample paths of a one-dimensional Brownian bridge between (r, a) and (t, b), where $r < t$ are times and $a, b \in \mathbb{R}$.

E6.5 Write a program to simulate paths of a geometric Brownian motion X until a time $T > 0$, for given parameters α, β, initial value X_0, and time horizon T. Test your program by generating plots of X for different values of α and β.

E6.6 Use Monte Carlo integration with a sample size of $N = 10^6$ to estimate the expectation of $X_t = \exp(B_t - t/2)$ for $t \geq 0$. By experiment, determine the range of t-values where this method results in reasonable estimates.

E6.7 Implement the Euler–Maruyama scheme for computing an approximate solution of the SDE (6.17). Test your program by generating plots of solutions X with drift

$$\mu(t, x_1, x_2) = \begin{pmatrix} x_2 \\ x_1(1 - x_1^2) \end{pmatrix}$$

and diffusion coefficient

$$\sigma(t, x_1, x_2) = s(x_1, x_2)\begin{pmatrix} 1 & 0 \\ 0 & 1 \end{pmatrix}$$

where

$$s(x_1, x_2) = \begin{cases} 0.2 & \text{if } x_1, x_2 > 0 \text{ and} \\ 0.03 & \text{otherwise.} \end{cases}$$

E6.8 Implement the Milstein scheme for computing approximate solutions to SDEs of the form (6.21). Test your program by simulating solutions of

$$dX_t = 0.1 \, dt + X_t \, dB$$

with initial condition $X_0 = 1$.

E6.9 Implement a Monte Carlo method to estimate the strong error for the Euler–Maruyama and Milstein schemes, when simulating paths of a geometric Brownian motion. Use the resulting program to recreate the plot in Figure 6.8.

E6.10 Implement a program to use Monte Carlo estimation to estimate the weak error of the Euler–Maruyama and Milstein methods, as described in example 6.19. Use your program to recreate the plot in Figure 6.9. Note, the resulting program will probably take a long time to run!

E6.11 Use Monte Carlo estimation to estimate the probability that the solution X of the SDE (6.29) exceeds the level $c = 2.5$ before time $T = 5$. Determine the approximate root-mean squared error of your estimate.

E6.12 Use importance sampling to estimate the probability that the solution X of the SDE (6.29) exceeds the level $c = 2.5$ before time $T = 5$. Determine an approximate confidence interval for the probability in question.

E6.13 Write a program to simulate paths $(S_t)_{t \in [0,T]}$ and $(V_t)_{t \in [0,T]}$ from the Heston model (6.43).

E6.14 Write a program to compute Monte Carlo estimates for the price C of a call option with expiry time $T > 0$ and strike price $K > 0$, where the underlying stock is described by the Heston model (6.43).

 Test your program by determining an estimate for C for the case where $S_0 = 1$ and $V_0 = 0.16$, and where the parameters in the Heston model are $r = 1.02$, $\lambda = 1$, $\sigma = 0.5$, $\xi = 1$ and $\rho = -0.5$. Ignoring the bias, estimate the mean squared Monte Carlo error of your estimate.

E6.15 Write a program to compute multilevel Monte Carlo estimates for the price C of a call option with expiry time $T > 0$ and strike price $K > 0$, where the underlying stock is described by the Heston model (6.43). Test your program by comparing the result with the result from exercise E6.14.

Appendix A

Probability reminders

In this text we assume that the reader is familiar with the basic results of probability. For reference, and to fix notation, this chapter summarises some important concepts and results.

A.1   Events and probability

The basic objects in probability theory are events and random variables. Typically *events* are denoted by capital letters such as A, B and C and we write $P(A)$ for the probability that an event A occurs. *Random variables* are typically denoted by upper case letters such as X, Y, Z and they can take values either in the real numbers $\mathbb{R}$, in the Euclidean space $\mathbb{R}^d$ or even in more general spaces. Random variables are often used to construct events; for example $\{X < 3\}$ denotes the event that X takes a value which is smaller than 3 and we write $P(X < 3)$ for the probability of this event. If X is a random variable taking values in some set, and if f is a function on this set, then $f(X)$ is again a random variable.

Each random variable has a *distribution* or *probability distribution*, which completely describes its probabilistic behaviour. Special probability distributions are often designated by calligraphic upper case letters, sometimes with parameters given in brackets, for example $\mathcal{N}(\mu, \sigma^2)$ for the normal distribution with mean μ and variance σ^2 or $\mathcal{U}[a, b]$ for the uniform distribution on the interval $[a, b]$. General distributions are often designated by P or μ. We write

$$X \sim P$$

to state that a random variable X has distribution P. For real-valued random variables, the distribution can always be completely described by a distribution function.

An Introduction to Statistical Computing: A Simulation-based Approach, First Edition. Jochen Voss.
© 2014 John Wiley & Sons, Ltd. Published 2014 by John Wiley & Sons, Ltd.

Definition A.1   The *cumulative distribution function* (CDF) of a random variable X on $\mathbb{R}$ is given by

$$F(a) = P(X \leq a)$$

for all $a \in \mathbb{R}$.

Often the CDF of a random variable is simply referred to as the 'distribution function', omitting the term 'cumulative' for brevity. Distribution functions are normally denoted by capital letters such as F or F_X (where the subscript denotes which random variable this is the CDF of), occasionally Greek letters such as Φ are used. We sometimes write $X \sim F$ (slightly abusing notation) to indicate that the distribution of X has distribution function F.

Definition A.2   A random variable X has *probability density* f, if

$$P(X \in A) = \int_A f(x)\, dx \tag{A.1}$$

for every set A.

Often the probability density of a random variable X is referred to just as the *density* of X. While every random variable on $\mathbb{R}$ has a distribution function F, a density f may or may not exist. If f exists, then it can be found as the derivative of the CDF, that is $f = F'$. Probability densities are typically denoted by Roman or Greek letters such as f, g, p, φ and ψ.

One important property of probability densities is that they are not uniquely defined. If the density φ is only changed on sets which are small enough to not affect the value of the integral in (A.1), for example in a finite number of points, the changed density will still describe the same distribution.

Example A.3   The *uniform* distribution $\mathcal{U}[0, 1]$ on the interval $[0, 1]$ has density

$$f(x) = \begin{cases} 1 & \text{if } x \in [0, 1] \text{ and} \\ 0 & \text{otherwise.} \end{cases}$$

Since we can change f at individual points without changing the distribution, we can equivalently use densities such as

$$\tilde{f}(x) = \begin{cases} 1 & \text{if } x \in (0, 1) \text{ and} \\ 0 & \text{otherwise} \end{cases}$$

to describe the distribution $\mathcal{U}[0, 1]$. The corresponding distribution function is

$$F(a) = \int_{-\infty}^{a} f(x)\, dx = \begin{cases} 0 & \text{if } a < 0 \\ a & \text{if } 0 \leq a < 1 \text{ and} \\ 1 & \text{otherwise.} \end{cases}$$

Note that the CDF F is uniquely determined; we get the same result if we compute F using the alternative density $\tilde{f}$.

Example A.4   The *standard normal distribution* $\mathcal{N}(0, 1)$ has density

$$f(x) = \frac{1}{\sqrt{2\pi}} e^{-x^2/2}$$

and distribution function

$$F(x) = \frac{1}{\sqrt{2\pi}} \int_{-\infty}^{x} e^{-y^2/2} \, dy.$$

The integral in the CDF cannot be evaluated explicitly, but many programming languages provide functions to evaluate F numerically.

Example A.5   The *exponential distribution* $\text{Exp}(\lambda)$ has density

$$f(x) = \begin{cases} \lambda e^{-\lambda x} & \text{if } x \geq 0 \text{ and} \\ 0 & \text{if } x < 0. \end{cases}$$

The distribution function is

$$F(x) = \begin{cases} 1 - e^{-\lambda x} & \text{if } x \geq 0 \text{ and} \\ 0 & \text{if } x < 0. \end{cases}$$

Example A.6   The value $X \in \{1, 2, 3, 4, 5, 6\}$ of a single dice throw has no density. Its distribution function is

$$F(x) = \begin{cases} 0 & \text{for } x < 1 \\ 1/6 & \text{for } 1 \leq x < 2 \\ 2/6 & \text{for } 2 \leq x < 3 \\ 3/6 & \text{for } 3 \leq x < 4 \\ 4/6 & \text{for } 4 \leq x < 5 \\ 5/6 & \text{for } 5 \leq x < 6 \text{ and} \\ 1 & \text{for } 6 \leq x. \end{cases}$$

The most important properties of densities are given by the following characterisation: a function f is a probability density if and only if it satisfies the two properties:

(a) $f \geq 0$; and

(b) f is integrable with $\int f(x) \, dx = 1$.

Sometimes a density f is only known up to a constant Z, that is we know the function $g(x) = Zf(x)$ for all x, but we do not know f and the value of Z. In these cases, the second property listed can be used to find Z: if we let

$$Z = Z \int f(x)\,dx = \int Zf(x)\,dx = \int g(x)\,dx,$$

then

$$f(x) = \frac{1}{Z}g(x)$$

is a probability density. In this context, Z is called the *normalisation constant* for the unnormalized density g.

Definition A.7   Two random variables X and Y are *independent* of each other if

$$P(X \in A, Y \in B) = P(X \in A)P(Y \in B)$$

for all sets A and B. More generally, random variables $X_1, \ldots, X_n$ are independent if

$$P\big(X_i \in A_i \text{ for } i = 1, \ldots, n\big) = \prod_{i=1}^{n} P(X_i \in A_i)$$

for all sets $A_1, \ldots, A_n$.

If the random variable X_i has density f_i for $i = 1, 2, \ldots, n$, then we can characterise independence of the X_i via their densities: $X_1, \ldots, X_n$ are independent if and only if the joint density f of $X_1, \ldots, X_n$ is of the form

$$f(x_1, \ldots, x_n) = \prod_{i=1}^{n} f(x_i).$$

A.2   Conditional probability

The conditional probability of an event A, given another event B with $P(B) > 0$, is defined as

$$P(A|B) = \frac{P\big(A \cap B\big)}{P(B)}, \tag{A.2}$$

where $P(A \cap B)$ is the probability that both A and B occur simultaneously. The same relation multiplied by $P(B)$ is known as *Bayes' rule*:

$$P(A \cap B) = P(A|B)P(B). \tag{A.3}$$

Often, the event and condition in a conditional probability concern the distribution of a random variable X: if $P(X \in B) > 0$ we can consider $P(X \in A | X \in B)$. For fixed B, the *conditional distribution* $P_{X|X \in B}$ of X given $X \in B$, defined by

$$P_{X|X \in B}(A) = P(X \in A | X \in B), \tag{A.4}$$

is itself a probability distribution.

The random variable X here can take values on an arbitrary space. By taking X to be a vector, $X = (X_1, X_2) \in \mathbb{R}^2$ say, and by choosing $A = A_1 \times \mathbb{R}$ and $B = \mathbb{R} \times B_2$, we get

$$P(X \in A | X \in B) = P(X_1 \in A_1 | X_2 \in B_2).$$

Using this idea, results such as proposition 1.26 can also be applied to the distribution of a random variable X, conditioned on the values of a different random variable Y.

If the pair $(X, Y) \in \mathbb{R}^m \times \mathbb{R}^n$ has a joint density $f(x, y)$, it is also possible to consider X conditioned on the event $Y = y$. Since the event $Y = y$ has probability 0, definition (A.2) can no longer be used; instead, one defines the *conditional density* of X given $Y = y$ as

$$f_{X|Y}(x|y) = \begin{cases} \frac{f(x,y)}{f_Y(y)} & \text{if } f_Y(y) > 0 \text{ and} \\ \pi(x) & \text{otherwise,} \end{cases} \tag{A.5}$$

where

$$f_Y(y) = \int f(\tilde{x}, y) \, d\tilde{x}$$

is the density of Y and π is an arbitrary probability density. The choice of π in the definition does not matter, since it is used only for the case $f_Y(y) = 0$, that is when conditioning on cases which never occur. The function $f_{X|Y}(x|y)$ defined in this way satisfies

$$\int f_{X|Y}(x|y) \, dx = 1$$

for all y and

$$\int_A \int_B f_{X|Y}(x|y) f_Y(y) \, dy \, dx = \int_A \int_B f(x, y) \, dy \, dx = P(X \in A, Y \in B).$$

The latter relation is the analogue of Bayes' rule (A.3) for conditional densities.

A.3   Expectation

Real-valued random variables X often (but not always) have an *expectation*, which we denote by $\mathbb{E}(X)$. If X only takes finitely many values, say $x_1, \ldots, x_n$, then the expectation of X is given by

$$\mathbb{E}(X) = \sum_{i=1}^{n} x_i\, P(X = x_i).$$

If the distribution of X has a density f, the expectation of X can be computed as

$$\mathbb{E}(X) = \int x\, f(x)\, dx.$$

Similarly, if φ is a function, the expectation of the random variable $\varphi(X)$ can be computed as

$$\mathbb{E}\big(\varphi(X)\big) = \int \varphi(x)\, f(x)\, dx. \tag{A.6}$$

In cases where the integral on the right-hand side can be solved explicitly, this formula allows expectations to be computed analytically.

Probabilities of events involving X can be rewritten as expectations using the *indicator function* of the event: the indicator function of the event $\{X \in A\}$ is the random variable $\mathbb{1}_A(X)$ given by

$$\mathbb{1}_A(x) = \begin{cases} 1 & \text{if } x \in A \text{ and} \\ 0 & \text{otherwise.} \end{cases} \tag{A.7}$$

Using the definition of the expectation, we find

$$\mathbb{E}(\mathbb{1}_A(X)) = 1 \cdot P(X \in A) + 0 \cdot P(X \notin A) = P(X \in A)$$

and from (A.6) we get

$$P(X \in A) = \mathbb{E}\big(\mathbb{1}_A(X)\big) = \int \mathbb{1}_A(x)\, f(x)\, dx,$$

where f is the density of X. Similarly, if $\varphi(X)$ is a function of X, we have

$$P\big(\varphi(X) \in A\big) = \mathbb{E}\Big(\mathbb{1}_A\big(\varphi(X)\big)\Big) = \int \mathbb{1}_A\big(\varphi(x)\big)\, f(x)\, dx. \tag{A.8}$$

A.4   Limit theorems

In this section we cite, for reference, two well-known limit theorems which all Monte Carlo methods rely on heavily: the law of large numbers and the central limit theorem.

The strong law of large numbers allows to approximate an expectation by the average of a large i.i.d. sample. This approximation forms the basis of the Monte Carlo methods discussed in Chapter 3.

Theorem A.8   (strong law of large numbers). Let $(X_n)_{n \in \mathbb{N}}$ be a sequence of i.i.d. random variables with expectation μ. Then

$$\lim_{n \to \infty} \frac{1}{n} \sum_{i=1}^{n} X_i = \mu$$

with probability 1.

This theorem is due to Kolmogorov and it can be found in most textbooks about probability, for example as theorem 20.2 in Jacod and Protter (2000) or as 4.3S in Williams (2001).

From the law of large numbers we know that

$$\frac{1}{n} \sum_{i=1}^{n} (X_i - \mu) \longrightarrow 0$$

as $n \to \infty$. The central limit theorem, given below, is a refinement of this result: it describes the fluctuations around this limit. The central limit theorem can for example be used to analyse the error of numerical methods based on the law of large numbers.

Theorem A.9   (central limit theorem). Let $(X_n)_{n \in \mathbb{N}}$ be a sequence of independent random variables with expectation μ and finite variance $\sigma^2 > 0$. Then we have

$$\frac{1}{\sqrt{n}} \sum_{i=1}^{n} (X_i - \mu) \overset{d}{\longrightarrow} \mathcal{N}(0, \sigma^2), \tag{A.9}$$

where $\overset{d}{\longrightarrow}$ denotes convergence in law.

In the central limit theorem, the 'convergence in law' in equation (A.9) is equivalent to the statement

$$P\left(\frac{1}{\sqrt{n}} \sum_{i=1}^{n} (X_i - \mu) \in [a, b] \right) \longrightarrow \frac{1}{\sqrt{2\pi\sigma^2}} \int_a^b \exp\left(-\frac{x^2}{2\sigma^2} \right) dx$$

as $n \to \infty$ for all $a, b \in \mathbb{R}$ with $a \leq b$. Again, this result can be found in most textbooks about probability, for example as theorem 21.1 in Jacod and Protter (2000) or in Section 5.4 of Williams (2001).

A.5   Further reading

A concise and rigorous exposition of basic probability is given in Jacod and Protter (2000). A longer introduction, with a view towards applications in statistics, can be found in Williams (2001) and a classical text is the book by Feller (1968).

Appendix B

Programming in R

This appendix contains a short introduction to programming with the R programming language (R Development Core Team, 2011). R can be downloaded from the R homepage at `http://www.r-project.org/` and there are many online tutorials available which describe how to install and use the R system. Therefore, we assume that the reader has access to a working R installation and already knows how start the program and how to use the built-in help system. The presentation here will focus on aspects of the language which are relevant for statistical computing.

B.1   General advice

One of the most important points to understand when learning to program is the distinction between *learning to program* and *learning a programming language*. Learning to program involves understanding how to approach and structure a problem in order to turn it into an algorithm which a computer can execute. This process often requires that the problem is rephrased or broken down into smaller subproblems. Becoming a proficient programmer is a slow process which requires a lot of practice and which can take a long time. In contrast, once you already know how to program, learning a new programming language is relatively easy. Typically this just requires learning a small number (normally much less than 100) of commands and rules. This appendix gives an introduction to both, the basics of programming and of the programming language R.

While it may be possible to learn the mathematical contents of this book just by reading and understanding the text, programming can only be learned by practising a lot (just like learning to juggle or learning to play a musical instrument). One of the main reasons for this is that attention to detail is crucially important in computer programming. While a typo in a mathematical argument will normally just make the argument a bit more difficult to follow, a single typo in a computer program will typically cause the program to abort with an error or to return nonsensical results.

An Introduction to Statistical Computing: A Simulation-based Approach, First Edition. Jochen Voss.
© 2014 John Wiley & Sons, Ltd. Published 2014 by John Wiley & Sons, Ltd.

The text includes numerous exercises which allow you to practise your programming skills. Another way to train your programming skills is to read and understand other people's programs; but only after you have written the corresponding program yourself! For this purpose, the text includes answers for all of the exercises. But, again, it is not possible to learn to program by just reading programs (just as it is impossible to become a good violinist only by watching other people play the violin), so it is important that you try the exercises yourself before looking at the answers. Finally, and as already remarked, learning to program takes a long time, so make sure that you commit enough time to learning and practising.

B.2   R as a Calculator

In this section we discuss basic use of R and introduce some fundamental concepts. Many of the topics discussed here will be covered in more detail in the following sections.

You interact with the R system by entering textual commands, and the system then reacts on these commands. A simple example of a command is

```
3 + 4
```

This command asks R to add the numbers 3 and 4, the result 7 is printed to the screen. After you try the above command in R, you should see the following two lines:

```
> 3 + 4
[1] 7
```

Your command is preceded by > and the result is preceded by [1]. We use this distinction throughout the appendix: listings starting with the character > show both your commands and the resulting output. Listings not starting with the character > only show the commands without giving the output. You can (and should!) always run the commands yourself to see the result.

An example of a more complicated command is

```
plot(c(1,2,3,4), c(5,2,1,6), xlab="x", ylab="y", type="o")    (B.1)
```

Try this example yourself! If everything worked, a graph such as the one in Figure B.1 will appear on the screen.

R processes commands one at a time. Commands can be split over more than one line, for example a plot command could look as follows:

```
plot(c(1,2,3,4), c(5,2,1,6), type="o",
     xlab="really long label for the x-axis", ylab="y")
```

When you enter this command into R, nothing will happen when you enter the first line (since R notices that the first bracket is not yet closed, so the command cannot be

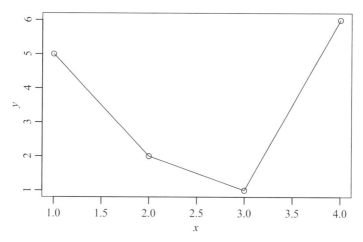

Figure B.1   The graph generated by the command in listing (B.1). The circles correspond to the x and y coordinates provided in the command, the axis labels are as specified.

complete). Only after the command is completed by the second line, the plot appears. For programming, we need to split the solution to a problem into steps which are small enough that each step can be performed in a single command.

B.2.1   Mathematical operations

Some of the simplest commands available are the ones which correspond directly to mathematical operations. In the first example of this section, we could just type 3 + 4 to compute the corresponding sum. Table B.1 lists the R equivalent of the most important mathematical operations.

B.2.2   Variables

In R you can use variables to store intermediate results of computations. The following transscript illustrates the use of variables:

```
> a <- 1
> b <- 7
> c <- 2
> root1 <- (-b + sqrt(b^2 - 4*a*c)) / (2*a)
> root2 <- (-b - sqrt(b^2 - 4*a*c)) / (2*a)
> root1
[1] -0.2984379
> root2
[1] -6.701562
> root1*root2
[1] 2
```

Table B.1   List of some commonly used mathematical operations in R.

Operation	Example	R code
Addition	$8 + 3$	8 + 3
Subtraction	$8 - 3$	8 - 3
Multiplication	$8 \cdot 3$	8 * 3
Division	$8/3$	8 / 3
Power	8^3	8 ^ 3
Modulus	$8 \bmod 3$	8 %% 3
Absolute value	$|x|$	abs(x)
Square root	$\sqrt{x}$	sqrt(x)
Exponential	e^x	exp(x)
Natural logarithm	$\log(x)$	log(x)
Sine	$\sin(2\pi x)$	sin(2 * pi * x)
Cosine	$\cos(2\pi x)$	cos(2 * pi * x)

You can freely choose names for your variables, consisting of letters, digits and the dot character (but starting with a letter), and you can assign values to variables using the assignment operator <-. After a value is assigned to a variable, the name of the variable can be used as a shorthand for the assigned value.

Since variable names in R cannot contain accented or greek characters, some creativity is required when translating a mathematical formula into R code: instead of X_k we could, for example, write Xk and instead of $\tilde{\alpha}$ we could, for example, write alpha.tilde.

There is a subtle difference between the use of variables in mathematics and in programming. While in mathematics expressions like $x = x + 1$ are not very useful, the corresponding expression x <- x+1 in R has a useful meaning:

```
> x <- 6
> x <- x + 1
> x
[1] 7
```

What happens here is that in the assignment x <- x+1, the right-hand side x + 1 is evaluated first: by the rules for the use of variables, x is replaced by its value 6, and then 6 + 1 is evaluated to 7. Once the value to be assigned is determined, this value (the 7) is assigned to x. Consequently, the effect of the command x <- x+1 is to increase the value stored in the variable x by 1.

An alternative notation for assigning a value to a variable in R is using the operator =. The R statements x <- 1 and x = 1 are equivalent. To avoid confusion with the equality sign from mathematics, we use the <- operator to assign values to variables throughout. One consequence of the fact that <- indicates an assignment, that is some action the computer will perform, is that the order of commands can matter when assignments are involved:

```
x <- 3
y <- x + 1
```

sets x to 3 and y to 4, whereas

```
y <- x + 1
x <- 3
```

also sets x to 3 but sets y to whatever value x previously had, plus 1. If the value of x has not been previously set, an error message to this effect will be shown when the line y <- x + 1 is executed.

The main use of variables is to store data and results of computations for later reference. This has several advantages: first, once a value is stored, the variable name can be used to refer to this value, potentially saving much typing and making the program shorter. If a descriptive name is chosen for the variable, this can also make the intention of a command much clearer to human readers. Secondly, if the result of a time-consuming computation is stored in a variable, the result from the variable can be reused without performing the computation again. Thus, sometimes the speed of a program can be greatly improved by storing results in variables instead of redoing the same computation over and over again.

B.2.3   Data types

Every object in R represents one of a handful of possible 'data types'. In the examples above we have already seen numbers and strings (short for 'character strings').

B.2.3.1   Numbers

The basic data type in R are numbers. Most of the time, numbers in R programs are written in the same way as that used in mathematics, but there are a few aspects to be aware of. These mostly relate to very small or very large numbers.

A special notation is used in R (and in many other programming languages) to denote multiplication with a power of 10: the R code xey, where x and y are ordinary numbers, stands for $x \cdot 10^y$. Thus, 1e6 is a shorthand notation for one million and 1.3e-5 stands for 0.000013. The number before the symbol e is allowed to be a decimal fraction, but the number after the e must be an integer. This notation is widely used and R uses it, for example, when it needs to print very large or very small numbers. For example the output of the following command tells us that e^{100} is approximately equal to $2.69 \cdot 10^{43}$:

```
> exp(100)
[1] 2.688117e+43
```

An important issue to keep in mind is the fact that R, like any programming language, has only limited precision when storing numbers: the internal representation of a number may be affected by *truncation error*, slightly changing the value of the number stored. As a result, very small positive numbers are rounded to 0 in the internal representation, and very large values are represented as ∞:

```
> exp(-10000)
[1] 0
> exp(10000)
[1] Inf
```

The number system in R can explicitly represent the numbers $+\infty$ and $-\infty$. The corresponding symbols in R are Inf and -Inf.

Truncation error in particular causes problems in situations where two large but nearly equal values are subtracted and nearly cancel. For example, if we subtract 10^{15} from $10^{15} + 0.2$ in R, we do not get the expected result 0.2, but 0.25 instead.

```
> (1e15 + 0.2) - 1e15
[1] 0.25
```

Similar problems can occur when very small, but approximately equal, positive numbers are divided. This type of cancellation can, for example, be encountered when evaluating the acceptance probability (4.3) of a Metropolis-Hastings algorithm. The best solution to problems caused by such cancellations is to rewrite the underlying mathematical expressions so that the cancellations in the program no longer occur. For the problem of evaluating the acceptance probability (4.3), this approach is illustrated in equation (4.19).

Finally, the special value NaN, short for 'not a number', is used to represent the outcome of calculation where the result is not mathematically defined.

```
> 0 / 0
[1] NaN
> log(-1)
[1] NaN
```

B.2.3.2   Vectors

Vector objects in R are useful to represent mathematical vectors in a program; they can also be used as a way to store data for later processing. An easy way to create vector objects is the function c (short for 'concatenate') which collects all its arguments into a vector. The vector

$$x = \begin{pmatrix} 1 \\ 2 \\ 3 \end{pmatrix}$$

can be represented in R as follows:

```
> c(1, 2, 3)
[1] 1 2 3
```

The elements of a vector can be accessed by using square brackets: if x is a vector, x[1] is the first element, x[2] the second element and so on:

```
> x <- c(7, 6, 5, 4)
> x[1]
[1] 7
> x[1] + x[2]
[1] 13
> x[1] <- 99
> x
[1] 99  6  5  4
```

The function c can also be used to append elements to the end of an existing vector, thus increasing its length:

```
> x <- c(1, 2, 3)
> x <- c(x, 4)
> x
[1] 1 2 3 4
```

In addition to the function c, there are several ways of constructing vectors: one can start with an empty vector and add elements one-by-one:

```
> x <- c()
> x[1] <- 1
> x[2] <- 1
> x[3] <- x[2] + x[1]
> x[4] <- x[3] + x[2]
> x
[1] 1 1 2 3
```

Vectors consisting of consecutive, increasing numbers can be created using the colon operator:

```
> 1:15
[1]  1  2  3  4  5  6  7  8  9 10 11 12 13 14 15
> 10:20
[1] 10 11 12 13 14 15 16 17 18 19 20
```

More complicated vectors can be generated using the function seq:

```
> seq(from=1, to=15, by=2)
[1]  1  3  5  7  9 11 13 15
> seq(from=15, to=1, by=-2)
[1] 15 13 11  9  7  5  3  1
```

Vectors in R programs are mostly used to store data sets, but they can also be used to store the components of mathematical vectors, that is of elements of the space $\mathbb{R}^d$. Mathematical operations on vectors work as expected:

```
> c(1, 2, 3) * 2
[1] 2 4 6
> c(1, 2, 3) + c(3, 2, 1)
[1] 4 4 4
```

Other useful functions on vectors include sum (to compute the sum of the vector elements), mean (the average), var (the sample variance), sd (the sample standard deviation), and length (the length of the vector):

```
> x <- c(1, 2, 3)
> (x[1] + x[2] + x[3]) / 3
[1] 2
> sum(x) / length(x)
[1] 2
> mean(x)
[1] 2
```

The operators and functions from Table B.1, when applied to a vector, operate on the individual elements:

```
> x <- c(-1, 0, 1, 2, 3)
> abs(x)
[1] 1 0 1 2 3
> x^2
[1] 1 0 1 4 9
```

This allows to efficiently operate on a whole data set with a single instruction.

B.2.3.3 Matrices

Matrices can be constructed in R using the function matrix. To represent the matrix

$$A = \begin{pmatrix} 1 & 2 & 3 \\ 4 & 5 & 6 \end{pmatrix}$$

in R, the following command can be used:

```
> A <- matrix(c(1, 2, 3,
+                4, 5, 6),
+             nrow=2, ncol=3, byrow=TRUE)
> A
     [,1] [,2] [,3]
[1,]    1    2    3
[2,]    4    5    6
```

The first argument to matrix is a vector giving the numbers to be stored in the matrix. The following two arguments set the number of rows/columns of the matrix, and the last argument states that we gave the entries row-by-row. The whole command can be given on one line, the line breaks are only inserted to increase readability. The $n \times n$ identity matrix can be generated by diag(n) and an $m \times n$ zero matrix by matrix(0, nrow=m, ncol=n).

Individual elements of a matrix can be accessed using square brackets, just as for vectors: if A is a matrix, A[1,1] denotes the top-left element of the matrix, A[1,] denotes the first row of the matrix (as a vector), and A[,1] denotes the first column of A:

```
> A <- matrix(c(1,2,3,4), nrow=2, ncol=2, byrow=TRUE)
> A[1,1] <- 9
> A
     [,1] [,2]
[1,]    9    2
[2,]    3    4
> A[1,]
[1] 9 2
> A[,1]
[1] 9 3
```

For complicated matrices it is sometimes useful to first create an empty matrix, and then use a short program to fill in the values. This can be achieved by using a command such as matrix(nrow=m, ncol=n), without specifying values for the matrix elements:

```
> A <- matrix(nrow=2, ncol=10)
> A
     [,1] [,2] [,3] [,4] [,5] [,6] [,7] [,8] [,9] [,10]
[1,]   NA   NA   NA   NA   NA   NA   NA   NA   NA    NA
[2,]   NA   NA   NA   NA   NA   NA   NA   NA   NA    NA
> A[1,] <- 1
> A[2,] <- 1:10
> A
```

```
     [,1] [,2] [,3] [,4] [,5] [,6] [,7] [,8] [,9] [,10]
[1,]    1    1    1    1    1    1    1    1    1     1
[2,]    1    2    3    4    5    6    7    8    9    10
```

The entries of the empty matrix are originally displayed as NA (for 'not available'), and we have to assign values to the elements of A before the matrix can be used.

The sum of matrices and the product of a matrix with a number can be computed using + and *, the matrix-matrix and matrix-vector products from linear algebra are given by %*%. (Note, A * A is not the matrix product, but the element-wise product!)

```
> A <- matrix(c(1, 2,
+                2, 3),
+              nrow=2, ncol=2, byrow=TRUE)
> A
     [,1] [,2]
[1,]    1    2
[2,]    2    3
> A %*% A
     [,1] [,2]
[1,]    5    8
[2,]    8   13
> x <- c(0, 1)
> A %*% x
     [,1]
[1,]    2
[2,]    3
> x %*% A %*% x
     [,1]
[1,]    3
```

The last command shows that vectors are automatically interpreted as row vectors or column vectors as needed: in R it is not required to transpose the vector x when evaluating expressions such as $x^\top A x$.

Many matrix operations are available, for example the transpose of a matrix A can be computed using t(A), the inverse by solve(A), the functions rowSums and colSums return the row and column sums as vectors, and rowMeans and colMeans return the row and column averages, respectively. The solution x to a system $Ax = b$ of linear equations can be computed using the command solve(A, b).

B.2.3.4   Strings

Strings are used in R to represent short texts, represented as a sequence of characters. Strings are used, for example, to specify the text used for the axis labels in a plot. Strings in R are enclosed in quotation marks.

Sometimes a bit of care is needed when using strings: the string "12" represents the text consisting of the digits one and two; this is different from the number 12:

```
> 12 + 1
[1] 13
> "12" + "1"
Error in "12" + "1" : non-numeric argument to binary operator
```

R complains that it cannot add "12" and "1", because these values are not numbers.

Strings can be stored in variables just as for numbers and the function paste can be used to concatenate strings:

```
> s <- paste("this", "is", "a", "test")
> paste(s, ": ", "a", "bra", "ca", "da", "bra", sep="")
[1] "this is a test: abracadabra"
> paste("x=", 12, sep="")
[1] "x=12"
```

The argument sep for the function paste specifies a 'separator' which is put between the individual strings. The default value is a single space character. If the arguments of paste are not strings, they are converted to a string before the concatenation:

```
> paste("x=", 12, sep="")
[1] "x=12"
```

B.2.3.5   Truth values

To represent truth values (also known as *boolean values*), R uses the special values TRUE and FALSE. The shorthand versions T for TRUE and F for FALSE can also be used. Truth values are most commonly encountered as the results of comparisons, using the operators from Table B.2:

```
> 1 < 2
[1] TRUE
> 3 < 2
[1] FALSE
```

Truth values can be stored in variables:

```
> result <- 1 < 2
> result
[1] TRUE
```

When used in a context where a numeric value is expected, truth values are automatically converted to numbers: TRUE is interpreted as 1 and FALSE is interpreted as 0. This behaviour is, for example, useful to count how many elements of a list

Table B.2   List of comparison operators in R. The values textttx and y can be replaced by arbitrary numeric expressions, the result of the comparison is either TRUE or FALSE.

Comparison	Example	R code
Strictly smaller	$x < y$	x < y
Smaller or equal	$x \leq y$	x <= y
Strictly bigger	$x > y$	x > y
Bigger or equal	$x \geq y$	x >= y
Equal	$x = y$	x == y
Not equal	$x \neq y$	x != y

satisfy a given condition; adding the truth values using sum gives the number of TRUE entries in a vector of booleans:

```
> x <- rnorm(10)
> x
 [1] -1.1497316  0.9251614 0.2357447 0.7422293  1.8712479
 [6] -0.7683445 -0.4054793 0.9313800 1.1917342 -0.7275102
> x > 0
 [1] FALSE TRUE TRUE TRUE TRUE FALSE FALSE TRUE TRUE FALSE
> sum(x > 0)
 [1] 6
```

B.3   Programming principles

In the previous section we have seen many examples of R commands. An *R program* is a sequence of commands, designed to solve a specific problem when executed in order. As an example, consider the Fibonacci sequence defined by $x_1 = x_2 = 1$ and $x_k = x_{k-1} + x_{k-2}$ for all $n > 2$. The following commands form an R program to compute x_6:

```
x <- c()
x[1] <- 1
x[2] <- 1
x[3] <- x[2] + x[1]
x[4] <- x[3] + x[2]
x[5] <- x[4] + x[3]
x[6] <- x[5] + x[4]
cat("the 6th Fibonacci number is", x[6], "\n")
```

(B.2)

When these commands are executed in R, one after another, the last command prints 'the 6th Fibonacci number is 8' to the screen (the \n starts a new line in the output). While this program works, it still has a number of shortcomings which we will address in the following sections.

B.3.1   Don't repeat yourself!

A fundamental principle in programming is to avoid duplication wherever possible. This principle applies on many different levels and in many different situations. It is sometimes called the *don't repeat yourself* (DRY) principle. In this section we will discuss some aspects of the DRY principle, using the program (B.2) as an example.

B.3.1.1   Loops

The way we compute x[6] in the program (B.2) involves a lot of repetition: to represent the equation $x_n = x_{n-1} + x_{n-2}$, we used four different lines in our program! Because of this, it will be impractical to use a similar program to compute x[100]. A second problem is that it can be difficult to spot mistakes caused by this repetition.

Another complication appears if we want to modify the program to use the equation $x_n = x_{n-1} - x_{n-2}$ instead: a single change to the mathematical formula requires multiple changes to our program.

The problem of this specific repetition can be solved using a 'loop' in the R program. Such a loop instructs R to execute a command repeatedly. We can use such a loop to write the command for the relation $x_k = x_{k-1} + x_{k-2}$ only once, and then let R repeat this command as needed:

```
n <- 10
x <- c()
x[1] <- 1
x[2] <- 1                                              (B.3)
for (k in 3:n) {
   x[k] <- x[k-1] + x[k-2]
}
cat("the ", n, "th Fibonacci number is ", x[n], "\n", sep="")
```

The loop is implemented by the for statement: for is followed by a group of one or more commands, enclosed in curly brackets { and }, which are executed repeatedly. The number of repetitions is determined by the vector 3:n, the commands are executed once for each element of this vector, that is $n - 2$ times in total. The variable k is set to the corresponding element of the vector before each iteration starts: the first time the loop is executed, k is set to 3. The executed statement is then x[3] <- x[3-1] + x[3-2]. Before the next iteration, k is set to 4 and, after substituting k, the executed statement is x[4] <- x[4-1] + x[4-2]. This is repeated until, in the last iteration of the loop, k is set to the value of n. Once the loop is completed, the program continues with the first instruction after the loop and the program outputs 'the 10th Fibonacci number is 55'.

The choice of the name k for the loop variable in the example above was arbitrary (it was chosen to match the index in the mathematical formula), any other variable name could have been used instead. Similarly, the vector 3:n can be replaced by any other vector, the elements are not required to be adjacent or increasing.

A further improvement of this program, compared with the first version, is the introduction of the variable n: with the new program, we can compute a different Fibonacci number by changing only a single line. Similarly, if we want to implement a different recursion relation, we can just replace the command x[k] <- x[k-1] + x[k-2] with a different one. By avoiding repetition in the program text, mistakes such as the one in exercise EB.5 where the required change to the output string was forgotten, are no longer possible.

There is a second kind of loop available, the while loop, which can be used when the required number of iterations is not known in advance. For example, the following program determines the first Fibonacci number which is bigger than or equal to 1000:

```
x <- c()
x[1] <- 1
x[2] <- 1
n <- 2
while (x[n] < 1000) {
  n <- n + 1
  x[n] <- x[n-1] + x[n-2]
}
cat("the ", n, "th Fibonacci number is ", x[n], "\n", sep="")
```

The while loop repeats its commands while the condition in the round brackets is satisfied. In the condition, all the usual comparison operators can be used (Table B.2). For this type of loop, no automatic assignment to a loop variable takes place, so we have to increment n ourselves using the command n <- n + 1. As soon as the condition is false, the loop ends and the program continues with the first command after the loop.

B.3.1.2   Conditional statements

Inside loops it is often useful to be able to execute different commands in different iterations of the loop. For example, to print all numbers $n = 1, 2, \ldots, 100$ satisfying $\sin(n) > \frac{1}{2}$ to the screen, we can use the following program:

```
for (n in 1:100) {
  if (sin(n) > 0.5) {
    cat(n, "\n")
  }
}
```

The if statement is followed by a block of one or more commands, enclosed in curly brackets { and }. These commands are only executed if the condition given inside the round brackets is true, otherwise the whole `if` block has no effect. As with the `while` statement, all the usual comparison operators (Table B.2) can be used in the condition. There is a second form of the `if` statement:

```
for (n in 1:10) {
  if (n %% 2 == 1) {
    cat(n, "is odd\n")
  } else {
    cat(n, "is even\n")
  }
}
```
(B.4)

Here, the commands in the first block of curly brackets are executed if the condition is true, and otherwise the commands in the second block are executed. Since $n \bmod 2$ equals 0 for even numbers and 1 for odd numbers, this program gives the correct output.

Finally, for cases where a result is computed using one of two alternative formulas, the `ifelse` statement can be used. Using `ifelse`, the loop from listing (B.4) can be rewritten as follows:

```
for (n in 1:10) {
  cat(n, ifelse(n %% 2 == 1, "is odd", "is even"), "\n")
}
```

The first argument of `ifelse` must be a truth value which determines which of the following two values should be used: if the first argument is true, the result is taken from the second argument, otherwise it is taken from the third argument. While the `ifelse` function can always be replaced by an `if` statement, use of `ifelse` sometimes allows to simplify programs.

B.3.1.3   Command scripts

The DRY principle applies also to repetition between different programs: if you have solved a problem once, it is better to reuse the old program, instead of writing a new one. Reuse of old programs does not only avoid unnecessary work, it also reduces the risk of mistakes being introduced in the code and allows to improve the program over time. To facilitate this, the following guidelines are useful:

- Save all your R programs in files (file names for such files typically end in `.R`), store these files somewhere safe, and keep the programs organised in a way which allows you to find them again when needed at a later time.

- When writing programs, take care to write them in a way which makes the code reusable. As we have seen, the DRY principle can help with this. Also, it is useful to write the program as clearly as possible, to use descriptive names for your variables, and to use indentation to make the structure of loops and `if` statements easy to follow.

- If the program uses any nonobvious constructions or clever tricks, it is often helpful to add comments with explanations to the program. Such comments can be very helpful when trying to reuse programs written more than a few weeks ago. Since R ignores every line of input which starts with the character #, such comments can be stored directly in the program. For example, when reading the (slightly cryptic) program

```
# compute Fibonacci numbers x[n]
# y = (x[n], x[n-1])
y <- c(1,1)
for (n in 3:17) {
  y <- c(sum(y), y[1])
}
print(y[1])
```

the comments make it a lot easier to figure out what the program does.

B.3.2 Divide and conquer!

In this section we will discuss a second fundamental programming principle, sometimes called the 'divide and conquer paradigm'. This principle is to break down a problem into smaller subproblems which can be solved individually. After solving the individual subproblems, the individual solutions can be combined to obtain a solution to the full problem. As for the DRY principle discussed above, the divide and conquer principle applies on many different levels and in many different situations. In this text we will focus on the most basic aspects of this principle.

The basic tool for isolating individual building blocks of a program is a 'function' which allows to use a simple name as an abbreviation for a list of commands. We will illustrate the concept of functions in R with the help of examples.

Example B.1 The program from listing (B.3) can be turned into an R function as follows:

```
fibonacci <- function(n) {
  x <- c()
  x[1] <- 1
  x[2] <- 1
  for (k in 3:n) {
    x[k] <- x[k-1] + x[k-2]
  }
  return(x[n])
}
```
(B.5)

When we execute the above lines in R, nothing seems to happen. The commands only define the name fibonacci as an abbreviation for the commands on the following lines, the commands are not yet executed. This step of assigning a name to the function is referred to as 'defining the function'.

After the function is defined, we can execute the commands in the function by using the assigned name fibonacci:

```
> fibonacci(5)
[1] 5
> res <- fibonacci(10)
> res
[1] 55
```
(B.6)

This step is referred to as 'calling the function'. The second call to fibonacci shows that we can assign the result of a function call to a variable and thus store it for later use.

The first line of listing (B.5) not only gives the name of the function but also, in the brackets after the keyword function, it indicates that the function has exactly one argument, called n. Every time the function is called, we have to specify a value for this argument n, for the first call of fibonacci in (B.6) we write fibonacci(5) to indicate that the value 5 should be used for n, for the second call n equals 10. The next six lines in listing (B.5), copied from listing (B.3), then compute the first n elements of the Fibonacci sequence, and the return statement indicates the result of the computation and ends the function.

Example B.2 The R equivalent of a mathematical function is often straightforward to implement. For example, to define an R function for computing the value of

$$f(x) = e^{-x^2},$$

for a given x, we can use the following R code:

```
f <- function(x) {
   return(exp(-x^2))
}
```

The idea of a function like fibonacci in listing (B.5) is that, following the DRY principle, we write the function only once; from then on we can use the function by calling it without having to think about the details of how Fibonacci numbers are computed. The function can be used as one of the building blocks for a bigger program, alongside the built-in R functions such as sqrt and plot.

To allow for easy reuse of existing functions, some of the effect of commands inside the function is hidden from the caller: while the variables n and x are used and modified inside the function, R takes care not to disturb any variables which may be used by the caller:

```
> n <- 3
> fibonacci(10)
[1] 55
> n
[1] 3
> x
Error: object 'x' not found
```

As we can see, the assignment n=10 used inside fibonacci does not affect the value 3 we stored in n before the call and, similarly, the variable x used inside the function is not visible to the caller. This isolation of the variables inside the function is the reason that we need to use return to pass the result of a function to the caller.

Sometimes, it is required to pass several argument values into a function. For example, the built-in function rnorm to generate normally distributed random variables could be defined as follows:

```
rnorm <- function(n, mean, sd) {
  # code for generating random numbers
  ...
}
```

This declares three arguments: n specifies how many random numbers should be generated, mean specifies the mean, and sd specifies the standard deviation. We can use the command rnorm(10, 0, 1) to get 10 standard normally distributed random variables. Since the special case of mean 0 and standard deviation 1 is very common, these value are used as 'default values'. The true declaration of rnorm is

```
rnorm <- function(n, mean=0, sd=1) {
  ...
}
```

The mean=0 tells R to use 0 for the mean, if no other value is given, and similarly for the standard deviation. Thus, rnorm(10, 5) generates 10 random values with mean 5 (since we specified the mean in the second argument), and standard deviation 1 (the default value is used because we did not give a third argument).

When calling a function which uses default values for some or all of its arguments, and if we want to specify only some of the arguments, we can give values for individual function arguments as in the following example:

```
X <- rnorm(10, sd=3)
```

This will generate 10 values from a normal distribution with mean 0 (the default value is used because we did not specify this argument) and standard deviation 3 (as specified by `sd=3`).

Example B.3   Consider the following function:

```
test <- function(a=1, b=2, c=3, d=4) {
  cat("a=", a, ", b=", b, ", c=", c, ", d=", d, "\n", sep="")
}
```

When experimenting with this function, we get the following output:

```
> test()
a=1, b=2, c=3, d=4
> test(5, 6)
a=5, b=6, c=3, d=4
> test(a=5, d=6)
a=5, b=2, c=3, d=6
> n <- 3
> test(a=n, b=n+1, c=n+2, d=n+3)
a=3, b=4, c=5, d=6
```

As we have seen, functions can be used to encapsulate parts of your program as a single unit. Functions help to follow both the 'divide and conquer' principle (because they allow to easily break down a program into smaller building blocks) and the DRY principle (because they are easily reused). We conclude this section with some general guidelines for writing functions:

- Sometimes there is a choice about which parts of a program could be split into functions. A good idea in such situations is to aim for functions whose purpose is easily explained: 'compute the nth Fibonacci number' is a task which will make a good function. In contrast, it seems less clear whether implementing a formula such as:

$$x_n = \begin{cases} x_{n-1}/2 & \text{if } x_{n-1} \text{ is even and} \\ 3x_{n-1} + 1 & \text{if } x_{n-1} \text{ is odd} \end{cases} \tag{B.7}$$

in a function is a good idea: the description of this function will likely be as long as the function itself, and reuse of the resulting function seems not very likely.
- Using descriptive names for functions is a good idea. Examples of well-chosen function names include `mean`, `sin` and `plot`; in all three cases it is easy to guess from the name what the function will do.

- The 'divide and conquer' principle becomes most powerful when functions make use of previously defined functions which implement more fundamental building blocks. The extreme case of this is when the definition of a function includes calls to itself (for example when sorting of a list of length n is reduced to the problem of sorting lists of length $n - 1$); this technique is called 'recursion'.

B.3.3   Test your code!

The final programming principle we will discuss in this text is the importance of testing programs: even for experienced programmers, it is almost impossible to get a non-trivial program right at the first attempt. Thus, the usual procedure is to complete a program bit by bit (following the approach in the previous section), and every time a part of the program is completed to systematically test for the presence of errors.

There are different kinds of errors which can occur in computer programs:

(a) 'Syntax errors' are errors caused by not following the restrictions and requirements of the programming language. In this case, R does not understand the program at all and complains immediately with an error message. Examples include excess commas and brackets:

```
> mean(c(1,2,,3))
Error in c(1, 2, , 3) : argument 3 is empty
> mean(c(1,2,3)))
Error: unexpected ')' in "mean(c(1,2,3)))"
```

Normally, the cause of these errors can be spotted immediately, and the problem is usually easy to resolve by following the hints given in the error message.

(b) 'Run-time errors' occur when the program is syntactically correct, but when R encounters an operation which cannot be performed during the execution of the program. Examples include programs which try to compute the sample variance of an empty data set. In this case, the program is stopped immediately and an error message is printed:

```
> print.var <- function(x) {
+       cat("the variance is", var(x), "\n")
+ }
> print.var(c(1,2,3))
the variance is 1
> print.var(c())
Error in var(x) : 'x' is NULL
```

This kind of error is often more difficult to detect and fix, because the error may not occur in every run of the program and because it may be difficult to find the exact location of the error in the program.

'Arithmetic errors' are special cases of run-time errors, where the program tries to evaluate expressions like 0/0 or to compute the square root of a negative number. In these cases, R just sets the result to NaN (short for 'not a number') and sometimes (but not always) produces a warning message:

```
> 0/0
[1] NaN
> 2/2 + 1/1 + 0/0
[1] NaN
> sqrt(-1)
[1] NaN
Warning message:
In sqrt(-1) : NaNs produced
```

(c) 'Semantic errors' are the errors caused by a program being a valid program, but one which does not implement the intended algorithm. These errors are cases where the program does what you *told* it to do instead of what you *meant* it to do. Such errors can be very difficult to spot and fix. As a simple example, consider the following function for computing the mean of a vector:

```
# something is wrong here!
average <- function(x) {
  n <- length(x)
  s <- sum(x)
  return(x/n)
}
```
(B.8)

This function does not work as intended (since the return statement erroneously uses x instead of s), but there is no error message. The best way to find this mistake is to test the function by trying to call it.

Errors in programs are sometimes called 'bugs', and the process of locating and fixing errors in a program is called 'debugging'. There are various aspects to debugging:

- A good start is to try the code in question for a few cases where the correct result is known. Make sure to try some boundary cases (such as very short data sets) as well as some typical cases. Test each function of a program separately, starting with the most fundamental ones.
- Carefully look out for any error messages or warnings. These messages often contain useful hints about what went wrong.
- To locate the position of an error, it is often helpful to temporarily insert a print or cat statement into a program to check whether the program actually

executes certain lines of code and to see whether variables at these locations still have the expected values.

- The functions debug and undebug can be used to watch the execution of a function step by step. See the R help text for debug [obtained by typing help(debug)] for usage instructions.

Example B.4 Assume we want to debug the function average given in listing (B.8). The first step is to try a few values:

```
> average(c(1, 2, 3))
[1] 0.3333333 0.6666667 1.0000000
```

Since the mean of $(1, 2, 3)$ is 2, we see immediately that something is wrong. Assuming that we do not spot the typo yet, we can try to modify the function by inserting a cat statement as follows:

```
average <- function(x) {
  n <- length(x)
  s <- sum(x)
  cat("n =", n, " s =", s, "\n")
  return(x/n)
}
```

If we rerun the function, we now get the following output:

```
> average(c(1, 2, 3))
n = 3  s = 6
[1] 0.3333333 0.6666667 1.0000000
```

Since the printed values are what we expect (the list has length $n = 3$, and the sum of the elements should be $s = 1 + 2 + 3 = 6$), we know that the mistake must be after the cat statement. Thus, we have narrowed down the location of the problem to a single line (the return statement), and looking at this line it is now easy to spot that x should be replaced with s.

B.4   Random number generation

R contains an extensive set of built-in functions for generating (pseudo-)random numbers of many of the standard probability distributions. There are also functions available to compute densities, CDFs and quantiles of these distributions. Since these functions will be used extensively for the exercises in the main text, we give a short introduction here.

Table B.3   Some probability distributions supported by R.

Distribution	Name in R
Binomial distribution	binom
χ^2 distribution	chisq
Exponential distribution	exp
Gamma distribution	gamma
Normal distribution	norm
Poisson distribution	pois
Uniform distribution	unif

The names of all these functions are constructed using the following scheme: the first letter is

r   for random number generators,
d   for densities (weights for the discrete case),
p   for the CDFs and
q   for quantiles.

The rest of the name determines the distribution; some possible distributions are given in Table B.3.

Example B.5   The function to generate normal distributed random numbers is rnorm and the density of the exponential distribution is dexp.

The functions starting with r, examples include runif and rnorm, can be used to generate random numbers. The first argument for each of these functions is the number n of random values required; the output is a vector of length n. The following arguments give parameters of the underlying distribution; often these arguments have the most commonly used parameter values as their default values. Details about how to use these functions and how to set the distribution parameters can be found in the R online help. Finally, the function set.seed is available to set the seed of the random number generator (see the discussion in section 1.1.3).

Example B.6   A vector of 10 independent, standard normally distributed random numbers can be obtained using the command rnorm(10). A single sample from a $\mathcal{N}(2, 9)$ distribution can be obtained using rnorm(1, 2, 3), where the last argument gives the standard deviation (not the variance) of the distribution.

An exception to the naming scheme for random number generators is the discrete uniform distribution: a sample $X_1, \ldots, X_n \sim \mathcal{U}\{1, 2, \ldots, a\}$ can be generated with the R command sample.int(a, n, replace=TRUE).

B.5   Summary and further reading

The best way to learn programming is to gain a lot of practice. For this reason, I recommend trying to solve as many of the exercises provided in this text as possible. Other interesting sources of programming challenges can be found online, for example on the web page of Project Euler.[1]

An extensive introduction to all aspects of R can be found online in Venables *et al.* (2011). A more in-depth description can be found in Venables and Ripley (2000). R source code for many of the methods discussed here can be found in Rizzo (2008). Information about specific R functions can be found using the built-in help system (accessed using the function `help`).

Exercises

EB.1 Use R to compute the following values:

(a) $1 + 2 + 3 + 4 + 5 + 6 + 7 + 8 + 9 + 10$

(b) 2^{16}

(c) 2^{100}

(d) $\dfrac{2}{1 + \sqrt{5}}$

EB.2 What are the values of x and y after the following commands are executed in R?

```
x <- 1
y <- 2
x <- x + y
y <- x + y
```

EB.3 Use R to compute the value of $1 + 2 + \cdots + 100$.

EB.4 Write an R function to compute the sample excess kurtosis

$$g_2 = \frac{\frac{1}{n}\sum_{i=1}^{n}(x_i - \bar{x})^4}{\left(\frac{1}{n}\sum_{i=1}^{n}(x_i - \bar{x})^2\right)^2} - 3,$$

for a given vector $x = (x_1, \ldots, x_n)$, where $\bar{x}$ is the average of the elements of x.

---

[1] Available from `http://projecteuler.net/`.

EB.5   The following code is an extension of the program from listing (B.2), changed to compute the Fibbonacci numbers until x[8] (instead of x[6]).

```
x <- c()
x[1] <- 1
x[2] <- 1
x[3] <- x[2] + x[1]
x[4] <- x[3] + x[2]
x[6] <- x[5] + x[4]
x[7] <- x[6] + x[5]
x[8] <- x[7] + x[6]
cat("the 6th Fibonacci number is", x[8], "\n")
```

Can you spot the mistake?

EB.6   Let $x_0 = 0$ and $x_n = \cos(x_{n-1})$ for all $n \in \mathbb{N}$. Use R to compute the value of x_{20}.

EB.7   Write an R program which uses a `for` loop to print the elements of the decreasing sequence $100, 99, \ldots, 0$ to the screen, one per line.

EB.8   Write an R program which uses a `while` loop to determine the smallest square number bigger than 5000.

EB.9   Write an R program which uses a `while` loop to determine the biggest square number smaller than 5000.

EB.10 Given $x_0 \in \mathbb{N}$, a sequence $(x_n)_{n \in \mathbb{N}}$ of integers can be defined by equation (B.7). Once this sequence reaches 1, it starts to cycle through the values $3 \cdot 1 + 1 = 4$, $4/2 = 2$, and $2/2 = 1$, but it is unknown whether this cycle is reached from every starting point x_0. Write an R program which prints the values of x_n, starting with $x_0 = 27$. The program should stop when the value 1 is reached for the first time.

EB.11 What does the following function compute?

```
f <- function(x) {
  n <- length(x)
  m <- mean(x)
  s <- 0
  for (i in 1:n) {
    s <- s + (x[i] - m)^2
  }
  return(s/(n-1))
}
```

EB.12 Continuing example B.3, feed the following commands into R:

```
test <- function(a=1, b=2, c=3, d=4) {
  cat("a=", a, ", b=", b, ", c=", c, ", d=", d, "\n", sep="")
}
```

```
b <- "a"
test(c=b)
```

Explain the result R prints to the screen.

EB.13   Section B.3.3 introduces three categories of programming errors. Which categories do the two (!) errors from exercise EB.5 fall into?

EB.14   When trying to solve exercise E3.7, one could try to implement the function $\hat{\rho}(X, Y)$ from equation (3.23) as follows:

```
# something is wrong here!
rxy <- function(X,Y) {
  mX <- mean(X)
  mY <- mean(Y)
  numerator <- sum((X - mX) * (Y - mY))
  denominator <- sqrt(sum(X-mX)^2 * sum(Y-mY)^2)
  return(numerator / denominator)
}
```

When testing this function, we quickly discover that something must be wrong:

```
> X <- c(1.6, -1.1, -1.2)
> Y <- c(0.1, 0.2, 0.1)
> rxy(X, Y)
[1] -Inf
```

One can check that $-1 \le \hat{\rho}(X, Y) \le 1$, so the value $-\infty$ clearly cannot be the correct result! Use the techniques explained in section B.3.3 to find the mistake.

EB.15   The following function is a (failed) attempt to compute

$$\sum_{i=1}^{n-1}(x_{i+1} - x_i)^2,$$

that is the sum of squared increments, in R:

```
SomethingWrong <- function(x) {
  n <- length(x)
  sum <- 0
  for (i in 1:n-1) {
    sum <- sum + (x[i+1] - x[i])^2
  }
  return(sum)
}
```

When we apply this function to the vector (1, 2, 3), we do not get the correct answer 2, but `numeric(0)` instead.

```
> SomethingWrong(c(1,2,3))
numeric(0)
```

What is the mistake in the function `SomethingWrong`?

EB.16 Use R to create plots of the densities and CDFs of the following distributions:

(a) $\mathcal{N}(0, 1)$ — the standard normal distribution;

(b) $\mathcal{N}(3, 4)$ — the normal distribution with mean $\mu = 3$ and variance $\sigma^2 = 4$;

(c) $\text{Exp}(1)$ — the exponential distribution with rate 1;

(d) $\Gamma(9, 0.5)$ — the gamma distribution with shape parameter $k = 9$ and scale parameter $\theta = 0.5$.

Appendix C

Answers to the exercises

This appendix contains solutions to the exercises found throughout the book. Most of the answers require use of a computer. Any programs used in the solutions here are written using the R programming language (R Development Core Team, 2011); of course, similar solutions can be written in different programming languages. For reference, a short introduction to programming in R can be found in Appendix B.

All R programs used for this book, both to generate the figures and for the answers of the exercises, can be downloaded from the accompanying web page at www.wiley.com/go/statistical_computing

C.1   Answers for Chapter 1

Solution E1.1   To implement the LCG in R, we can execute the commands from algorithm 1.2 in a loop.

```
LCG <- function(n, m, a, c, X0) {
  X <- numeric(length=n)
  Xn <- X0
  for (i in 1:n) {
    Xn <- (a*Xn + c) %% m
    X[i] <- Xn
  }
  return(X)
}
```

The resulting function LCG can be called as follows:

```
> LCG(10, 8, 5, 1, 0)
 [1] 1 6 7 4 5 2 3 0 1 6
```

An Introduction to Statistical Computing: A Simulation-based Approach, First Edition. Jochen Voss.
© 2014 John Wiley & Sons, Ltd. Published 2014 by John Wiley & Sons, Ltd.

The output matches the manually computed result from Example 1.3, so we can assume that the program is correct.

Solution E1.2   Using the function LCG from exercise E1.1, we can generate the graphs as follows:

```
par(mfrow=c(2,2))

X <- runif(1000)
plot(X[1:999], X[2:1000], asp=1, cex=.5,
     xlab=expression(X[i]), ylab=expression(X[i+1]))

m <- 81
a <- 1
c <- 8
seed <- 0
X <- LCG(1000, m, a, c, seed)/m
plot(X[1:999], X[2:1000], asp=1, cex=.5,
     xlab=expression(X[i]), ylab=expression(X[i+1]))

m <- 1024
a <- 401
c <- 101
seed <- 0
X <- LCG(1000, m, a, c, seed)/m
plot(X[1:999], X[2:1000], asp=1, cex=.5,
     xlab=expression(X[i]), ylab=expression(X[i+1]))

m <- 2^32
a <- 1664525
c <- 1013904223
seed <- 0
X <- LCG(1000, m, a, c, seed)/m
plot(X[1:999], X[2:1000], asp=1, cex=.5,
     xlab=expression(X[i]), ylab=expression(X[i+1]))
```

The output is shown in Figure 1.1.

Solution E1.3

(a) There are different approaches to picking out the middle two digits of a number. One method is to use algebraic operations:

```
middle.square <- function (x) {
  square <- x^2
  middle <- floor(square / 10) %% 100
  return(middle);
}
```

Another method is to convert between numbers and strings:

```
middle.square2 <- function(x) {
  square <- x^2
  padded <- formatC(square, width=4, flag="0")
  middle <- substring(padded, 2, 3)
  return(as.numeric(middle))
}
```

(b) The easiest method to find a loop is to generate elements of the sequence X_n, until a previously seen element is reached:

```
find.loop <- function (start) {
  seen <- c()

  # follow the sequence until it runs back into itself
  x <- start
  while (! is.element(x, seen)) {
    seen <- c(seen, x)
    x <- middle.square(x)
  }

  # get the complete loop
  loop <- c(x)
  y <- middle.square(x)
  while (y != x) {
    loop <- c(loop, y)
    y <- middle.square(y)
  }
  return(loop)
}
```

Using this function, we can systematically look for loops:

```
all.loops <- c()
for (x in 0:99) {
  loop <- find.loop(x)

  # To see whether we've found this loop already,
  # we keep track of the smallest element in each loop.
  smallest <- min(loop)
  if (! is.element(smallest, all.loops)) {
    cat("found a loop: ", loop, "\n");
    all.loops <- c(all.loops, smallest)
  }
}
```

This program generates the following output:

```
found a loop:   0
found a loop:   10
found a loop:   60
found a loop:   24 57
found a loop:   50
```

The function middle.square is illustrated in Figure C.1.

(c) Figure C.1 shows that the output starts repeating itself after at most 15 steps and that all loops are very short (length 1 or 2). Therefore, the method forms a rather poor PRNG.

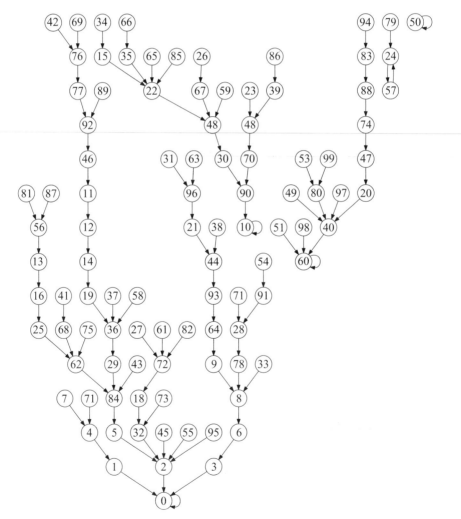

Figure C.1   The transition graph of the middle square method for two-digit numbers.

Solution E1.4   The first step in solving this question is to find the CDF of the given distribution. For $x < 1$ we have $f(x) = 0$ and thus $F(x) = 0$. For $x \geq 1$ we have $f(x) = 1/x^2$ and thus

$$F(a) = \int_1^a \frac{1}{x^2}\,dx = \left(-\frac{1}{x}\right)\Big|_{x=1}^a = \frac{1}{1} - \frac{1}{a} = 1 - \frac{1}{a}.$$

Next, we have to find the inverse of F. Since F is continuous, we can just compute the ordinary inverse:

$$u = F(x) = 1 - \frac{1}{x} \quad \Longleftrightarrow \quad \frac{1}{x} = 1 - u \quad \Longleftrightarrow \quad x = \frac{1}{1-u}.$$

By proposition 1.14, if $U \sim \mathcal{U}[0, 1]$, then $X = 1/(1 - U)$ has density f. The resulting method can be implemented in R as follows:

```
GenerateSample <- function(n) {
  U <- runif(n)
  return(1/(1-U))
}
```

To generate a histogram of $10\,000$ values, we can then use the following commands:

```
X <- GenerateSample(10000)
hist(X, freq=FALSE, breaks=seq(0, max(X)+1, 0.1),
     xlim=c(0,10), ylim=c(0,1),
     main=NULL, col="gray80", border="gray20")
```

Some care is needed, because X occasionally takes very big values:

```
> summary(X)
   Min.  1st Qu.   Median     Mean  3rd Qu.      Max.
  1.000    1.336    1.971    8.073    3.899  2746.000
```

To make the histogram useful we have to choose sufficiently narrow bars (using the `breaks` argument) and to restrict the displayed horizontal coordinate range (using the `xlim` argument). Also, since we want to compare the output with the density, we have to switch to a density plot instead of a frequency plot (using the argument `freq=FALSE`). The density can be added to the plot using the `lines` command:

```
x <- seq(0, 10, 0.01)
f <- ifelse(x <= 1, 0, 1/x^2)
lines(x, f, lw=1)
```

The resulting plot is shown in Figure C.2. The plot shows a good match between the density function and the histogram.

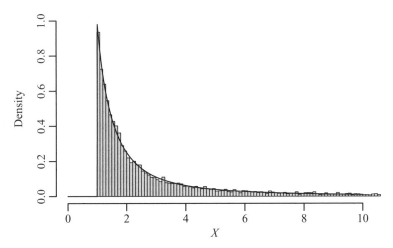

Figure C.2 A histogram of 10 000 samples from the distribution given in exercise E1.4, together with the corresponding density function (solid line).

Solution E1.5   Let $Y_n = 1$, if the proposal X_n is accepted and $Y_n = 0$ otherwise. Then we can write $K_n = \sum_{i=1}^{n} Y_i$. From equation (1.4) in the proof of proposition 1.20 we know that each of the proposals is accepted with probability Z and thus we have

$$\mathbb{E}(Y_n) = 1 \cdot P(Y_n = 1) + 0 \cdot P(Y_n = 0) = Z$$

for all $n \in \mathbb{N}$. Finally, by the strong law of large numbers (theorem A.8), we find

$$\lim_{n\to\infty} \frac{1}{n} K_n = \lim_{n\to\infty} \frac{1}{n} \sum_{i=1}^{n} Y_i = \mathbb{E}(Y_1) = Z.$$

This completes the proof.

Solution E1.6   We can implement the rejection algorithm as follows:

```
f <- function(x) {
  return((x> 0) * 2 * dnorm(x,0,1))
}
g <- function(x) { return(dexp(x,1)) }
c <- sqrt(2 * exp(1) / pi)

rhalfnormal <- function(n) {
  res <- numeric(length=n)
  i <- 0
```

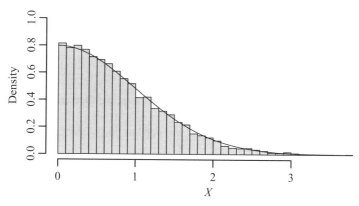

Figure C.3   A histogram of the output of the algorithm from exercise E1.6. The histogram is generated from 10 000 samples from the half-normal distribution. For comparison, the solid line gives the exact density from equation (1.5).

```
while (i<n) {
  U <- runif(1, 0, 1)
  X <- rexp(1, 1)
  if (c * g(X) * U <= f(X)) {
    i <- i+1
    res[i] <- X;
  }
}
return(res)
}
```

To test the function `rhalfnormal`, we generate a histogram as follows:

```
X <- rhalfnormal(10000)
hist(X, breaks=50, prob=TRUE, ylim=c(0,1),
     main=NULL, col="gray80", border="gray20")
curve(f, min(X), max(X), n=500,
      ylim=c(0,1), ylab="f", add=TRUE)
```

The output of this program is shown in Figure C.3.

Solution E1.7   In this question we are asked to construct a rejection algorithm to convert the $\text{Exp}(\lambda)$-distributed proposal into $\mathcal{N}(0, 1)$-distributed samples. We construct the algorithm in two steps: first we generate half-normal distributed samples with density

$$f(x) = \frac{2}{\sqrt{2\pi\sigma^2}} \exp\left(-\frac{x^2}{2\sigma^2}\right)$$

for all $x \geq 0$. Since the corresponding samples are positive, we can use the density of the $\text{Exp}(\lambda)$-distribution, that is

$$g(x) = \lambda \exp(-\lambda x)$$

for all $x \geq 0$ as the proposal density. In a second step, we then attach a random sign to get the full normal distribution.

For the rejection sampling algorithm, we need to find a constant c with

$$
\begin{aligned}
f(x) &\leq c \cdot g(x) \\
\Longleftrightarrow \quad & \frac{2}{\sqrt{2\pi\sigma^2}} \exp\left(-\frac{x^2}{2\sigma^2}\right) \leq c \cdot \lambda \exp(-\lambda x) \\
\Longleftrightarrow \quad & \frac{2}{\lambda\sqrt{2\pi\sigma^2}} \exp\left(-\frac{x^2}{2\sigma^2} + \lambda x\right) \leq c \\
\Longleftrightarrow \quad & \frac{2}{\lambda\sqrt{2\pi\sigma^2}} \exp\left(-\frac{1}{2\sigma^2}(x - \lambda\sigma^2)^2\right) \exp\left(\lambda^2\sigma^2/2\right) \leq c
\end{aligned}
$$

for all $x \in \mathbb{R}$. Since $(x - \lambda\sigma^2)^2 \geq 0$, this condition is satisfied for all values of c with

$$c \geq \frac{2}{\lambda\sqrt{2\pi\sigma^2}} \exp\left(\lambda^2\sigma^2/2\right).$$

Since the method is most efficient for small values of c, the best choice is to choose c equal to this bound. The resulting acceptance condition is

$$
\begin{aligned}
c \cdot g(X)U &\leq f(X) \\
\Longleftrightarrow \quad & \frac{2}{\lambda\sqrt{2\pi\sigma^2}} \exp\left(\lambda^2\sigma^4\right) \lambda \exp(-\lambda x) U \leq \frac{2}{\sqrt{2\pi\sigma^2}} \exp\left(-\frac{x^2}{2\sigma^2}\right) \\
\Longleftrightarrow \quad & U \leq \exp\left(-\frac{1}{2\sigma^2}(x - \lambda\sigma^2)^2\right).
\end{aligned}
$$

This leads to the following algorithm:

(a) Generate $X \sim \text{Exp}(\lambda)$.
(b) Generate $U \sim \mathcal{U}[0, 1]$.
(c) Accept X if $U \leq \exp\left(-(X - \lambda\sigma^2)^2/2\sigma^2\right)$, otherwise reject X.
(d) For accepted samples, return $-X$ or X randomly, each with probability $1/2$.

We can optimise the method by choosing the value of λ which minimises

$$c(\lambda) = \frac{2\exp\left(\lambda^2\sigma^2/2\right)}{\lambda\sqrt{2\pi\sigma^2}}.$$

Setting the derivative of $c(\lambda)$ equal to zero, we find the following necessary condition for an extremum:

$$
\begin{aligned}
0 &= c'(\lambda) \\
&= \frac{2}{\sqrt{2\pi\sigma^2}} \cdot \frac{\lambda\sigma^2 \exp\left(\lambda^2\sigma^2/2\right) \cdot \lambda - 1 \cdot \exp\left(\lambda^2\sigma^2/2\right)}{\lambda^2} \\
&= \left(\lambda^2\sigma^2 - 1\right) \frac{2\exp\left(-\lambda^2\sigma^4\right)}{\lambda^2\sqrt{2\pi\sigma^2}}.
\end{aligned}
$$

This condition is only satisfied for $\lambda = \lambda^* = 1/\sigma$ and, since $c(\lambda)$ converges to ∞ for $\lambda \downarrow 0$ and $\lambda \to \infty$, the value λ^* is a minimum of $c(\lambda)$. Thus, the optimal value of λ is λ^* and the corresponding, optimal value of c is

$$
c(\lambda^*) = \frac{\sigma\, 2\exp(1/2)}{\sqrt{2\pi\sigma^2}} = \sqrt{\frac{2e}{\pi}} \approx 1.315.
$$

This shows that, when the algorithm is optimally tuned, we need on average 1.315 proposals to generate one output sample.

Solution E1.8   Since the proposals are Exp(1)-distributed, the proposal density is given by

$$
g(x) = \exp(-x)
$$

and we need to find a constant $c > 0$ with $f(x) \leq cg(x)$ for all $x \geq 0$. We have

$$
\begin{aligned}
f(x) &\leq c \cdot g(x) \\
&\Longleftrightarrow \frac{1}{\sqrt{x}} \exp\left(-y^2/2x - x\right) \leq c\exp(-x) \\
&\Longleftrightarrow \frac{1}{\sqrt{x}} \exp\left(-y^2/2x\right) \leq c,
\end{aligned}
$$

for all $x \geq 0$ and thus we need to find the maximum of the left-hand side of this equation. Setting the derivative equal to zero, gives the condition

$$
\begin{aligned}
0 &= \left(\frac{1}{\sqrt{x}} \exp\left(-y^2/2x\right)\right)' \\
&= \frac{\frac{y^2}{2x^2}\exp(-y^2/2x)\sqrt{x} - \frac{1}{2\sqrt{x}}\exp(-y^2/2x)}{x} \\
&= \left(\frac{y^2}{x} - 1\right)\frac{\exp(-y^2/2x)}{2x^{3/2}},
\end{aligned}
$$

which is satisfied for $x = y^2$; it is easy to check that this value corresponds to a maximum. Thus, the optimal choice for c is

$$c = \frac{1}{\sqrt{y^2}} \exp\left(-y^2/2y^2\right) = \frac{1}{|y|} \exp(-1/2).$$

This is the required result.

Solution E1.9   The inequality $c \geq 1$ follows from

$$1 = \int f(x)\,dx \leq \int cg(x)\,dx = c \int g(x)\,dx = c.$$

Now assume $c = 1$. Then we have $f \leq g$. For $\varepsilon > 0$ let $A_\varepsilon = \{x \mid f(x) \leq g(x) - \varepsilon\}$. Then we have

$$1 = \int f(x)\,dx = \int_{A_\varepsilon} f(x)\,dx + \int_{A_\varepsilon^{\mathsf{C}}} f(x)\,dx$$

$$\leq \int_{A_\varepsilon} (g(x) - \varepsilon)\,dx + \int_{A_\varepsilon^{\mathsf{C}}} g(x)\,dx = \int g(x)\,dx - \varepsilon|A_\varepsilon| = 1 - \varepsilon|A_\varepsilon|$$

and thus $\varepsilon|A_\varepsilon| \leq 0$. Since $\varepsilon > 0$, this is only possible if $|A_\varepsilon| = 0$. Therefore we find

$$\left|\{x \mid f(x) < g(x)\}\right| = \left|\bigcup_{\varepsilon > 0} A_\varepsilon\right| = \lim_{\varepsilon \downarrow 0} |A_\varepsilon| = 0.$$

This completes the proof.

Solution E1.10   Since we have $X \sim \mathcal{U}[a, b]$ and $Y \sim \mathcal{U}[c, d]$, the density of X is given by $\mathbb{1}_{[a,b]}(x)/(b - a)$ and the density of Y is $\mathbb{1}_{[c,d]}(y)/(d - c)$, where

$$\mathbb{1}_{[a,b]}(x) = \begin{cases} 1 & \text{if } x \in [a, b] \text{ and} \\ 0 & \text{otherwise.} \end{cases}$$

Since X and Y are independent, the joint density of (X, Y) is

$$f(x, y) = \frac{\mathbb{1}_{[a,b]}(x)}{b - a} \cdot \frac{\mathbb{1}_{[c,d]}(y)}{d - c} = \frac{\mathbb{1}_{[a,b]}(x)\mathbb{1}_{[c,d]}(y)}{(b - a)(d - c)}.$$

Consequently,

$$P\big((X, Y) \in A\big) = \int \int \mathbb{1}_A(x, y) f(x, y)\,dy\,dx$$

$$= \int \int \mathbb{1}_A(x, y) \frac{\mathbb{1}_{[a,b]}(x)\mathbb{1}_{[c,d]}(y)}{(b - a)(d - c)}\,dy\,dx$$

$$= \frac{1}{(b - a)(d - c)} \int_a^b \int_c^d \mathbb{1}_A(x, y)\,dy\,dx$$

where

$$\mathbb{1}_A(x, y) = \begin{cases} 1 & \text{if } (x, y) \in A \text{ and} \\ 0 & \text{otherwise.} \end{cases}$$

Since the rectangle $R = [a, b] \times [c, d]$ satisfies $|R| = (b - a)(d - c)$ and any set A satisfies

$$|A \cap R| = \int_a^b \int_c^d \mathbb{1}_A(x, y)\, dy\, dx$$

we find

$$P\big((X, Y) \in A\big) = \frac{|A \cap R|}{|R|}.$$

This is the probability from the definition of the uniform distribution on R and therefore the proof is complete.

Solution E1.11   The set A consists of two disjoint components. The probability for a sample to be in the left component is $1/3$ and the probability of being in the right component is $2/3$. Using lemma 1.31 we know that, conditioned on being in one of the components, the sample should be distributed uniformly on this component. Consequently we can generate samples $X_i \sim \mathcal{U}(A)$ as follows:

(a) Let $I \in \{1, 2\}$ with $P(I = 1) = 1/3$ and $P(I = 2) = 2/3$.
(b) If $I = 1$ generate $X \sim \mathcal{U}\big([0, 1] \times [0, 1]\big)$. Otherwise generate $X \sim \mathcal{U}\big([2, 4] \times [0, 1]\big)$.

The method can be implemented in R as follows:

```
r.two.squares <- function(n) {
  U <- runif(n)
  X <- ifelse(U < 1/3, runif(n), runif(n, 2, 4))
  Y <- runif(n)
  return(cbind(X, Y))
}

Z <- r.two.squares(1000)
plot(Z[,1], Z[,2], xlab="X", ylab="Y", cex=.5, asp=1)
```

The resulting plot is shown in Figure C.4.

Solution E1.12   We can proceed as in exercise E1.11, but some care is needed since the two squares composing the set B overlap. To avoid this problem, we write $B = B_1 \cup B_2 \cup B_3$ where $B_1 = [0, 2] \times [0, 1]$, $B_2 = [0, 3] \times [1, 2]$ and

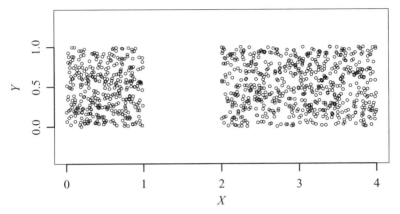

Figure C.4   A sample of 1000 points from the uniform distribution on set A from exercise E1.11.

$B_3 = [1, 3] \times [2, 3]$. The corresponding probabilities are $P(X \in B_1) = |B_1|/|B| = 2/7$, $P(X \in B_2) = |B_2|/|B| = 3/7$ and $P(X \in B_3) = |B_3|/|B| = 2/7$.
The method can be implemented in R as follows:

```
r.three.rectangles <- function(n) {
  U <- runif(n)
  X <- ifelse(U < 2/7, runif(n, 0, 2),
              ifelse(U < 5/7, runif(n, 0, 3), runif(n, 1, 3)))
  Y <- ifelse(U < 2/7, runif(n, 0, 1),
              ifelse(U < 5/7, runif(n, 1, 2), runif(n, 2, 3)))
  return(cbind(X, Y))
}
```

A plot of a sample generated by this function is shown in Figure C.5.

Solution E1.13

(a) We can use the following procedure: let $X_n \sim \mathcal{U}[-1, 1]$ and $Y_n \sim \mathcal{U}[0, 1]$ be independent. Accept $U_n = (X_n, Y_n)$ if $X_n^2 + Y_n^2 \leq 1$. Then the accepted points are uniformly distributed on the semicircle. The acceptance probability is

$$
\begin{aligned}
P\big((X_n, Y_n) \text{ accepted}\big) &= P\big((X_n, Y_n) \text{ in semicircle}\big) \\
&= \frac{|\text{semicircle}|}{|\text{rectangle}|} = \frac{\pi/2}{2} = \frac{\pi}{4} \approx 0.7854.
\end{aligned}
$$

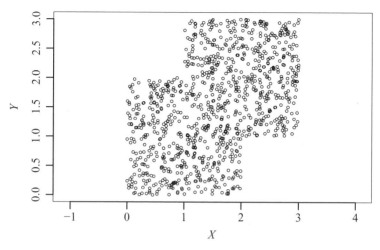

Figure C.5 A sample of 1000 points from the uniform distribution on set B from exercise E1.12.

(b) Since we do not know in advance how many proposals are required to generate a given number of samples, we use a `while` loop in our program:

```
rsemicircle <- function(n) {
  res = c()
  k <- 0
  while (length(res) < 2*n) {
    k <- k + 1
    X <- runif(1, -1, 1)
    Y <- runif(1, 0, 1)
    if (X^2 + Y^2 < 1) {
      res <- rbind(res, c(X, Y))
    }
  }
  cat(k, "proposals ->", n, "samples\n")
  return(res)
}

Z <- rsemicircle(1000)
plot(Z[,1], Z[,2], xlab="X", ylab="Y", cex=.5, asp=1)
```

The resulting plot is shown in Figure C.6.

Solution E1.14   To obtain a rejection method with acceptance probability of at least 80% we need to find a region for the proposals which contains the semicircle, but which has an area of at most $\pi/(2 \cdot 0.8)$. Here we use the area depicted in Figure C.7.

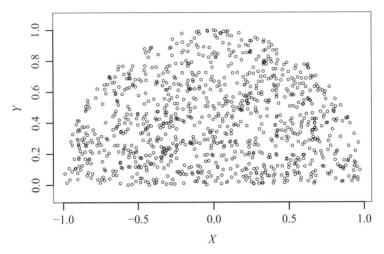

Figure C.6   A sample of 1000 points from the uniform distribution on the semicircle (see exercise E1.13). A total of 1274 proposals was used in the rejection algorithm to generate the 1000 samples.

The proposal region A has area $|A| = 2 - 2 \cdot 1/4^2 = 30/16$ (6/16 for the upper region and 24/16 for the lower region) and thus the acceptance probability is $\frac{\pi/2}{|A|} \approx$ 0.8378. We can sample from the uniform distribution on A as in exercise E1.12:

```
proposal <- function() {
 U <- runif(1)
 X <- ifelse(U < 6/30, runif(1, -0.75, 0.75), runif(1, -1, 1))
 Y <- ifelse(U < 6/30, runif(1, 0.75, 1), runif(1, 0, 0.75))
 return(c(X,Y))
}
```

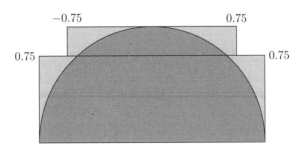

Figure C.7   Proposal region for the rejection sampling algorithm proposed in the solution to exercise E1.14.

Using rejection sampling to obtain the uniform distribution on the semicircle is now straightforward:

```
rsemicircle <- function(n) {
  res = c()
  k <- 0
  while (length(res) < 2*n) {
    k <- k + 1
    Z <- proposal()
    if (sum(Z^2) < 1) {
      res <- rbind(res, Z)
    }
  }
  return(res)
}
```

Solution E1.15   To find the density of X we can use lemma 1.33: since X is distributed uniformly on the area between the x-axis and the graph of the function $\sqrt{1-x^2}$, the vector $(X, \frac{2}{\pi}Y)$ is uniformly distributed on

$$A = \left\{(x, y) \in \mathbb{R}^2 \mid 0 \le y < f(x)\right\}$$

where

$$f(x) = \begin{cases} \frac{2}{\pi}\sqrt{1 - x^2} & \text{if } x \in [-1, 1] \text{ and} \\ 0 & \text{otherwise.} \end{cases}$$

Since $f \ge 0$ and $\int f(x)\,dx = \frac{2}{\pi} \cdot \frac{\pi}{2} = 1$, the function f is a probability density, and by lemma 1.33, the random variable X has density f.

To find the density of Y we use a direct calculation:

$$P(Y \in A) = \frac{\int_0^1 \mathbb{1}_A(y)\langle\text{width of semicircle at level } y\rangle\,dy}{\langle\text{area of semicircle}\rangle}$$

$$= \int_0^1 \mathbb{1}_A(y)\frac{2\sqrt{1 - y^2}}{\pi/2}\,dy$$

$$= \int_{\mathbb{R}} \mathbb{1}_A(y)\,g(y)\,dy$$

where

$$g(y) = \frac{4}{\pi}\sqrt{1 - y^2}\,\mathbb{1}_{[0, 1]}(y).$$

Thus, Y is distributed with density g.

Solution E1.16   Denote the density of cX by g and define $\varphi(x) = cx$ for all $x \in \mathbb{R}^d$. By theorem 1.34, if X has density

$$f(x) = g(\varphi(x)) \cdot |\det D\varphi(x)|,$$

then $\varphi(X)$ has density g. Since $\varphi(x)_i = cx_i$ for all $i = 1, 2, \ldots, d$, we find

$$\det D\varphi(x) = \det \begin{pmatrix} c & 0 & \cdots & 0 \\ 0 & c & \cdots & 0 \\ \vdots & \vdots & \ddots & \vdots \\ 0 & 0 & \cdots & c \end{pmatrix} = c^d$$

and thus $f(x) = g(cx)|c^d|$ for all $x \in \mathbb{R}^d$. We can use the substitution $y = cx$ to solve this equation for g, the result is

$$g(y) = \frac{1}{|c^d|} f(y/c).$$

This is the required density of cX.

Solution E1.17   We have $Y = \varphi(X)$ with $\varphi(x) = (x^2 - 1)/2$ for all $x \in \mathbb{R}$. We would like to apply theorem 1.34 but, since φ is not bijective, the theorem does not apply directly. To work around this problem we first consider the random variable $Z = |X|$. Since $X^2 = Z^2$, Y can also be written as $Y = \varphi(Z)$ where $\varphi: [0, \infty) \to \mathbb{R}$ is bijective. We have

$$P(Z \in A) = P(X \in A) + P(-X \in A) = \frac{2}{\sqrt{2\pi}} \int_A \exp(-x^2/2) \, dx$$

for all $A \subseteq [0, \infty)$ and thus Z has density

$$f(z) = \begin{cases} \frac{2}{\sqrt{2\pi}} \exp(-z^2/2) & \text{if } z \geq 0 \text{ and} \\ 0 & \text{otherwise.} \end{cases}$$

Now we can apply theorem 1.34 by considering φ to be a map from $A = (0, \infty)$ to $B = (-1/2, \infty)$. Denoting the density of Y by g we get

$$\frac{2}{\sqrt{2\pi}} \exp(-z^2/2) = g(\varphi(z))|\varphi'(z)| = g(\varphi(z))z$$

for all $z > 0$. Choosing $z = \varphi^{-1}(y) = \sqrt{2y + 1}$ we get

$$\frac{2}{\sqrt{2\pi}} \exp(-y - 1/2) = g(y)\sqrt{2y + 1}$$

and thus

$$g(y) = \begin{cases} \frac{\exp(-y-1/2)}{\sqrt{\pi}(y+1/2)} & \text{if } y \geq -1/2 \text{ and} \\ 0 & \text{otherwise.} \end{cases}$$

This is the required density of Y.

Solution E1.18   Following the steps presented in example 1.40, we can use the following R program:

```
rcauchy.ratio <- function(n) {
  X0 <- c()
  X1 <- c()
  while (length(X0) < 2*n) {
    x <- runif(1, 0, 1)
    y <- runif(1, -1, 1)
    if (x^2 + y^2 < 1) {
      X0 <- c(X0, x)
      X1 <- c(X1, y)
    }
  }
  return(X1 / X0)
}
```

C.2   Answers for Chapter 2

Solution E2.1   Following algorithm 2.9 we can sample from the mixture distribution as follows:

```
mu <- c(1, 2, 5)
sigma <- sqrt(c(0.01, 0.5, 0.02))
theta <- c(0.1, 0.7, 0.2)

Y <- sample(3, 10000, replace=TRUE, prob=theta)
X <- rnorm(10000, mu[Y], sigma[Y])
```

The required histogram can be plotted using the `hist` command.

```
hist(X, breaks=50, prob=TRUE,
     main=NULL, col="gray80", border="gray20")
```

Finally, from lemma 2.8 we know that the density of the mixture distribution is the mixture of the component densities. Using the function `rnorm` for the components, we can overlay the mixture density over the plot as follows:

```
curve(theta[1] * dnorm(x, mu[1], sigma[1])
      + theta[2] * dnorm(x, mu[2], sigma[2])
      + theta[3] * dnorm(x, mu[3], sigma[3]),
      min(X), max(X), n=1000, add=TRUE)
```

The resulting plot is shown in Figure 2.1.

Solution E2.2   To show that X is not a Markov chain, we need to find a case where the Markov chain condition (2.2) is violated. Example: if $X_{j-1} = a$ and $X_{j-2} = b$ then $X_j \sim \mathcal{N}(a + b, 1)$. Thus, for small $\varepsilon > 0$,

$$P\left(X_j \in [-3, 3] \,\Big|\, X_{j-1} \in [-\varepsilon, +\varepsilon], X_{j-2} \in [10 - \varepsilon, 10 + \varepsilon]\right) \approx 0$$

but

$$P\left(X_j \in [-3, 3] \,\Big|\, X_{j-1} \in [-\varepsilon, +\varepsilon], X_{j-2} \in [-\varepsilon, +\varepsilon]\right) \approx 1.$$

If X was a Markov chain, the two probabilities would be equal, but here they are not. Thus, X is not a Markov chain.

Solution E2.3   (a) To sample paths from the Markov chain we can use a program such as:

```
MC <- function(N, initial, P) {
  Xj <- sample(length(initial), 1, prob=initial)
  res <- c(Xj)

  for (j in 1:N) {
    p <- P[Xj,]
    Xj <- sample(length(p), 1, prob=p)
    res <- c(res, Xj)
  }

  return(res)
}

# transition matrix
P <- matrix(c(2/3, 1/3, 0, 0,
              .1, .9, 0, 0,
              .1, 0, .9, 0,
              .1, 0, 0, .9),
            nrow = 4, ncol = 4, byrow=TRUE)
```

```
# initial distribution
mu <- c(.25, .25, .25, .25)

MC(10, mu, P)
```

(b) To numerically estimate the distribution of X_{10} we can use a histogram:

```
data <- c()
for (k in 1:10000) {
  X <- MC(10, mu, P)
  data <- c(data, X[11])
}
hist(data, breaks=seq(0.5,4.5), probability=TRUE,
     main=NULL, col="gray80", border="gray20")
```

The histogram allows us to find estimated probabilities for the different states of the Markov chain: from the plot (not shown here) we read off the values $P(X_{10} = 1) \approx 0.22$, $P(X_{10} = 2) \approx 0.6$, $P(X_{10} = 3) \approx 0.09$, and $P(X_{10} = 4) \approx 0.09$.

(c) We have to compute $\mu^\top P^{10}$:

```
> mu %*% P %*% P %*% P %*% P %*% P %*% P %*% P %*% P %*% P %*% P
     [,1]       [,2]      [,3]       [,4]
[1,] 0.2308349 0.5948259 0.08716961 0.08716961
```

The result coincides well with the values we read off the histogram.

Solution E2.4   (a) This follows directly from the properties of a stochastic matrix:

$$(Pv)_y = \sum_{x \in S} p_{xy} v_y = \sum_{x \in S} p_{xy} 1 = 1 = 1 \cdot v_y$$

for all $y \in S$. Thus, v is an eigenvector with eigenvalue 1.

(b) Assume $Av = \lambda v$. Then $A(\alpha v) = \alpha Av = \alpha(\lambda v) = \lambda(\alpha v)$.

(c) We find the eigenvalues of P as follows:

```
> E <- eigen(t(P))
$values
[1] 1.0000000 0.9000000 0.9000000 0.5666667

$vectors
      [,1]           [,2]          [,3]          [,4]
[1,] -0.2873479 -3.925231e-17 -3.925231e-17 -0.7071068
[2,] -0.9578263 -7.071068e-01 -7.071068e-01  0.7071068
[3,]  0.0000000  7.071068e-01  0.000000e+00  0.0000000
[4,]  0.0000000  0.000000e+00  7.071068e-01  0.0000000
```

Looking at the first column of this matrix (corresponding to the eigenvalue 1.0000000), we see that

$$v = (-0.2873479, -0.9578263, 0.0000000, 0.0000000)$$

is the (only) eigenvector of $P^\top$ with eigenvalue 1. Thus, by normalising this vector to be a probability vector, we find the stationary distribution of P:

```
> v <- E$vectors[,1]
> v / sum(v)
[1] 0.2307692 0.7692308 0.0000000 0.0000000
```

Solution E2.5   To simulate a Poisson process with constant intensity, we can directly follow the steps described in example 2.38. In R, we can use the following function:

```
PoissonProcessConst <- function(a, b, lambda) {
  N <- rpois(1, lambda * (b-a))
  X <- runif(N, a, b)
  return(X)
}
```

If we want the function to return the points in increasing order, we could replace the return statement by `return(sort(X))`.

Solution E2.6   To implement the required function in R, we can directly follow the steps described in example 2.39:

```
PoissonProcessStep <- function(t, lambda) {
  stopifnot(length(t) == length(lambda)+1)
  X <- c()
  for (i in 1:length(lambda)) {
    N <- rpois(1, lambda * (t[i+1]-t[i]))
    X <- c(X, runif(N, t[i], t[i+1]))
  }
  return(X)
}
```

The function can be used as follows:

```
t <- c(1.4, 3,   4, 6,   6.2, 7.9,  8,   9.5)
lambda <- c(1, 4.5, 2, 1.4,   0,   1.1, 0.8)
X <- PoissonProcessStep(t, lambda)
```

The output of a call such as this has been used to determine the position of the marked points in Figure 2.3.

C.3   Answers for Chapter 3

Solution E3.1   We can use the rejection algorithm suggested in example 3.6, that is we apply the rejection sampling algorithm 1.22 with target density $f(x) = \exp(-y^2/2x - x)/\sqrt{x}$, proposal density $g(x) = \exp(-x)$ and $c = \exp(-1/2)/|y|$. The condition $cg(X)U \le f(X)$ from the algorithm takes the following form:

$$cg\,(X)U \;\le\; f(X)$$
$$\Longleftrightarrow \quad \frac{1}{|y|}\exp(-1/2)\exp(-x)U \;\le\; \frac{1}{\sqrt{x}}\exp\left(-y^2/2x - x\right)$$
$$\Longleftrightarrow \quad U \;\le\; \frac{|y|}{\sqrt{x}}\exp\left(-y^2/2x + 1/2\right).$$

This gives the following algorithm for generating samples from the conditional distribution of X given $Y = y$:

(a) Generate $X \sim \mathrm{Exp}(1)$.
(b) Generate $U \sim \mathcal{U}[0, 1]$.
(c) Accept X if $U \le \dfrac{|y|}{\sqrt{x}}\exp\left(-y^2/2x + 1/2\right)$.

We can implement this rejection algorithm in R as follows:

```
GeneratePosteriorSamples <- function(n, y) {
  res <- c()
  while (length(res) < n) {
    X <- rexp(1)
    U <- runif(1)
    if (U <= abs(y) * exp(-y^2/(2*X) + 0.5) / sqrt(X)) {
      res <- c(res, X)
    }
  }
  return(res)
};
```

Finally, Monte Carlo estimation can be used to get approximations for the mean $\mathbb{E}(X\,|\,Y = 4)$ and the variance $\mathrm{Var}(X\,|\,Y = 4)$:

```
> X <- GeneratePosteriorSamples(10000, 4)
> mean(X)
[1] 3.329493
> var(X)
[1] 1.934953
```

This shows that the posterior expectation is approximately 3.33 and the posterior variance is approximately 1.93.

Solution E3.2   To compute a single Monte Carlo estimate for $\mathbb{E}\left(\sin(X)^2\right)$ we can use the following function:

```
GetMCEstimate <- function(N) {
  X <- rnorm(N)
  return(mean(sin(X)^2))
}
```

This function takes N as its only parameter and computes one instance of the Monte Carlo estimate Z_N^{MC}. An estimate for $\mathbb{E}(\sin(X)^2)$ can be obtained with the following command:

```
> GetMCEstimate(1000000)
[1] 0.4318852
```

We want to compare the variances of the Monte Carlo estimates for $N = 1000$ and $N = 10\,000$. To do so, we first generate estimates repeatedly and then plot a histogram of the resulting values:

```
estimates <- replicate(10000, GetMCEstimate(1000))
range <- c(min(estimates), max(estimates))
hist(estimates, breaks=seq(range[1], range[2], length.out=50),
     xlab=expression(Z[1000]), xlim=range, ylim=c(0,2100),
     main=NULL, col="gray80", border="gray20")
```

By taking a huge value of N we can generate a single estimate which is close to the true value of the expectation. This value can be marked in the histogram using the following commands:

```
good.estimate <- GetMCEstimate(1000000)
abline(v=good.estimate)
```

The resulting plot is shown in Figure 3.1(a). The figure shows that the exact value for the expectation is approximately 0.43, the estimates for $N = 1000$ range from about 0.40 to about 0.46. The width of this range is more than 10% of the exact value and thus a single estimate for $N = 1000$ would not be a very accurate approximation for the mean.

Finally, for the second part of the question, we repeat the experiment with Monte Carlo sample size $N = 10\,000$:

```
estimates2 <- replicate(10000, GetMCEstimate(10000))
hist(estimates2, breaks=seq(range[1], range[2], length.out=50),
     xlab=expression(Z[10000]), xlim=range, ylim=c(0,2100),
     main=NULL, col="gray80", border="gray20")
abline(v=good.estimate)
```

The new histogram is shown in Figure 3.1(b). As expected from proposition 3.14, the variance of the samples in the second histogram is significantly lower than in the first histogram.

Solution E3.3   The Monte Carlo estimate for $\mathbb{E}\big(\cos(X)\big)$ is given by

$$Z_N^{\text{MC}} = \frac{1}{N} \sum_{j=1}^{N} \cos(X_j),$$

where $X_1, \ldots, X_N \sim \mathcal{N}(0, 1)$ are i.i.d. The difficulty lies in the choice of N.

The question asks us to obtain and estimate with 'three digits of accuracy'. This statement is open to interpretation and to answer the question, we need to choose which interpretation of 'three digits of accuracy' we choose. Here we interpret 'three digits of accuracy' to mean $\text{MSE}(Z_N^{\text{MC}}) \leq 0.001^2$. (The square on the right-hand side is needed, since the mean squared error is a squared quantity.) Since

$$\text{Var}\big(\cos(X)\big) = \mathbb{E}\big(\cos(X)^2\big) - \mathbb{E}\big(\cos(X)\big)^2 \leq 1 - 0 = 1,$$

we can use the estimate

$$\text{MSE}\left(Z_N^{\text{MC}}\right) \leq \frac{\text{Var}\left(\cos(X)\right)}{N} \leq \frac{1}{N}.$$

Using this estimate we see that $N = 10^6$ is enough to achieve $\text{MSE}(Z_N^{\text{MC}}) \leq 10^{-6} = 0.001^2$. A computer experiment can be used to get a numerical value for the expectation:

```
> N <- 1e6
> X <- rnorm(N)
> mean(cos(X))
[1] 0.6064109
```

Thus, our estimate is $\mathbb{E}(\cos(X)) \approx 0.606$.

Solution E3.4   From equation (3.5) we get

$$\begin{aligned}
P(X \leq a) &= \int_{-\infty}^{a} \frac{1}{\sqrt{2\pi}} e^{-x^2/2}\, dx \\
&= \frac{1}{2} + \frac{1}{\sqrt{2\pi}} \int_{0}^{a} e^{-x^2/2}\, dx \\
&= \frac{1}{2} + \frac{a}{\sqrt{2\pi}} \mathbb{E}\big(e^{-a^2 U^2/2}\big)
\end{aligned}$$

where $U \sim \mathcal{U}[0, 1]$. Since the estimate $\tilde{p}_N$ is obtained by replacing the expectation on the right-hand side with the corresponding Monte Carlo estimate, we know what $\tilde{p}_N \to p$ as $N \to \infty$ and that $\tilde{p}_N$ is unbiased.

Since both estimates are unbiased, we have

$$\text{MSE}(p_N) = \text{Var}(p_N) = \frac{1}{N}\text{Var}\big(\mathbb{1}_{[a,\infty)}(X)\big)$$

and

$$\text{MSE}(\tilde{p}_N) = \text{Var}(\tilde{p}_N) = \text{Var}\bigg(\frac{1}{2} + \frac{a}{\sqrt{2\pi}N} \sum_{j=1}^{N} e^{-a^2 U^2/2}\bigg)$$

$$= \frac{1}{N} \cdot \frac{a^2}{2\pi}\text{Var}\big(e^{-a^2 U^2/2}\big).$$

To estimate the MSE of the estimator p_N we can use the following R commands:

```
> a <- 1
> X <- rnorm(1000000)
> var(X >= a)
[1] 0.1332084
```

Thus, we have $\text{MSE}(p_N) \approx 0.1332/N$. For the estimator $\tilde{p}_N$ we get:

```
> U <- runif(1000000)
> var(exp(-a^2*U^2/2)) * a^2 / (2*pi)
[1] 0.002345373
```

that is we have $\text{MSE}(\tilde{p}_N) \approx 0.0023/N$. Thus the estimator $\tilde{p}_N$ has much smaller error than the estimator p_N does.

Solution E3.5   The key observation which allows to compute $\hat{\sigma}^2$ using only a constant amount of memory is that $\hat{\sigma}^2$ can be written as

$$\hat{\sigma}^2 = \frac{1}{N-1} \sum_{j=1}^{N} \Big(f(X_j)^2 - 2f(X_j)Z_N^{\text{MC}} + \big(Z_N^{\text{MC}}\big)^2\Big)$$

$$= \frac{1}{N-1} \sum_{j=1}^{N} f(X_j)^2 - 2\frac{Z_N^{\text{MC}}}{N-1} \sum_{j=1}^{N} f(X_j) + \frac{N}{N-1}\big(Z_N^{\text{MC}}\big)^2$$

$$= \frac{1}{N-1} \sum_{j=1}^{N} f(X_j)^2 - \frac{N}{N-1}(Z_N^{\text{MC}})^2$$

$$= \frac{1}{N-1} \sum_{j=1}^{N} f(X_j)^2 - \frac{1}{N(N-1)}\bigg(\sum_{j=1}^{N} f(X_j)\bigg)^2.$$

Using this formula, we can compute Z_N^{MC} and $\hat{\sigma}^2$ simultaneously, using the following steps:

```
1: s ← 0
2: t ← 0
3: for j = 1, 2, ..., N do
4:    generate X_j
5:    s ← s + f(X_j)
6:    t ← t + f(X_j)²
7: end for
8: return Z_N^MC = s/N and σ̂² = (t − s²/N)/(N − 1).
```

This algorithm requires a constant amount of memory, since only the variables s and t need to be stored between iterations of the loop.

Solution E3.6   We consider importance sampling estimates of the probability $p = P(X \in A)$, where $X \sim \mathcal{N}(0, 1)$ and $A = [3, 4]$. If samples $Y_j \sim \psi$ are used, the corresponding importance sampling estimate for p is given by

$$\hat{p} = \frac{1}{N} \sum_{j=1}^{N} \mathbb{1}_A(Y_j) \frac{\varphi(Y_j)}{\psi(Y_j)},$$

where

$$\varphi(x) = \frac{1}{\sqrt{2\pi}} \exp\left(-\frac{x^2}{2}\right)$$

is the density of X.

(a) An importance sampling estimate $\hat{p}$, together with the sample variance $\hat{\sigma}^2$ can be computed using the following R code

```
GetISEstimate <- function(N, gen.Y, psi) {
  Y <- gen.Y(N)
  phi <- function(x) dnorm(x, 0, 1);
  weighted.samples <- (Y >= 3 & Y <= 4) * phi(Y) / psi(Y)
  return(list(p=mean(weighted.samples),
              var=var(weighted.samples)))
}
```

The R function `GetISEstimate` takes three arguments: `N` is the sample size, `gen.Y` is an R function to generate the samples $Y_1, \ldots, Y_N$ and `psi` is an R function to compute the density ψ. The function `GetISEstimate` can be called as follows:

```
> gen.Y1 <- function(N) { rnorm(N, 1, 1) }
> psi1 <- function(x) { dnorm(x, 1, 1) }
> GetISEstimate(10000, gen.Y1, psi1)
$p
[1]  0.001243597
```

The output of GetISEstimate is a list which contains the estimate for p in the field $p and the estimated variance of the weighted samples in $var.

To get a list of estimated variances for all four methods, we have to call GetISEstimate repeatedly. One way of doing so is to define the following helper functions:

```
PrintISVariance <- function(name, gen.Y, psi) {
  est <- GetISEstimate(10000, gen.Y, psi)
  cat(name, "  -->  Var = ", est$var, "\n", sep="")
  return(est$var)
}

gen.Y2 <- function(N) { rnorm(N, 2, 1) }
psi2 <- function(x) { dnorm(x, 2, 1) }
gen.Y3 <- function(N) { rnorm(N, 3.5, 1) }
psi3 <- function(x) { dnorm(x, 3.5, 1) }
gen.Y4 <- function(N) { rexp(N) + 3 }
psi4 <- function(x) { dexp(x-3) }

PrintAllVariances <- function() {
  v1 <- PrintISVariance("N(1,1)  ", gen.Y1, psi1)
  v2 <- PrintISVariance("N(2,1)  ", gen.Y2, psi2)
  v3 <- PrintISVariance("N(3.5,1)", gen.Y3, psi3)
  v4 <- PrintISVariance("Exp(1)+3", gen.Y4, psi4)
  return(c(v1, v2, v3, v4))
}
```

These functions determine the variances for all four estimators. The result is as follows:

```
> res <- PrintAllVariances()
N(1,1)      -->    Var = 7.882264e-05
N(2,1)      -->    Var = 1.400159e-05
N(3.5,1)    -->    Var = 6.666196e-06
EXP(1)+3    -->    Var = 1.897206e-06
```

The results show that the method which uses $Y_j \sim \mathrm{Exp}(1) + 3$ has the smallest variance and thus gives the most accurate results for given sample size N. We store the list of variances in the variable res for use in (c).

(b) We use the most efficient method, identified in the previous part of the question, to obtain a good estimate for p:

```
> N <- 10000
> est <- GetISEstimate(N, gen.Y4, psi4)
> est$p
[1] 0.001329137
> 1.96 * sqrt(est$var) / sqrt(N)
[1] 2.70939e-05
```

The output shows that the computed value for the estimate is $\hat{p} = 0.001329137$ and that a 95% confidence interval for p around $\hat{p}$ has a width of approximately $2 \cdot 2.7 \cdot 10^{-5} = 0.000054$. Since the width of the confidence interval is much smaller than the value itself, we expect $\hat{p}$ to be an accurate estimate.

(c) To get an estimate with an error of 1% we choose N such that $\mathrm{MSE}(\hat{p}) = \varepsilon^2$, where $\varepsilon = 0.01\,\hat{p}$. Solving (3.18) for N we find that the required condition for N is given by

$$N \geq \frac{\hat{\sigma}^2}{\varepsilon^2} = \frac{\hat{\sigma}^2}{(0.01\,\hat{p})^2}.$$

Using the variances from (a) we get the following results:

```
> eps <- 0.01 * est$p
> ceiling(res / eps^2)
[1]   446182   79257   37735   10740
```

For comparison we note that from equation (3.11) we know that the corresponding lower bound for a basic Monte Carlo estimate is $N \geq p(1-p)/\varepsilon^2$:

```
> est$p * (1 - est$p) / eps^2
[1]   7513676
```

We see that computation of the basic Monte Carlo estimate will take approximately $7513676/10740 \approx 700$ times as long as computing the best of the importance sampling estimates considered here.

Solution E3.7   We study how Monte Carlo estimation can be used to estimate the bias of the estimator $\hat{\rho}(X, Y)$ from equation (3.23).

(a) The estimator $\hat{\rho}(X, Y)$ can be implemented as an R function as follows:

```
rxy <- function(x, y) {
  x.bar <- mean(x)
  y.bar <- mean(y)
  numerator <- mean((x-x.bar)*(y-y.bar))
  denominator <- sqrt(mean((x-x.bar)^2)*mean((y-y.bar)^2))
  return(numerator/denominator)
}
```

Using this function, the bias of $\hat{\rho}(X, Y)$ can be estimated by the following program: we generate random pairs (X, Y) with given length n and given correlation ρ as in the previous question. The following function generates samples $\hat{\rho}(X_j, Y_j)$ for $j = 1, 2, \ldots, N$:

```
rhohat.sample <- function(N, n, rho) {
  res <- c()
```

```
for (j in 1:N) {
  X <- rnorm(n)
  eta <- rnorm(n)
  Y <- rho*X + sqrt(1-rho^2)*eta
  res[j] <- rxy(X, Y)
}
return(res)
}
```

Using the output of this function, a Monte Carlo estimate for the bias can be computed as follows:

```
EstimateBias <- function(N, n, rho) {
  return(mean(rhohat.sample(N, n, rho)) - rho)
}
```

Since $\hat{\rho}(X, Y) \in [-1, 1]$, we have the upper bound $\text{Var}(\hat{\rho}(X, Y)) \leq 1$ and thus the variance of the estimate for the bias is smaller than $1/N$. Consequently, for $N = 10\,000$, the standard deviation of the estimate is guaranteed to be smaller than 0.01 (and in reality will be much smaller than this). Since correlations lie in the range $[-1, +1]$, bringing the error below 0.01 seems sufficient; thus we will use $N = 10\,000$ in (b) and (c).

(b) We can call the function EstimateBias in a loop in order to plot the estimated bias as a function of ρ. Assuming $n = 10$, we can use the following program:

```
n <- 10
N <- 10000
rho.values <- seq(-1, +1, by=0.05)
bias.values <- c()
for (rho in rho.values) {
  cat("rho =", rho, "\n")
  bias.values <- c(bias.values, EstimateBias(N, n, rho))
}
plot(rho.values, bias.values,
     xlab=expression(rho), ylab="estimated bias")
```

The resulting data points in the plot are still affected by random noise, caused by the Monte Carlo error for finite N. To make the picture clearer, we can add a smoothed curve through the estimated values:

```
s <- smooth.spline(rho.values, bias.values, spar=0.6)
lines(s)
```

The resulting plot is shown in Figure 3.2.

(c) We can compute $\hat{\rho}(X, Y)$ for the given data as follows:

```
X <- c(0.218, 0.0826, 0.091, 0.095, -0.826,
       0.208, 0.600, -0.058, 0.602, 0.620)
```

```
Y <- c(0.369, 0.715, -1.027, -1.499, 1.291,
       -0.213, -2.400, 1.064, -0.367, -1.490)
print(rxy(X,Y))
```

The result is -0.6917216. From Figure 3.2 we know that the bias for ρ-values of this magnitude is about 0.02, that is on average $\hat{\rho}(X, Y)$ will over-estimate ρ by about 0.02. Consequently, $-0.69 - 0.02$ is an approximately unbiased estimate for ρ.

Solution E3.8   We can use the following R code to get the Monte Carlo estimate of the confidence coefficients:

```
EstimateConfidenceCoefficient <- function(n, lambda, N=100000){
  p.alpha <- qt(0.975, df=n-1)
  k <- 0
  for (j in 1:N) {
    X <- rpois(n, lambda)
    m <- mean(X)
    sigma.hat <- sd(X)
    d <- p.alpha * sigma.hat / sqrt(n)
    if (m - d <= lambda && lambda <= m + d) {
      k <- k + 1
    }
  }
  return(k / N)
}
```

To call this function for a range of different λ, we can use a loop such as the following:

```
lambda.list <- c()
prob.list <- c()
width.list <- c()
for (lambda in seq(0.025, 1, length.out=40)) {
  prob <- EstimateConfidenceCoefficient(10, lambda)
  print(c(lambda, prob))
  lambda.list <- c(lambda.list, lambda)
  prob.list <- c(prob.list, prob)
}
```

The results of the estimation are stored in the vector prob.list, the corresponding values of λ are stored in lambda.list. Finally, we create a plot of the results:

```
plot(lambda.list, prob.list, ylim=c(0,1),
     xlab=expression(lambda), ylab="confidence coefficient")
```

The plot, shown in Figure 3.3, clearly shows that for Poisson-distributed data the interval (3.26), at least when λ is allowed to be arbitrarily small, has an extremely low confidence coefficient and thus is not a useful confidence interval.

C.4   Answers for Chapter 4

Solution E4.1   To implement the random walk Metropolis method described in example 4.10, we can use the following R function:

```
GenerateMCMCSample <- function(N, sigma) {
  X <- numeric(length=N)
  Xj <- runif(1, -1, 1)
  for (j in 1:N) {
    Yj <- rnorm(1, Xj, sigma)
    if (abs(Yj) <= 3 * pi) {
      alpha <- (Xj * sin(Yj) / (Yj * sin(Xj))) ^ 2
    } else {
      alpha <- 0
    }
    U <- runif(1)
    if (U < alpha) {
      Xj <- Yj
    }
    X[j] <- Xj
  }
  return(X)
}
```

Some care is needed with the choice of the initial value X_0: since, at least in our implementation, $\alpha(x, y)$ is undefined for $x = 0$, we cannot start the process with $X_0 = 0$. Instead we start with a random initial point $X_0 \sim \mathcal{U}[-1, 1]$. This choice avoids the problem at 0 and, at the same time, guarantees that X_0 falls into a region where π is large. Figure 4.1 shows paths resulting from three runs of the algorithm, for different values of σ.

Solution E4.2   To implement the random walk Metropolis method in R we can use the following functions:

```
pi <- function(x, log=FALSE) {
  dnorm(x, 100, 1, log=log)
}

p <- function(x, y, log=FALSE) {
  dnorm(y, x, log=log)
}
```

```
alpha <- function(x, y) {
  return(exp(pi(y, log=TRUE) + p(y, x, log=TRUE)
             - pi(x, log=TRUE) - p(x, y, log=TRUE)))
}

MCMC <- function(X0, N) {
  X <- numeric(length=N)
  Xj <- X0
  for (j in 1:N) {
    Yj <- rnorm(1, Xj)
    U <- runif(1)
    if (U < alpha(Xj, Yj)) {
      Xj <- Yj
    }
    X[j] <- Xj
  }
  return(c(X0, X))
}
```

These functions can be used as follows:

```
N <- 1000
j <- 0:N
X <- MCMC(0, N)
plot(j, X, type="l", ylab=expression(X[j]))
```

The resulting plot is shown in Figure 4.2. Experiments show that the process always moves towards the value 100 with approximately constant speed. Once the process is close to 100, it starts fluctuating around this value. A good choice for the burn-in period is, for example the time until the process reaches the interval [99, 101]. In this simple example it would be even better to always start with $X_0 = 100$ instead of using a burn-in period.

Solution E4.3 In exercise E4.1 we have already written an R-function to generate paths from the random walk Metropolis algorithm with target density π. Using this function, named `GenerateMCMCSample` we can generate plots of the estimated autocorrelations as follows:

```
PlotAutocorrelation <- function(N, sigma, ...) {
  X <- GenerateMCMCSample(N, sigma)
  a <- acf(X^2, lag.max=200, ci=0, ylab=expression(rho[k]), ...)
  text(195, 0.92, bquote(sigma==.(sigma)))
  # a$acf stores the autocorrelation coefficients,
  # a$acf[1] = rho_0 = 1, a$acf[2] = rho_1, ...
```

```
eff <- 1/(1 + 2 * sum(a$acf[-1]))
print(c(eff, 1/eff))
}
```

Here we use the built-in R function `acf` to estimate the autocorrelation of the samples. The function `PlotAutocorrelation` can be used as follows:

```
> PlotAutocorrelation(1000000, 6)
[1] 0.1520343
```

Solution E4.4 There are two different ways we can estimate the required acceptance probability. We can either use the ratio of accepted samples to proposed samples in the algorithm, or we can take the average of the acceptance probabilities $\alpha(X_{j-1}, Y_j)$ for $j = 1, 2, \ldots, N$. Here we use the second of these methods:

```
AverageAcceptanceRate <- function(N, sigma) {
  asum <- 0;
  Xj <- runif(1, -1, 1)
  for (j in 1:N) {
    Yj <- rnorm(1, Xj, sigma)
    if (abs(Yj) <= 3 * pi) {
      alpha <- min((Xj * sin(Yj) / (Yj * sin(Xj))) ^ 2, 1)
    } else {
      alpha <- 0
    }
    U <- runif(1)
    if (U < alpha) {
      Xj <- Yj
    }
    asum <- asum + alpha
  }
  return(asum / N)
}

xx <- c()
yy <- c()
for (p in seq(-2, 3, 0.2)) {
  sigma <- 10^p
  cat("sigma = ", sigma, "\n", sep="")
  xx <- c(xx, sigma^2)
  yy <- c(yy, AverageAcceptanceRate(100000, sigma))
}

plot(xx, yy, type="l", log="x",
  xlab=expression(sigma^2), ylab="average acceptance probability")
```

```
abline(v=1, lty=3)
abline(v=6^2, lty=3)
abline(v=36^2, lty=3)
```

The output of this program is shown in Figure 4.4.

Solution E4.5   The random walk Metropolis algorithm for the posterior distribution can be implemented as follows:

```
GeneratePosteriorSample <- function(N, x, eta) {
  n <- length(x)

  mu <- mean(x)
  sigma <- sd(x)
  theta <- matrix(nrow=N, ncol=2)
  for (j in 1:N) {
    eps <- rnorm(2, 0, eta)
    mu.tilde <- mu + eps[1]
    sigma.tilde <- sigma + eps[2]

    if (-10 <= mu.tilde && mu.tilde <= 10 && sigma.tilde > 0) {
      a <- n * (log(sigma) - log(sigma.tilde))
      b <- sum((x - mu.tilde)^2) / (2 * sigma.tilde) + sigma.tilde
      c <- sum((x - mu)^2) / (2 * sigma) + sigma
      alpha <- min(exp(a - b + c), 1)
    } else {
      alpha <- 0
    }

    U <- runif(1)
    if (U < alpha) {
      mu <- mu.tilde
      sigma <- sigma.tilde
    }

    theta[j,1] <- mu
    theta[j,2] <- sigma
  }
  return(theta)
}
```

The function can be used as follows:

```
n <- 100
x <- rnorm(n, 5, 2)

N <- 10000
theta <- GeneratePosteriorSample(N, x, 0.1)
```

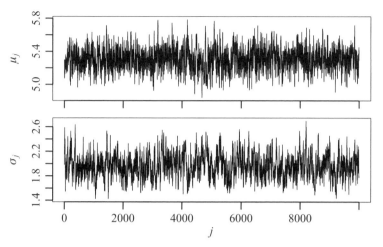

Figure C.8   Output of the program from exercise E4.5.

The path of the Markov chain is now stored in `theta`. The value `theta[j,1]` corresponds to μ_j and the entry `theta[j,1]` stores σ_j. These data can now be used for Monte Carlo estimates. Here we restrict ourselves to plotting the data:

```
j = 1:N
plot(j, theta[j,1], type="l", ylab=expression(mu[j]))
plot(j, theta[j,2], type="l", ylab=expression(sigma[j]))
```

The resulting plot is shown in Figure C.8.

Solution E4.6   The Gibbs sampler described in section 4.4.2 can be implemented in R as follows:

```
GeneratePosteriorSample <- function(N, burn.in=1000) {
  muj <- matrix(runif(2*k, -10, 10), nrow=k, ncol=2)
  Yj <- sample(1:k, n, replace=TRUE)
  for (j in 1:(burn.in + N)) {
    if (j %% 2 == 1) {
      for (a in 1:k) {
        ind <- Yj == a
        na <- sum(ind)
        if (na > 0) {
          m <- c(mean(x[ind,1]), mean(x[ind,2]))
          s <- sigma / sqrt(na)
          xi <- c(999, 999)  # this is re-set in the loop below
          while (xi[1] < -10 || xi[1] > 10
                 || xi[2] < -10 || xi[2] > 10) {
```

```
            xi = rnorm(2, m, s)
          }
          if (j > burn.in) {
            points(xi[1], xi[2], pch=a+1, cex=0.6)
          }
        } else {
          xi <- runif(2, -10, 10)
        }
        muj[a,] <- xi
      }
    } else {
      for (i in 1:n) {
        q <- numeric(length=k)
        for (a in 1:k) {
          q[a] <- (dnorm(x[i,1], muj[a,1], sigma)
                  * dnorm(x[i,2], muj[a,2], sigma))
        }
        Yj[i] <- sample(1:k, 1, prob=q)
      }
    }
  }
}
```

The function assumes that the number of modes is stored in the global variable k, the number of observations is stored in n and the observations are stored in an $n \times 2$ matrix x. For simplicity we plot the generated samples inside this function (using the points command) instead of returning the values.

To test the function we generate a synthetic data set from the model:

```
k <- 5
n <- 100
sigma <- 1

true.mu <- matrix(runif(2*k, -10, 10), nrow=k, ncol=2)
true.y <- sample(1:k, n, replace=TRUE)
x <- matrix(nrow=n, ncol=2)
x[,1] <- rnorm(n, true.mu[true.y, 1], sigma)
x[,2] <- rnorm(n, true.mu[true.y, 2], sigma)
```

A sample from the posterior distribution of the cluster means $\mu_1, \ldots, \mu_n$ given the observations stored in x can now be obtained by calling the function GeneratePosteriorSample as follows:

```
plot(x[,1], x[,2], xlim=c(-10,10), ylim=c(-10,10),
      xlab=expression(X[list(i,1)]), ylab=expression(X[list(i,2)]),
```

```
        col="gray60", asp=1)
    GeneratePosteriorSample(100)
```

The results of two different runs of this program are shown in Figure 4.5 (a typical output) and Figure 4.6 (an exceptional output, where the Markov chain has not yet converged).

Solution E4.7   Since the Gibbs sampler for the Ising model, as given by algorithm 4.31, can require a large number of iterations to reach equilibrium, we take some care to provide an efficient implementation:

- To avoid special-casing the pixels on the boundaries of the grid in step 4, we add extra rows/columns of zeros around the image, resulting in an extended grid size of $(L + 2) \times (L + 2)$.
- Since the variable d computed in step 4 can only take the five possible values $-4, -2, 0, 2,$ and 4, we can pre-compute the resulting probability weights

$$\pi_{X_m | X_{\neg m}}(+1 | x_{\neg m}) = \frac{\exp(+\beta d)}{\exp(+\beta d) + \exp(-\beta d)} = \frac{1}{1 + \exp(-2\beta d)},$$

to avoid evaluating the exponential function inside the main loop of the algorithm.

The resulting method can be implemented in R as follows:

```
RunGibbs <- function(M, beta, L) {
  X <- matrix(0, nrow=L+2, ncol=L+2)
  pixels <- sample(c(-1,+1), size=L*L, replace=TRUE)
  X[2:(L+1),2:(L+1)] <- matrix(pixels, nrow=L, ncol=L)
  p.table <- numeric(length=5)
  for (d in seq(-4, 4, by=2)) {
    p.table[(d+6)/2] <- 1 / (1 + exp(-2 * beta * d))
  }
  for (sweep in 1:M) {
    n <- matrix(0, nrow=L, ncol=L)
    for(ix in 2:(L+1)) {
      for (iy in 2:(L+1)) {
        d <- X[iy-1, ix] + X[iy+1, ix] + X[iy, ix-1] + X[iy, ix+1]
        p <- p.table[(d+6)/2]
        X[iy, ix] <- sample(c(-1, +1), size=1, prob=c(1-p, p))
      }
    }
  }
  return(X[2:(L+1),2:(L+1)])
}
```

For given inverse temperature β and given M, the program performs $N = M \cdot L^2$ steps of algorithm 4.31, that is it performs M 'sweeps' over the complete picture, and returns the final state $X^{(N)} \in S$. This state can then be plotted using R commands such as `image(1:L, 1:L, X)`. The results of different runs of this function are shown in Figure 4.7.

It is tempting to attempt to implement a much faster version of the program which would update all pixels in parallel. Unfortunately step 4 of algorithm 4.31 references pixels which have only been updated during the current 'sweep' over the image ($X_{m_1-1,m_2}^{(j-1)}$ and $X_{m_1,m_2-1}^{(j-1)}$), making parallel updates difficult.

Solution E4.8   For implementing algorithm 4.32 in R, we follow the steps described in the solution to exercise E4.7, and only modify the expression for sampling from the posterior distribution $p_{X_m|X_{-m}}$ to include the given observation Y of the original image:

```
RunGibbs <- function(M, beta, Y, sigma, burn.in=100) {
  L <- ncol(Y)
  count <- matrix(0, nrow=L, ncol=L)

  X <- matrix(0, nrow=L+2, ncol=L+2)
  pixels <- sample(c(-1,+1), size=L*L, replace=TRUE)
  X[2:(L+1),2:(L+1)] <- matrix(pixels, nrow=L, ncol=L)
  for (sweep in 1:(burn.in + M)) {
    for (ix in 2:(L+1)) {
      for (iy in 2:(L+1)) {
        d <- X[iy-1, ix] + X[iy+1, ix] + X[iy, ix-1] + X[iy, ix+1]
        p <- 1 / (1 + exp(-2 * (beta * d + Y[iy-1, ix-1] / sigma^2)))
        X[iy, ix] <- sample(c(-1, +1), size=1, prob=c(1-p, p))
      }
    }
    if (sweep > burn.in) {
      count <- count + (X[2:(L+1),2:(L+1)] == 1)
    }
  }
  return(count / M)
}
```

The return value of this function is a matrix of the same size as Y, giving the posterior probability that the corresponding pixel in the original image was black. The output of this function can be plotted using commands such as the following.

```
prob <- RunGibbs(N, beta, Y, sigma)
image(1:L, 1:L, prob)
```

The result of three different runs of this function is shown in Figure 4.8.

Solution E4.9   We start by implementing the mixture model from equation (4.47) in R. Samples from the mixture distribution can be generated using the following function:

```
model.clusters.mean <- 4  # the average number of clusters
model.r.min <- 0.5  # smallest possible stddev for a cluster
model.r.max <- 2.5  # largest possible stddev for a cluster

GenerateMixtureSample <- function(n) {
  # Sample the parameters:
  k <- rpois(1, lambda=model.clusters.mean-1) + 1
  mu <- matrix(nrow=2, ncol=k)
  mu[1,] <- runif(k, -10, 10)
  mu[2,] <- runif(k, -10, 10)
  sigma <- runif(k, model.r.min, model.r.max)

  # Generate a sample of size n from the mixture distribution:
  a <- sample.int(k, n, replace=TRUE)
  x_1 <- rnorm(n, mu[1,a], sigma[a])
  x_2 <- rnorm(n, mu[2,a], sigma[a])
  return(rbind(x_1, x_2))
}
```

We will need a function to evaluate the posterior distribution as given in equation (4.51). This is implemented in the function GetLogPosterior:

```
GetLogPrior <- function(mu, r) {
  # Since both, x and y coordinates of the mean have the same
  # distribution, we can merge them into one list:
  log.p.mu <- sum(dunif(as.vector(mu), -10, 10, log=TRUE))

  log.p.r <- sum(dunif(r, model.r.min, model.r.max, log=TRUE))

  k <- length(r)
  log.p.k <- dpois(k-1, lambda=model.clusters.mean-
1, log=TRUE)

  return(log.p.mu + log.p.r + log.p.k)
}

GetLogMixDens <- function(mu, r, x) {
  mix.dens <- mean(dnorm(mu[1,], x[1], r) * dnorm(mu[2,], x[2], r))
  return(log(mix.dens))
}

GetLogPosterior <- function(mu, r, obs) {
```

```
  log.prior <- GetLogPrior(mu, r)
  # Avoid calling GetLogMixDens for negative variances:
  if (log.prior == -Inf) return(-Inf)

  log.p.obs <- 0
  for (i in 1:ncol(obs)) {
    tmp <- GetLogMixDens(mu, r, obs[,i])
    log.p.obs <- log.p.obs + tmp
  }
  return(log.p.obs + log.prior)
}
```

To avoid problems caused by rounding errors, this function computes $\log \pi(k, x)$ instead of $\pi(k, x)$. Some care is needed, since the function GetLogPosterior may be called with 'impossible' parameter values, for example some components of r may be negative; in these cases, $\log 0 = -\infty$ must be returned.

Using the function GetLogPosterior to define the target density and using the different move types described in section 4.5.2, we can now implement algorithm 4.36.

The RJMCMC method can be implemented by the following R code. We start by introducing names for the different move probabilities:

```
MCMC.prob.mu.move <- 4 / 10
MCMC.prob.r.move <- 4 / 10
MCMC.prob.k.move <- 1 / 10
MCMC.prob.rot.move <- 1 / 10

MCMC.move.names = c("mu", "r", "k", "rot")
MCMC.move.probabilities <- c(MCMC.prob.mu.move,
                             MCMC.prob.r.move,
                             MCMC.prob.k.move,
                             MCMC.prob.rot.move)
names(MCMC.move.probabilities) <- MCMC.move.names
```

Moves of types m_μ and m_r depend on the parameters σ_μ and σ_r, respectively. Experimentation shows that the following values lead to reasonable acceptance rates:

```
MCMC.sigma.mu <- 0.28
MCMC.sigma.r <- 0.22
```

Next we define functions to compute the acceptance probabilities given in equation (4.52), equation (4.53), equation (4.55), equation (4.56), equation (4.57) and equation (4.58).

```
GetMuMoveAlpha <- function(GetLogPi, mu1, mu2, r) {
```

```
    return(exp(GetLogPi(mu2, r) - GetLogPi(mu1, r)))
}

GetRadiusMoveAlpha <- function(GetLogPi, mu, r1, r2) {
    return(exp(GetLogPi(mu, r2) - GetLogPi(mu, r1)))
}

GetLogPsi <- function(mu, r) {
    log.p.mu <- sum(dunif(mu, -10, 10, log=TRUE))
    log.p.r <- dunif(r, model.r.min, model.r.max, log=TRUE)
    return(log.p.mu + log.p.r)
}

GetAddMoveAlpha <- function(GetLogPi, mu1, mu2, r1, r2) {
    k.new <- length(r2)
    a <- exp(GetLogPi(mu2, r2)
            - GetLogPi(mu1, r1)
            - GetLogPsi(mu2[,k.new], r2[k.new]))
    return(ifelse(k.new == 2, a / 2, a))
}

GetRemoveMoveAlpha <- function(GetLogPi, mu1, mu2, r1, r2) {
    k.old <- length(r1)
    a <- exp(GetLogPi(mu2, r2)
            + GetLogPsi(mu1[,k.old], r1[k.old])
            - GetLogPi(mu1, r1))
    return(ifelse(k.old == 2, 2 * a, a))
}
```

With all these preparations in place, we can now implement algorithm 4.36:

```
RunRJMCMC <- function(GetLogPi, N, count.every=1) {
    types <- length(MCMC.move.probabilities)
    count.proposals <- rep(0, types)
    names(count.proposals) <- MCMC.move.names
    count.accepts <- rep(0, types)
    names(count.accepts) <- MCMC.move.names

    path.k <<- c()
    for (j in 1:N) {
        move.type = sample.int(types, 1,
                                prob=MCMC.move.probabilities)
        count.proposals[move.type] <- count.proposals[move.type] + 1

        k2 <- k
```

```
Mu2 <- Mu
R2 <- R
if (move.type == 1) {  # try a mean move
  a <- sample.int(k, 1)
  Mu2[,a] <- rnorm(2, Mu[,a], MCMC.sigma.mu)
  alpha <- GetMuMoveAlpha(GetLogPi, Mu, Mu2, R)
} else if (move.type == 2) {  # try a radius move
  a <- sample.int(length(R), 1)
  R2[a] <- rnorm(1, R[a], MCMC.sigma.r)
  alpha <- GetRadiusMoveAlpha(GetLogPi, Mu, R, R2)
} else if (move.type == 3) {
  if (k == 1 || runif(1) < 0.5) { # try to add a component
    k2 <- k + 1
    Mu2 <- cbind(Mu, runif(2, -10, 10))
    R2 <- c(R, runif(1, model.r.min, model.r.max))
    alpha <- GetAddMoveAlpha(GetLogPi, Mu, Mu2, R, R2)
  } else {  # try to remove a component
    k2 <- k - 1
    # prevent R from converting Mu2 to a vector if k2==1:
    Mu2 <- as.matrix(Mu[,-k], nrow=k2, ncol=2)
    R2 <- R[-k]
    alpha <- GetRemoveMoveAlpha(GetLogPi, Mu, Mu2, R, R2)
  }
} else {  # cyclic right shift of components by a:
  if (k > 1) {
    a <- sample.int(length(R)-1, 1)
    Mu2[,1:a] <- Mu[,(k-a+1):k]
    Mu2[,(a+1):k] <- Mu[,1:(k-a)]
    R2[1:a] <- R[(k-a+1):k]
    R2[(a+1):k] <- R[1:(k-a)]
    alpha <- 1
  }
}

if (runif(1) < alpha) {
  # accept the move
  count.accepts[move.type] <- count.accepts[move.type] + 1

  # store the updated values in the global namespace
  k <<- k2
  Mu <<- Mu2
  R <<- R2
}
if (j %% count.every == 0) {
  path.k <<- c(path.k, k)
```

```
      }
    }

    cat("acceptance rates for each move type:\n")
    print(count.accepts / count.proposals, digits=3)
    cat("\n")
  }
```

The argument GetLogPi to the function RunRJMCMC must be a function which returns the logarithm $\log \pi(k, \mu, r)$. Since k can be found by inspecting μ, the function GetLogPi just takes Mu and R as arguments. The function uses the global variables k, Mu and R to store the current state of the Markov chain. These variables need to be set before RunRJMCMC is called, and they contain the final state of the Markov chain on return from this function.

To test the function RunRJMCMC, we first run the RJMCMC algorithm without any observations, that is in the case where π is the prior distribution:

```
  k <- rpois(1, lambda=model.clusters.mean-1) + 1
  Mu <- matrix(runif(2*k, -10, 10), nrow=2, ncol=k)
  R <- runif(k, model.r.min, model.r.max)

  RunRJMCMC(GetLogPrior, 10000)   # burn-in
  RunRJMCMC(GetLogPrior, 1000000, count.every=100)

  h <- hist(path.k, breaks=seq(-0.5, max(path.k) + 1, by=1),
            prob=TRUE, xlab="k", ylab="probability",
            main=NULL, col="gray80", border="gray20")
  p <- c(0, dpois(0:9, lambda=model.clusters.mean-1))
  points(0:10, p)
```

The resulting histogram is shown in Figure C.9.

Finally, we can run the function RunRJMCMC on data generated from the model:

```
  k <- rpois(1, lambda=model.clusters.mean-1) + 1
  Mu <- matrix(runif(2*k, -10, 10), nrow=2, ncol=k)
  R <- runif(k, model.r.min, model.r.max)

  obs <- GenerateMixtureSample(80)
  GetLogPi <- function(mu, r) {
    return(GetLogPosterior(mu, r, obs))
  }
  RunRJMCMC(GetLogPi, 5000)   # burn-in
  RunRJMCMC(GetLogPi, 50000, count.every=100)
```

To visualise the results of the run, we define two auxiliary functions for plotting:

```
  PlotObservations <- function(obs) {
    plot(obs[1,], obs[2,], xlim=c(-13,13), ylim=c(-11,11),
```

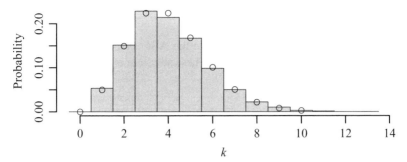

Figure C.9   A histogram showing the distribution of k in the RJMCMC algorithm from exercise E4.9, in the case where there are no observations and the target distribution coincides with the prior distribution π. The bars in the histogram give the frequencies observed in one run of the algorithm, the circles denote exact probabilities expected for the stationary distribution. Both sets of values coincide well, giving confidence that the implementation of the algorithm works correctly.

```
        asp=1, cex=0.5, xlab="", ylab="")
   rect(-10, -10, 10, 10)
}

PlotState <- function(mu, r) {
   symbols(mu[1,], mu[2,], circles=r,
        inches=FALSE, add=TRUE)
   symbols(mu[1,], mu[2,], circles=1.5*r,
        inches=FALSE, add=TRUE)
   symbols(mu[1,], mu[2,], circles=2*r,
        inches=FALSE, add=TRUE)
}
```

Using these functions, we can plot the final state of the RJMCMC Markov chain, together with a histogram of K_j.

```
par(mai=c(0.35, 0.35, 0.05, 0.05))
PlotObservations(obs)
PlotState(Mu, R)

par(mai=c(0.65, 0.75, 0.15, 0.05))
hist(path.k, breaks=seq(-0.5, max(path.k) + 0.5, by=1), prob=TRUE,
     xlab="K", ylab="probability",
     main=NULL, col="gray80", border="gray20")
```

The resulting plot is shown in Figure 4.9.

C.5   Answers for Chapter 5

Solution E5.1   The function to generate the ABC samples is modelled directly after the steps described in example 5.4. These steps are executed in a loop until N samples are accumulated.

```
S <- function(x) {
  return(c(sum(x), sum(x^2)) / n)
}

ABC.basic <- function(N, n, s, delta) {
  res.theta <- matrix(nrow=N, ncol=2) # col. 1: mu, col. 2: sigma
  res.s <- matrix(nrow=N, ncol=2) # row j = s_j
  j <- 1
  k <- 1
  while (k <= N) {
    mu <- runif(1, -10, 10)
    sigma <- rexp(1)
    X <- rnorm(n, mu, sigma)
    sj <- S(X)
    if (sum((sj-s)^2) <= delta^2) {
      res.theta[k,] <- c(mu, sigma)
      res.s[k,] <- sj
      k <- k + 1
    }
    j <- j + 1
  }
  cat("delta=", delta, ": ", j/N, " proposals/sample\n",
  sep="")
  return(list(theta=res.theta, s=res.s))
}
```

The newly defined function ABC.basic does not only return the samples θ_{j_k}, but also the values s_{j_k}. The values s_j are sometimes useful, for example in methods which weight the samples according to the discrepancy $s_j - s^*$. The function ABC.basic can be used as follows:

```
n <- 20
s.obs <- c(6.989, 52.247)
X <- ABC.basic(100, n, s.obs, 0.1)
```

Figure 5.1 shows histograms of the output values X$theta[,1] (corresponding to μ) and X$theta[,2] (corresponding to σ) for different values of δ.

Solution E5.2   Using the function `ABC.basic` from exercise E5.1, we can implement algorithm 5.6 as follows:

```
ABC.regression <- function(N, n, s, delta) {
  samples <- ABC.basic(N, n, s, delta)
  S <- cbind(samples$s, 1)
  C <- t(S) %*% samples$theta
  B.hat <- solve(t(S) %*% S, C)
  beta.hat <- B.hat[1:2, 1:2]
  correction <- t((s - t(samples$s))) %*% beta.hat
  theta.tilde <- samples$theta + correction
  return(list(theta=theta.tilde))
}
```

Here, we return a list with just one element `theta` so that `ABC.regression` can be used as a drop-in replacement for `ABC.basic`. Different from the function in exercise E5.1, we do not return the s_j, since the correction makes these values largely meaningless for the returned values. The function `ABC.regression` can be called as follows:

```
n <- 20
s.obs <- c(6.989, 52.247)
samples <- ABC.basic(100, n, s.obs, 0.1)
```

Solution E5.3   Algorithm 5.15 can be implemented in R as follows:

```
bias.boot <- function(theta.hat, x, N=10000) {
  n <- length(x)
  theta.star <- c()
  for (j in 1:N) {
    X.star <- sample(x, n, replace=TRUE)
    theta.star[j] <- theta.hat(X.star)
  }
  cat("RMSE* = ", sd(theta.star)/sqrt(N), "\n", sep="")
  return(mean(theta.star) - theta.hat(x))
}
```

The first argument of this function must be an implementation of the estimator in question. For the estimator $\hat{\sigma}^2$ from the question we can use the following function.

```
sigma.squared <- function(x) {
  x.bar <- mean(x)
  return(mean((x - x.bar)^2))
}
```

The bootstrap estimate for the bias can then be computed with calls such as:

```
bias.boot(sigma.squared, x)
```

To test the function, we use a generated data set x:

```
x <- rnorm(100)
```

The resulting estimate of the bias is then

```
> bias.boot(sigma.squared, x)
RMSE* = 0.001194376
[1] -0.009388906
```

Since this result is close to the theoretical value $-\mathrm{Var}(X)/\sqrt{n} = -0.1$ for the bias, we can assume that our implementation works correctly.

The Monte Carlo sample size N affects the size of error in the approximation (5.22). To justify our choice of N we first note that, by proposition 3.14, the magnitude of this error is given by the root-mean-square error

$$\mathrm{RMSE} = \frac{\mathrm{stdev}^*\left(f(X^*)\right)}{\sqrt{N}}.$$

The function `bias.boot` prints an estimate for the RMSE; for the example the value is approximately 0.0012. With 95% probability the magnitude of the Monte Carlo error is less than $1.96 \cdot \mathrm{RMSE} \approx 0.0023$. Thus, we expect the error introduced by the Monte Carlo integration to be approximately 25%. Since this ratio is still quite large, we increase the value of N from 10 000 to 100 000 by specifying the optional, third argument to `bias.boot`:

```
> bias.boot(sigma.squared, x, 100000)
RMSE* = 0.0003801055
[1] -0.0107199
```

This brings the Monte Carlo error down to about 7% which seems acceptable, since the result is only approximate even in the absence of Monte Carlo error.

Solution E5.4   If we interpret each head as $X = 1$ and each tail as $X = 0$, then the probability p of heads can be written as $p = \mathbb{E}(X)$, that is p is the mean of the distribution of X, and our estimate is $\hat{p} = \sum_{i=1}^{n} X_i/n = \bar{X}$. The bootstrap estimate of the standard error of the mean is

$$\widehat{\mathrm{se}}^*(\hat{p}) = \sqrt{\frac{1}{n^2} \sum_{i=1}^{n} (X_i - \bar{X})^2}.$$

Each of the 1600 terms in the sum where $X_i = 1$ contributes $(1 - 16/25)^2 = 9^2/25^2$. Each of the 900 terms in the sum where $X_i = 0$ contributes $(0 - 16/25)^2 = 16^2/25^2$. Thus we get

$$\widehat{se}^*(\hat{p}) = \sqrt{\frac{1}{n^2}\left(1600 \cdot \frac{9^2}{25^2} + 900 \cdot \frac{16^2}{25^2}\right)}$$

$$= \sqrt{\frac{1}{n^2}\frac{100 \cdot 16 \cdot 9 \cdot (9 + 16)}{(16 + 9)^2}}$$

$$= \frac{10 \cdot 4 \cdot 3}{n \cdot 5}$$

$$= \frac{24}{2500}.$$

This is the bootstrap estimate of the standard error of $\hat{p}$.

Solution E5.5   We can implement algorithm 5.20 in R as follows:

```
GetSimpleBootstrapCI <- function(x, theta.hat, N, alpha=0.05) {
  # generate the bootstrap samples
  n <- length(x)
  theta.star <- c()
  for (j in 1:N) {
    X.star <- sample(x, n, replace=TRUE)
    theta.star[j] <- theta.hat(X.star)
  }

  # construct the confidence interval
  t <- theta.hat(x)
  l <- ceiling(0.5 * alpha * N)
  u <- ceiling((1 - 0.5 * alpha) * N)
  theta.star <- sort(theta.star)
  return(c(2 * t - theta.star[u], 2 * t - theta.star[l]))
}
```

In order to test GetSimpleBootstrapCI, we consider confidence intervals for the mean. Algorithm 5.20 requires the plug-in estimator for the parameter in question; from example 5.13 we know that the plug-in estimator for the mean is the average $\bar{x}$ of the sample x. This estimator, in form of the built-in R function mean, can be used as the parameter theta.hat of GetSimpleBootstrapCI. To check our function, we compare the output to the exact confidence interval. Assuming the variance is not known, the exact confidence interval is given by

$$[U, V] = \left[\bar{x} - \frac{t_{1-\alpha/2, n-1} \hat{\sigma}}{\sqrt{n}}, \bar{x} + \frac{t_{1-\alpha/2, n-1} \hat{\sigma}}{\sqrt{n}} \right],$$

where $\hat{\sigma}^2 = \sum_{i=1}^{n} (x_i - \bar{x})^2 / (n - 1)$ and $t_{1-\alpha/2, n-1}$ denotes the $1 - \alpha/2$ quantile of the t-distribution with $n - 1$ degrees of freedom. We can compute the exact confidence interval using the following R code.

```
GetExactCI <- function(x, alpha=0.05) {
  n <- length(x)
  m <- mean(x)
  s <- sd(x)
  c <- qt(1 - 0.5*alpha, n-1)
  return(c(m - c * s / sqrt(n), m + c * s / sqrt(n)))
}
```

Finally, in order to easily perform repeated comparisons, we write a function which, for a given sample size, prints the output of both GetSimpleBootstrapCI and GetExactCI to the screen.

```
Test <- function(n, alpha=0.05) {
  X <- rnorm(n)
  ci.bootstrap <- GetSimpleBootstrapCI(X, mean, 1000)
  cat("bootstrap CI: [", ci.bootstrap[1], ", ",
      ci.bootstrap[2], "]\n", sep="")
  ci.exact <- GetExactCI(X)
  cat("exact CI:     [", ci.exact[1], ", ",
      ci.exact[2], "]\n", sep="")
}
```

Using this function, we can perform tests as follows:

```
> Test(10)
bootstrap CI: [-0.5593357, 0.206685]
exact CI:     [-0.6447165, 0.3395043]
> Test(50)
bootstrap CI: [-0.08583986, 0.448177]
exact CI:     [-0.1023353, 0.4421098]
> Test(100)
bootstrap CI: [-0.1195222, 0.2811508]
exact CI:     [-0.1205862, 0.2828386]
```

The output shows that there is reasonably good agreement between the bootstrap confidence intervals and the theoretically exact confidence intervals. Thus, we can assume that our implementation of algorithm 5.20 works correctly.

Solution E5.6   First we generate the test data set:

```
X <- rnorm(25)
```

From example 5.14 we know that the plug-in estimate for the variance is given by $\hat{\sigma}_n(x) = \sum_{i=1}^{n}(x_i - \bar{x})^2/n$. Since the built-in R function var uses $n - 1$ instead of n, we replace var by our own version:

```
PlugInVar <- function(x) {
  return(mean((x - mean(x))^2))
}
```

Next, we write a function to estimate the standard deviation of the boundary points and the average width of the confidence interval for the variance, always using the data set X we created above:

```
PrintStdDevAndMean <- function(N) {
  lower <- c()
  upper <- c()
  width <- c()
  for (i in 1:1000) {
    ci <- GetSimpleBootstrapCI(X, PlugInVar, N);
    lower[i] <- ci[1]
    upper[i] <- ci[2]
    width[i] <- ci[2] - ci[1]
  }
  cat("lower bound stddev: ", sd(lower), "\n", sep="")
  cat("upper bound stddev: ", sd(upper), "\n", sep="")
  cat("average width: ", mean(width), "\n", sep="")
}
```

Finally, we call this function for different values of N:

```
> PrintStdDevAndMean(1000)
lower bound stddev: 0.01664069
upper bound stddev: 0.0102032
average width: 0.6364756
> PrintStdDevAndMean(2000)
lower bound stddev: 0.01174421
upper bound stddev: 0.007007113
average width: 0.6358301
> PrintStdDevAndMean(4000)
lower bound stddev: 0.008348285
upper bound stddev: 0.005002864
average width: 0.6371888
```

As expected, the standard deviation of and thus the error in the boundaries decreases. But even for $N = 4000$ the standard deviation of either boundary is still approximately 1% of the width of the interval, so the error is still significant.

Solution E5.7   The BC_a algorithm 5.21 can be implemented in R as follows:

```
GetBCaBootstrapCI <- function(x, theta.hat, N, alpha=0.05) {
  # generate the bootstrap samples
  n <- length(x)
  theta.star <- c()
  for (j in 1:N) {
    X.star <- sample(x, n, replace=TRUE)
    theta.star[j] <- theta.hat(X.star)
  }

  # compute z.hat and a.hat
  z.hat <- qnorm(sum(theta.star < theta.hat(x)) / N)
  theta.jack <- c()
  for (i in 1:n) {
    theta.jack[i] <- theta.hat(x[-i])
  }
  y <- theta.jack - mean(theta.jack)
  a.hat <- (sum(y^3) / (6 * sum(y^2)^1.5))

  # pick out the BCa quantiles
  p <- c(0.5 * alpha, 1 - 0.5 * alpha)
  w <- z.hat + qnorm(p)
  q <- pnorm(z.hat + w / (1 - a.hat * w))
  idx <- ceiling(q * N)
  theta.star <- sort(theta.star)
  return(theta.star[idx])
}
```

Solution E5.8   The Monte Carlo estimate for the coverage probabilities can be implemented in R as follows:

```
TestConfInt <- function(get.ci, gen.sample, n, theta, theta.hat) {
  res <- c(0, 0, 0)
  names(res) <- c("left", "inside", "right")
  for (i in 1:1000) {
    x <- gen.sample(n, theta)
    ci <- get.ci(x, theta.hat, 5000)
    if (theta < ci[1]) {
      res[1] <- res[1] + 1
```

```
    } else if (theta > ci[2]) {
      res[3] <- res[3] + 1
    } else {
      res[2] <- res[2] + 1
    }
  }
  return(res / sum(res))
}
```

Using this function, we can now compare the two different methods for estimating confidence intervals.

```
CompareConfInts <- function(gen.sample, n, theta, theta.hat) {
  res.simple <- TestConfInt(GetSimpleBootstrapCI, gen.sample,
                            n, theta, theta.hat)
  res.BCa <- TestConfInt(GetBCaBootstrapCI, gen.sample,
                          n, theta, theta.hat)
  res <- rbind(res.simple, res.BCa)
  rownames(res) <- c("simple", "BCa")
  return(res)
}


# test 1: mean of standard normal random variables
print(CompareConfInts(rnorm, 10, 0, mean))
print(CompareConfInts(rnorm, 50, 0, mean))
print(CompareConfInts(rnorm, 100, 0, mean))


# test 2: mean of exponentially distributed random variables
gen2 <- function(n, mu) {
  return(rexp(n, 1/mu))
}
print(CompareConfInts(gen2, 10, 1, mean))
print(CompareConfInts(gen2, 50, 1, mean))
print(CompareConfInts(gen2, 100, 1, mean))


# test 3: standard deviation of normal random variables
gen3 <- function(n, sigma) {
  return(rnorm(n, 0, sigma))
}
theta.hat3 <- function(x) {
  x.bar <- mean(x)
  return(sqrt(sum((x - x.bar)^2)/length(x)))
}
print(CompareConfInts(gen3, 10, 1, theta.hat3))
print(CompareConfInts(gen3, 50, 1, theta.hat3))
print(CompareConfInts(gen3, 100, 1, theta.hat3))
```

The resulting output of one run of this program is summarised in Table 5.1. The results shown in the table indicate that for the test problems considered here, the performance of both methods is comparable. The data also show that the quality of the confidence intervals improves as the sample size n increases.

C.6   Answers for Chapter 6

Solution E6.1   We construct the solution iteratively, starting at time $t = 0$ and then computing B_{ih} from $B_{(i-1)h}$ for $i = 1, 2, \ldots, n$ by adding $\mathcal{N}(0, h)$-distributed random increments as described in algorithm 6.6. In R, this can be implemented as follows:

```
BrownianMotion1d <- function(t) {
  s.prev <- 0  # initial time
  B <- 0   # initial value
  res <- c()
  for (s in t) {
    h <- s - s.prev
    dB <- rnorm(1, 0, sqrt(h))
    B <- B + dB
    res <- c(res, B)
    s.prev <- s
  }
  return(res)
}
```

Since R has a built-in function cumsum for computing cumulative sums, we can use this function to write a shorter and more efficient implementation of the same algorithm:

```
BrownianMotion1d <- function(t) {
  n <- length(t)
  h <- t - c(0, t[-n])
  dB <- rnorm(n, 0, sqrt(h))
  B <- cumsum(dB)
  return(B)
}
```

(C.1)

We test the new function BrownianMotion1d by creating a plot of one path of a Brownian motion until time $T = 10$:

```
t <- seq(0, 100, by=0.01)
B <- BrownianMotion1d(t)
plot(t, B, type="l", xlab="t", ylab=expression(B[t]))
```

The resulting plot is shown in Figure 6.1.

Solution E6.2   We construct the solution iteratively, starting at time $t = 0$ and then computing B_{ih} from $B_{(i-1)h}$ for $i = 1, 2, \ldots, n$ by adding $\mathcal{N}(0, h)$-distributed random increments as described in algorithm 6.6. In R, this can be implemented as follows:

```
BrownianMotion <- function(t, d=1) {
  n <- length(t)
  sqrt.h <- sqrt(t - c(0, t[-n]))
  B <- c()
  for (i in 1:d) {
    dB <- rnorm(n, 0, sqrt.h)
    B <- cbind(B,cumsum(dB))
  }
  return(B)
}
```

We test the new function BM by creating a plot of one path of a two-dimensional Brownian motion until time $T = 10$:

```
n <- 32000
t <- seq(0, 10, length.out=n+1)
B <- BrownianMotion(t, d=2)
plot(B[,1], B[,2], asp=1, type="l", lwd=0.1,
     xlab=expression(B[t]^(1)), ylab=expression(B[t]^(2)))
col <- gray(n:0/n*0.8)
segments(B[1:(n-1),1], B[1:(n-1),2], B[2:n,1], B[2:n,2],
         asp=1, col=col)
```

The resulting plot is shown in Figure 6.2.

Solution E6.3   A function B to return samples from a single Brownian path can be implemented in R as follows:

```
t.known <- 0;  # times sampled so far
B.known <- 0;  # values of B corresponding to the times
in t.known

B <- function(times) {
  res <- c()
  for (s in times) {
    i <- max(c(which(t.known < s), 1))
    r <- t.known[i]
    if (s == r) {
      Bs <- B.known[i]
    } else {
```

```
      if (i < length(t.known)) {
        t <- t.known[i+1]
        mu <- B.known[i]*(t-s)/(t-r)+B.known[i+1]*(s-r)/(t-r)
        sigma <- sqrt((t-s)*(s-r)/(t-r))
      } else {
        mu <- B.known[i]
        sigma <- sqrt(s-r)
      }
      Bs <- rnorm(1, mu, sigma)
      t.known <<- append(t.known, s, after=i)
      B.known <<- append(B.known, Bs, after=i)
    }
    res <- c(res, Bs)
  }
  return(res)
}
```

The function works by storing all values simulated so far in the global variable B.known and the corresponding times in the global variable t.known.

The function B can be used like an ordinary function, the argument can be either a single time or a vector of times. The following commands illustrate how to use B:

```
t <- seq(0, 20, by=2)
plot(t, B(t), type="l", ylim=c(min(B(t))-0.5, max(B(t))+0.5),
    ylab=expression(B[t]), col="gray", lwd=1.4)

t <- seq(0, 20, by=0.02)
lines(t, B(t))
```

The plot created by these commands is shown in Figure 6.4.

Solution E6.4   To generate a Brownian bridge, we first generate a Brownian motion and then use equation (6.2) to convert this into a Brownian bridge. Using the function BrownianMotion1d from listing (C.1), we can implement this method in R as follows:

```
BrownianBridge <- function(s, r=0, a=0, t=1, b=0) {
  s.shifted <- c(s-r, t-r)  # make sure we sample B_{t-r}
  B <- BrownianMotion1d(s.shifted)
  Bt <- B[length(B)]
  X <- B + a - s.shifted / (t - r) * (Bt - b + a)
  return(X[-length(X)])
}
```

Solution E6.5   We can use the following function to simulate a path of a geometric Brownian motion, based on the function `BrownianMotion1d` from listing (C.1):

```
GeometricBM <- function(t, x0, alpha, beta, B) {
  if (missing(B)) {
    B <- BrownianMotion1d(t)
  }
  X <- x0 * exp(alpha * B + beta * t)
  return(X)
}
```

As in the solution to exercise E6.1, we use the `cumsum` function to efficiently simulate the Brownian motion B. We can plot paths of the simulated geometric Brownian motion with commands such as:

```
t <- seq(0, 10, by=0.01)
X <- GeometricBM(t, x0=1, alpha=1, beta=-0.1)
plot(t, X, type="l", xlab="t", ylab=expression(X[t]))
```

The plot resulting from one run of this program is shown in Figure 6.5.

Solution E6.6   From lemma 6.9 we know that $X_t = \exp(B_t - t/2)$ has expectation

$$\mathbb{E}(X_t) = 1 \cdot \exp\big((1/2 - 1/2)t\big) = \exp(0) = 1.$$

We can use the following R code to generate Monte Carlo estimates for this value with $t = 0, 1, 2, \ldots, 100$ and to generate a plot of the resulting estimates:

```
N <- 1e6
tt <- seq(0, 100, by=1)
xx <- c()
for (t in tt) {
  Bt <- rnorm(N, 0, sqrt(t))
  Xt <- exp(Bt - 0.5*t)
  xx <- c(xx, mean(Xt))
}
plot(tt, xx, ylim=c(0,3), xlab="t", ylab=expression(E(X[t])))
abline(h=1)
```

The plot resulting from one run of this program is shown in Figure 6.6. The figure shows that for $t \le 15$ the Monte Carlo estimates are accurate and for $15 < t \le 40$ the estimates have high variance but may be still acceptable. For $t > 40$ most of the estimates are close to 0 instead of being close to the exact value 1, and thus for this range of times t the sample size $N = 10^6$ is much too small.

Solution E6.7   To solve SDE (6.17) with the given drift and diffusion coefficient, we implement the Euler-Maruyama method from algorithm 6.12. This can be done by using an R function such as:

```
EulerMaruyama1d <- function(t, x0, mu, sigma, B) {
  n <- length(t)
  h <- t - c(0, t[-n])
  if (missing(B)) {
    dB <- rnorm(n, 0, sqrt(h))
  } else {
    if (length(B) != n) {
      stop("lengths of t and B don't match (",
          n, " vs. ", length(B), ")")
    }
    dB <- B - c(0, B[-n])
  }

  path <- c()
  s = 0
  X = x0
  for (i in 1:n) {
    X <- X + mu(s, X) * h[i] + sigma(s, X) * dB[i]
    s <- t[i]
    path[i] <- X
  }
  return(path)
}
```

A function which works for $d > 1$ or $m > 1$ can be implemented as follows:

```
EulerMaruyama <- function(t, x0, mu, sigma, B) {
  n <- length(t)

  ## dimension of the solution (from the initial condition)
  d <- length(x0)

  ## dimension of the noise (from the columns of sigma)
  m <- ncol(sigma(0, x0))
  if (! is.numeric(m)) m <- 1;

  h <- t - c(0, t[-n])
  if (missing(B)) {
    dB <- matrix(rnorm(n*m, 0, sqrt(h)), nrow=n, ncol=m)
  } else {
    B <- as.matrix(B)
```

```
    if (nrow(B) != n) {
      stop("lengths of t and B don't match (",
           n, " vs. ", nrow(B), ")")
    }
    if (ncol(B) != m) {
      stop("dimensions of sigma and B don't match (",
           m, " vs. ", ncol(B), ")")
    }
    dB = B - rbind(rep(0, d), as.matrix(B[-n,]))
  }

  path <- matrix(nrow=n, ncol=d)
  s = 0
  X = x0
  for (i in 1:n) {
    X <- X + mu(s, X) * h[i] + sigma(s, X) %*% dB[i,]
    s <- t[i]
    path[i,] <- X
  }
  return(path)
}
```

For one-dimensional SDEs, `EulerMaruyama1d` takes about half the time that `EulerMaruyama` does to solve the same SDE.

In order to use the function `EulerMaruyama`, we have to provide implementations of the drift μ and of the diffusion coefficient σ as R functions. For the given functions, this can be done as follows:

```
mu <- function(t, x) { c(x[2], x[1]*(1-x[1]^2)) }
sigma <- function(t, x) {
  ifelse(x[1]> 0 && x[2]> 0, 0.2, 0.03) * diag(2)
}

x0 <- c(-0.03, -0.49)
t <- seq(0, 15, by=0.001)
X <- EulerMaruyama(t, x0, mu, sigma)
```

Then the path of X in $\mathbb{R}^2$ can be plotted as follows:

```
plot(X[,1], X[,2], type="l")
```

The output, together with arrows visualising the drift μ, is shown in Figure 6.7.

Solution E6.8   The Milstein scheme from algorithm 6.15 can be implemented in R as shown in the following. Here we allow the functions `mu` and `sigma` to depend on the time t, in order to make the function more compatible with the `EulerMaruyama1d` function from exercise E6.7.

```
Milstein1d <- function(t, x0, mu, sigma, sigma.prime, B) {
  n <- length(t)
  h <- t - c(0, t[-n])
  if (missing(B)) {
    dB <- rnorm(n, 0, sqrt(h))
  } else {
    if (length(B) != n) {
      stop("lengths of t and B don't match (",
           n, " vs. ", length(B), ")")
    }
    dB <- B - c(0, B[-n])
  }

  path <- c()
  s = 0
  X = x0
  for (i in 1:n) {
    sig <- sigma(t, X)
    X <- (X
          + mu(s, X) * h[i]
          + sigma(s, X) * dB[i]
          + 0.5 * sig * sigma.prime(s, X) * (dB[i]^2 - h[i]))
    s <- t[i]
    path[i] <- X
  }
  return(path)
}
```

The function can be used as follows:

```
mu <- function(t, x) { 0.1 }
sigma <- function(t, x) { x }
sigma.prime <- function(t, x) { 1 }
t <- seq(0, 20, length.out=10001)
X <- Milstein1d(t, 1, mu, sigma, sigma.prime)
plot(t, X, type="l")
```

Solution E6.9   In order to determine the strong error using Monte Carlo integration, we will need to solve SDE (6.20) many times, the and efficiency of the method becomes important. Since the strong error in (6.23) only depends on the final point X_T

of the path, we introduce a specialised, faster version of the function `EulerMaruyama1d` (see Exercise 6.7), which returns only the endpoint instead of the whole path.

```
EulerMaruyama1dEndpoint <- function(t, x0, mu, sigma, B) {
  n <- length(t)
  h <- t - c(0, t[-n])
  if (missing(B)) {
    dB <- rnorm(n, 0, sqrt(h))
  } else {
    if (length(B) != n) {
      stop("lengths of t and B don't match (",
           n, " vs. ", length(B), ")")
    }
    dB <- B - c(0, B[-n])
  }

  s = 0
  X = x0
  for (i in 1:n) {
    X <- X + mu(s, X) * h[i] + sigma(s, X) * dB[i]
    s <- t[i]
  }
  return(X)
}
```

As a result of the simplified code, a call to `EulerMaruyama1dEndpoint` takes only approximately 30% of the time `EulerMaruyama1d` does. A similar simplification can be applied to `Milstein1d` from exercise E6.8:

```
Milstein1dEndpoint <- function(t, x0, mu, sigma, sigma.prime, B) {
  n <- length(t)
  h <- t - c(0, t[-n])
  if (missing(B)) {
    dB <- rnorm(n, 0, sqrt(h))
  } else {
    if (length(B) != n) {
      stop("lengths of t and B don't match (",
           n, " vs. ", length(B), ")")
    }
    dB <- B - c(0, B[-n])
  }

  s = 0
  X = x0
  for (i in 1:n) {
```

```
    sig <- sigma(t, X)
    X <- (X
          + mu(s, X) * h[i]
          + sigma(s, X) * dB[i]
          + 0.5 * sig * sigma.prime(s, X) * (dB[i]^2 - h[i]))
    s <- t[i]
  }
  return(X)
}
```

Next, we define the drift and diffusion coefficient.

```
alpha <- 1
beta <- 0.4
mu <- function(t,x) { (0.5*alpha^2 + beta) * x }
sigma <- function(t,x) { alpha * x }
sigma.prime <- function(t,x) { alpha }

x0 <- 1  # initial value
T <- 5  # time horizon
```

Now we have to simulate solutions of the SDE repeatedly, for different grid sizes, and to compute the strong error from the results. This can be done as follows:

```
n.max <- 1024  # largest discretisation parameter to try
n.min <- 64  # smallest discretisation parameter to try
K <- 9  # number of resolution steps
n <- round(exp(seq(log(n.min), log(n.max), length.out=K)))

N <- 20000  # number of runs to average over
samples.euler <- matrix(nrow=N, ncol=K)
samples.milstein <- matrix(nrow=N, ncol=K)
progress <- txtProgressBar(min=0, max=N, style=3)
t <- seq(0, T, length.out=n.max+1)
for (j in 1:N) {
  B <- c(0, cumsum(rnorm(n.max, 0, sqrt(t[2:(n.max+1)]-
  t[1:n.max]))))
  X.exact <- x0 * exp(alpha*B + beta*t)
  Xt.exact <- X.exact[n.max+1]

  for (k in 1:K) {
    I <- round(seq(1, n.max+1, length.out=n[k]+1))

    Xt.euler <- EulerMaruyama1dEndpoint(t[I], x0, mu, sigma, B[I])
    samples.euler[j,k] <- abs(Xt.euler - Xt.exact)
```

```
   Xt.milstein <- Milstein1dEndpoint(t[I], x0, mu, sigma,
                                     sigma.prime, B[I])
   samples.milstein[j,k] <- abs(Xt.milstein - Xt.exact)
 }
 setTxtProgressBar(progress, j)
}
close(progress)

error.euler <- colMeans(samples.euler)
error.milstein <- colMeans(samples.milstein)
```

Finally, we use the resulting data to create a plot:

```
h <- T / n
plot(h, error.euler, pch="x",
     xlab="h", ylab="strong error",
     ylim=range(c(error.euler, error.milstein)),
     log="xy")
points(h, error.milstein, pch="o")
```

The resulting plot, together with confidence intervals for each measurement, is shown in Figure 6.8.

Solution E6.10   First, we have to implement the Euler-Maruyama scheme and the Milstein scheme for SDE (6.20). To speed up the computation, we perform N simulations at the same time, thus reducing the required number of loops in the R program:

```
alpha <- 1
beta <- 0.4
x0 <- 1  # initial value
T <- 5  # time horizon

EulerMaruyama <- function(n, N) {
  h <- T / n
  X <- rep(x0, times=N)
  for (i in 1:n) {
    dB <- rnorm(N, 0, sqrt(h))
    X <- (X
          + (0.5 * alpha^2 + beta) * X * h
          + alpha * X * dB)
  }
  return(X)
```

```
}
Milstein <- function(n, N) {
  h <- T / n
  X <- rep(x0, times=N)
  for (i in 1:n) {
    dB <- rnorm(N, 0, sqrt(h))
    X <- (X
          + (0.5 * alpha^2 + beta) * X * h
          + alpha * X * dB
          + 0.5 * alpha^2 * X * (dB^2 - h))
  }
  return(X)
}
```

Next, we implement the test function f:

```
prob <- 0.6
threshold <- x0 * exp(alpha*qnorm(prob, 0, sqrt(T)) + beta*T)
f <- function(x) {
  as.numeric(x <= threshold)
}
```

Now we have to solve the SDE repeatedly for different grid sizes and to compute the weak error. This can be done as follows:

```
n.max <- 1024  # largest discretisation parameter to try
n.min <- 64  # smallest discretisation parameter to try
K <- 9  # number of resolution steps
n <- round(exp(seq(log(n.min), log(n.max), length.out=K)))

N <- 10000000  # number of runs to average over
mean.euler <- c()
mean.milstein <- c()
sd.euler <- c()
sd.milstein <- c()
progress <- txtProgressBar(min=0, max=2*K, style=3)
for (k in 1:K) {
  Xt.euler <- f(Euler--Maruyama(n[k], N))
  mean.euler[k] <- mean(Xt.euler)
  sd.euler[k] <- sd(Xt.euler)
  setTxtProgressBar(progress, 2*k-1)

  Xt.milstein <- f(Milstein(n[k], N))
  mean.milstein[k] <- mean(Xt.milstein)
  sd.milstein[k] <- sd(Xt.milstein)
```

```
  setTxtProgressBar(progress, 2*k)
}
close(progress)

exact <- prob
error.euler <- abs(mean.euler - exact)
error.milstein <- abs(mean.milstein - exact)
```

Finally, we use the resulting data to create a plot:

```
h <- T / n
plot(h, error.euler, pch="x",
     xlab="h", ylab="weak error",
     ylim=range(c(error.euler, error.milstein)),
     log="xy")
points(h, error.milstein, pch="o")
```

This completes the answers.

Solution E6.11   A basic Monte Carlo estimate for the probability that X exceeds a given level c can be based on the following R code.

```
f <- function(X, c) {
  return(as.numeric(max(X) > c))
}

x0 <- 0
T <- 5
t <- seq(0, T, by=0.01)

mu <- function(t, x) { -x }
sigma <- function(t, x) { 1.0 }

MC.sample <- function(N, c) {
  res <- c()
  for (i in 1:N) {
    X <- EulerMaruyama1d(t, x0, mu, sigma)
    res[i] <- f(X, c)
  }
  return(res)
}
```

For $c = 2.5$, we can use the function MC.sample as follows:

```
N <- 10000
```

```
Z <- MC.sample(N, 2.5)

m <- mean(Z)
s <- sd(Z)/sqrt(N)
cat("estimate for p: ", m, "\n", sep="")
cat("estimate for RMSE: ", sd(Z)/sqrt(N), "\n", sep="")
cat("estimate for CI: [", m - 1.96*s, ", ", m + 1.96*s, "]\n",
    sep="")
```

In the estimation of the root-mean squared error we neglected the discretisation error of the Euler-Maruyama method and only use the Monte Carlo error. The true error will be slightly bigger than the obtained estimate, due to the bias introduced by the discretisation error. The output of one run of the program is as follows:

```
estimate for p: 0.007
estimate for RMSE: 0.0008337683
CI = [0.005365814, 0.008634186]
```

This completes our solution of exercise E6.11.

Solution E6.12   An importance sampling estimate for the probability that X exceeds a given level c can be based on the following R code.

```
q <- 0.5
mu.tilde <- function(t, x) { -q*x }

phi.over.psi <- function(t, Y) {
  n <- length(t)
  h <- t[-1] - t[-n]
  dY <- Y[-1] - Y[-n]
  stoch.int <- sum(Y[-n] * dY)
  int <- sum(Y[-n]^2 * h)
  return(exp(-(1-q)*stoch.int - .5*(1-q*q)*int))
}

IS.sample <- function(N, c) {
  res <- c()
  for (i in 1:N) {
    Y <- EulerMaruyama1d(t, x0, mu.tilde, sigma)
    res[i] <- f(Y, c) * phi.over.psi(t, Y)
  }
  return(res)
}
```

For $c = 3.4$ and $N = 200\,000$, we can use the function IS.sample as follows:

```
N <- 200000
```

```
Z <- IS.sample(N, 3.2)

m <- mean(Z)
s <- sd(Z)/sqrt(N)
cat("estimate for p: ", m, "\n", sep="")
cat("estimate for RMSE: ", sd(Z)/sqrt(N), "\n", sep="")
cat("estimate for CI: [", m - 1.96*s, ", ", m + 1.96*s, "]\n",
    sep="")
```

The output of one run of the program is as follows:

```
estimate for p: 4.241444e-05
estimate for RMSE: 3.296227e-06
estimate for CI: [3.595384e-05, 4.887505e-05]
```

Solution E6.13   Algorithm 6.29 can be implemented as shown in the following. In the program we use the absolute value inside the square roots $\sqrt{|V_t|}$ in order to avoid problems when V_t temporarily takes negative values caused by discretisation error.

```
Heston <- function(t, S0, V0, r, lambda, sigma, xi, rho) {
  n <- length(t)
  h <- t - c(0, t[-n])
  dB <- matrix(rnorm(2*n, 0, sqrt(h)), nrow=n, ncol=2)
  sigma.square <- sigma^2
  rho.prime <- sqrt(1 - rho^2)
  path <- matrix(nrow=n, ncol=2)
  S <- S0
  V <- V0
  for (i in 1:n) {
    sqrt.V = sqrt(abs(V))
    S <- S + r * S * h[i] + S * sqrt.V * dB[i,1]
    V <- (V + lambda * (sigma.square - V) * h[i]
            + xi * sqrt.V * (rho * dB[i,1] + rho.prime * dB[i,2]))
    path[i,1] <- S
    path[i,2] <- V
  }
  return(path)
}
```

We pre-compute `sigma^2` and `sqrt(1 - rho^2)` outside the loop in order to avoid unnecessary, repeated evaluations of the squares and the square root. The function can be used as follows:

```
t <- seq(0, 5, by=0.001)
```

```
X <- Heston(t, S0=1, V0=0.16, r=1.02,
            lambda=1, sigma=1, xi=1, rho=-0.5)

plot(t, X[,1], type="l")
plot(t, X[,2], type="l")
```

The resulting plot is shown in Figure 6.13.

Solution E6.14   The Monte Carlo estimate requires us to solve SDE (6.43) many times in a row. To reduce the time required for computing these estimates, we slightly modify the code from exercise E6.13 to make it faster: since we are only interested in the final value S_T, there is no need to store the whole paths of S and V, and we can also assume a fixed step size $h = T/n$ for the Euler scheme. The resulting, simplified function is as follows:

```
HestonEndpointS <- function(n) {
  h <- T/n
  dB <- matrix(rnorm(2*n, 0, sqrt(h)), nrow=n, ncol=2)
  sigma.square <- sigma^2
  rho.prime <- sqrt(1 - rho^2)
  S <- S0
  V <- V0
  for (i in 1:n) {
    sqrt.V = sqrt(abs(V))
    S <- S + r * S * h + S * sqrt.V * dB[i,1]
    V <- (V + lambda * (sigma.square - V) * h
          + xi * sqrt.V * (rho * dB[i,1] + rho.prime * dB[i,2]))
  }
  return(S)
}
```

To compute the Monte Carlo estimate, we have to call this function repeatedly, with the given parameters, and we have to apply f to the results.

```
r <- 1.02
lambda <- 1
sigma <- 0.5
xi <- 1
rho <- -0.5

S0 <- 1
V0 <- 0.16

T <- 1
K <- 3
```

```
f <- function(s) {
  exp(-r*T) * max(s - K, 0)
}

MC.estimate <- function(n, N) {
  C <- replicate(N, f(HestonEndpointS(n)))
  est <- mean(C)
  mse <- var(C) / N
  return(list(est=est, mse=mse))
}
```

Estimates can now be obtained using calls such as:

```
> MC.estimate(100, 10000)
C=0.1147109, MSE=5.295385e-06
```

For the specified parameters, our estimate for the option price is $C = 0.1139534$, with an estimated mean squared error of $5.3 \cdot 10^{-6}$.

Solution E6.15   To implement the multilevel algorithm 6.26, we modify the function `HestonEndpointS` from exercise E6.14 to compute the solution of the SDE for discretisation parameters n and n/m simultaneously.

```
MLMCSample <- function(n, m) {
  if (n == m) return(f(HestonEndpointS(n)))
  if (n %% m != 0 || n <= m) stop("invalid grid size")
  h <- T/n
  dB <- matrix(rnorm(2*n, 0, sqrt(h)), nrow=n, ncol=2)
  sigma.square <- sigma^2
  rho.prime <- sqrt(1 - rho^2)
  S <- S0
  V <- V0
  Sm <- S0
  Vm <- V0
  dBm <- c(0, 0)
  for (i in 1:n) {
    sqrt.V = sqrt(abs(V))
    S <- S + r * S * h + S * sqrt.V * dB[i,1]
    V <- (V + lambda * (sigma.square - V) * h
          + xi * sqrt.V * (rho * dB[i,1] + rho.prime * dB[i,2]))
    dBm <- dBm + dB[i,]
    if (i %% m == 0) {
      sqrt.Vm = sqrt(abs(Vm))
      Sm <- Sm + r * Sm * m * h + Sm * sqrt.Vm * dBm[1]
```

```
        Vm <- (Vm + lambda * (sigma.square - Vm) * m * h
            + xi * sqrt.Vm * (rho * dBm[1] + rho.prime * dBm[2]))
        dBm <- c(0, 0)
    }
  }
  return(f(S)-f(Sm))
}
```

This function gives us the difference $Y_i^{(i,j)} - Y_{i-1}^{(i,j)}$ in the multilevel Monte Carlo estimate (6.41). The full estimate can be computed by combining the individual steps as follows:

```
MLMC.estimate <- function(n, m, L) {
  k = round(log(n) / log(m))
  if (n != m^k) stop("n=", n, " is not a power of k=", k)

  est <- 0
  mse <- 0
  n <- 1
  N <- L * m^k
  for (i in 1:k) {
    n <- n * m
    N <- N / m
    Ci <- replicate(N, MLMCSample(n, m))
    est <- est + mean(Ci)
    mse <- mse + var(Ci) / N
  }
  return(list(est=est, mse=mse))
}
```

C.7   Answers for Appendix B

Solution EB.1   As the following transcript of an R session shows, the first three expressions are straightforward:

```
> 1+2+3+4+5+6+7+8+9+10
[1] 55
> 2^16
[1] 65536
> 2^100
[1] 1.267651e+30
```

The sum of the first 10 natural numbers is 55 and $2^{16} = 65536$. The final answer 1.267651e+30 uses a shorthand notation: the returned result stands for $1.267651 \cdot 10^{30}$

(e is short for 'exponent'). R chooses this representation because the result is very big and would be difficult to read when written in standard notation.

For the final part of the exercise, computing $2/(1 + \sqrt{5})$, we can use the fact that brackets in R can be used exactly as they are used in mathematical expressions: using the function sqrt (see Table B.1), we can write

```
> 2/(1+sqrt(5))
[1] 0.618034
```

to get the correct (up to six decimal digits) result 0.618034.

Solution EB.2   The value of x is 3, the value of y is 5.

Solution EB.3   The sum can be computed by first constructing a vector of the numbers 1, 2, ..., 100, and then using the R function sum to add all elements of the vector:

```
> x <- 1:100
> x
  [1]    1   2   3   4   5   6   7   8   9  10  11  12  13  14  15
 [16]   16  17  18  19  20  21  22  23  24  25  26  27  28  29  30
 [31]   31  32  33  34  35  36  37  38  39  40  41  42  43  44  45
 [46]   46  47  48  49  50  51  52  53  54  55  56  57  58  59  60
 [61]   61  62  63  64  65  66  67  68  69  70  71  72  73  74  75
 [76]   76  77  78  79  80  81  82  83  84  85  86  87  88  89  90
 [91]   91  92  93  94  95  96  97  98  99 100
> sum(x)
[1] 5050
```

Thus, the required sum is 5050. A shorter solution is to merge the two steps into one command by just typing sum(1:100). This results in the same answer.

Solution EB.4   There are various ways to implement expressions such as the one for g_2 in R. To get a readable and short solution, it is important to make good use of the built-in R functions such as sum and mean. To compute g_2, we can use the following R code:

```
excessKurtosis <- function(x) {
  x.bar <- mean(x)
  numerator <- mean((x - x.bar)^4)
  denominator <- mean((x - x.bar)^2)^2
  return(numerator / denominator - 3)
}
```

This function computes the value g_2 for any sample x.

To test the function, we try large samples of distributions with a known kurtosis. First, the theoretical excess kurtosis of normal distributed values is 0. If we use the estimator g_2 for large samples, we should get results close to 0:

```
> excessKurtosis(rnorm(1000000))
[1] -0.00452515
> excessKurtosis(rnorm(1000000))
[1] 0.0002580271
> excessKurtosis(rnorm(1000000))
[1] -7.535562e-05
```

As a second test, we estimate the kurtosis of an exponential distribution; in this case, the theoretical excess kurtosis equals 6:

```
> excessKurtosis(rexp(1000000))
[1] 5.945269
> excessKurtosis(rexp(1000000))
[1] 5.991549
> excessKurtosis(rexp(1000000))
[1] 6.09649
```

Both sets of tests give the results close to the theoretical values, so we have reason to assume that the function excessKurtosis works.

Solution EB.5   The program contains at least two (!) mistakes: the line to assign a value to x[5] is missing, and the output claims wrongly that x[8] is the sixth (not eighth) Fibonacci number.

Solution EB.6   Since we need to repeat the operation of computing the cosine 20 times, we use a for loop in R:

```
X <- 0
for (i in 1:20) X <- cos(X)
print(X)
```

The output of these commands, and thus the value of x_{20}, is 0.7389378.

Solution EB.7   We first use seq to construct a vector which contains the numbers from 100 to 0, and then use this vector to control the for loop:

```
for (i in seq(from=100, to=0, by=-1)) {
  cat(i, "\n")
}
```

The command `print(i)` can be used instead of `cat(i, "\n")`, but the output is not quite as tidy (try this yourself!).

Solution EB.8   We can use the following program:

```
n <- 1
while (n^2 <= 5000) {
  n <- n + 1
}
print(n)
```

The result is $n = 71$.

Solution EB.9   One solution is the following program:

```
n <- 1
while ((n+1)^2 < 5000) {
  n <- n + 1
}
print(n)
```

Alternatively we can use the program from EB.8, and print `n-1` instead of `n` [we also need to check that `(n-1)^2` is not exactly equal to 5000; otherwise we would have to use `n-2` instead]. Using either method, the result is 70.

Solution EB.10   One solution is given in the following program.

```
x <- 27
n <- 0
while (x != 1) {
  cat("x[", n, "] = ", x, "\n", sep="")
  n = n + 1
  if (x %% 2 == 0) {
    x = x / 2
  } else {
    x = 3*x + 1
  }
}
cat("x[", n, "] = ", x, "\n", sep="")
```

This program violates the DRY principle: the `cat` command is duplicated to output the final value. One way to fix this problem is to stop the loop one iteration later, that is when the previous value equals one. This idea is implemented in the following version of the program:

```
n <- 0
x <- 27
old <- 0
while (old != 1) {
  old <- x
  cat("x[", n, "] = ", x, "\n", sep="")
  n = n + 1
  if (x %% 2 == 0) {
    x = x / 2
  } else {
    x = 3*x + 1
  }
}
```

Solution EB.11   The function computes the sample variance of the elements of x.

Solution EB.12   We set the variable b to the string "a". Since variables inside the function are independent of the names used by the caller, the function call test(c=b) is equivalent to test(c="a"), that is inside the function the value of the variable c is the string "a". Finally, the function cat does not include the enclosing quotation marks into the output when processing strings and thus the output is 'a=1, b=2, c=a, d=4'.

Solution EB.13   The missing line, where x[5] should have been set, causes a run-time error: the command x[6] <- x[5] + x[4] tries to read the uninitialised value x[5] and consequently x[6] and all the following values are set to NA. The mistake in the message printed via cat is a semantic error.

Solution EB.15   To find the mistake in the given R function, we first add a series of print commands to the function, to see at which step the function deviates from our expectations:

```
SomethingWrong <- function(x) {
  n <- length(x)
  cat("x =", x, ", n =", n, "\n")
  sum <- 0
  for (i in 1:n-1) {
    cat("i =", i, ", sum =", sum, "\n")
    sum <- sum + (x[i+1] - x[i])^2
```

```
  }
  return(sum)
}
```

This results in the following output:

```
> SomethingWrong(c(1,2,3))
x = 1 2 3 ,n = 3
i = 0 , sum = 0
i = 1 , sum =
i = 2 , sum =
numeric(0)
```

The first line of the output is what we expect: the input data are 1 2 3 and the length of the input data are 3. The second line already shows a problem: we asked R to use the values $1, \ldots, n-1$ for i, but the first iteration of the loop has i = 0. This is the cause of the problem, since the loop evaluates x[0] which is not the value we intended to use.

To find out why i took the value 0, we inspect the loop statement: we have i in 1:n-1 and we know that at this point n equals 3. Thus the range of i is 1:3-1:

```
> 1:3-1
[1] 0 1 2
> 1:2
[1] 1 2
> 1:(3-1)
[1] 1 2
```

The experiments shown above clearly point out the problem: R interprets 1:n-1 as $(1, 2, \ldots, n) - 1$, that is the computer subtracts 1 from every value in the sequence 1:n. This is not what we intended, and we can fix this by introducing brackets around n-1. The following version of the function fixes the error:

```
SomethingWrong <- function(x) {
  n <- length(x)
  sum <- 0
  for (i in 1:(n-1)) {
    sum <- sum + (x[i+1] - x[i])^2
  }
  return(sum)
}
```

Finally, we test the new version of the function:

```
> SomethingWrong(c(1,2,3))
[1] 2
```

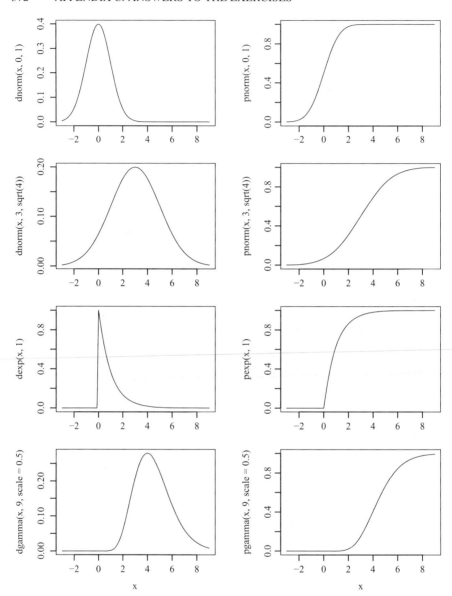

Figure C.10 Output of the R script from exercise EB.16. The graphs show the densities (left column) and distribution functions (right column) of the distributions $\mathcal{N}(0, 1)$, $\mathcal{N}(3, 4)$, $\mathrm{Exp}(1)$ and $\Gamma(9, 0.5)$ (top to bottom).

Since this is the expected result, we can assume that we have correctly identified and fixed the problem.

Solution EB.16   The function dnorm gives the density and pnorm gives the CDF of the normal distribution. Similarly, we can use the functions dexp and pexp for the exponential distribution and dgamma and pgamma for the gamma distribution. Thus, we can plot the required graphs as follows:

```
par(mfrow=c(4,2))

curve(dnorm(x,0,1), -3, 9)
curve(pnorm(x,0,1), -3, 9, ylim=c(0,1))

curve(dnorm(x,3,sqrt(4)), -3, 9)
curve(pnorm(x,3,sqrt(4)), -3, 9, ylim=c(0,1))

curve(dexp(x,1), -3, 9)
curve(pexp(x,1), -3, 9, ylim=c(0,1))

curve(dgamma(x,9,scale=.5), -3, 9)
curve(pgamma(x,9,scale=.5), -3, 9, ylim=c(0,1))
```

The resulting plot is shown in Figure C.10.

References

M. A. Beaumont, W. Zhang, and D. J. Balding. Approximate Bayesian Computation in population genetics. *Genetics*, 162(4):2025–2035, 2002.

M. Blum and O. François. Non-linear regression models for Approximate Bayesian Computation. *Statistics and Computing*, 20:63–73, 2010.

A. N. Borodin and P. Salminen. *Handbook of Brownian Motion — Facts and Formulae.* Probability and its Applications. Birkhäuser, 1996.

G. E. P. Box and M. E. Muller. A note on the generation of random normal deviates. *Annals of Mathematical Statistics*, 29(2):610–611, 1958.

G. Casella and R. L. Berger. *Statistical Inference*. Duxbury Press, second edition, 2001.

C. Chatfield. *The Analysis of Time Series*. Chapman & Hall/CRC Texts in Statistical Science Series. Chapman & Hall/CRC, sixth edition, 2004.

J. S. Dagpunar. *Simulation and Monte Carlo, with Applications in Finance and MCMC*. John Wiley & Sons, Ltd, 2007.

A. C. Davison and D. V. Hinkley. *Bootstrap Methods and their Application*. Cambridge University Press, 1997.

T. J. DiCiccio and B. Efron. Bootstrap confidence intervals. *Statistical Science*, 11(3):189–212, 1996.

B. Efron and R. J. Tibshirani. *An Introduction to the Bootstrap*. Chapman & Hall, 1993.

P. Fearnhead and D. Prangle. Constructing summary statistics for approximate Bayesian computation: semi-automatic approximate Bayesian computation. *Journal of the Royal Statistical Society: Series B*, 74(3):419–474, 2012.

W. Feller. *An Introduction to Probability Theory and Its Applications*, volume I. John Wiley & Sons, Ltd, third edition, 1968.

P. H. Garthwaite, I. T. Jolliffe, and B. Jones. *Statistical Inference*. Oxford University Press, 2002.

A. Gelman and D. B. Rubin. Inference from iterative simulation using multiple sequences. *Statistical Science*, 7(4):457–472, 1992.

J. E. Gentle, W. Härdle, and Y. Mori, editors. *Handbook of Computational Statistics*. Springer, 2004.

M. B. Giles. Multilevel Monte Carlo path simulation. *Operations Research*, 56(3):607–617, 2008a.

M. B. Giles. Improved multilevel Monte Carlo convergence using the Milstein scheme. In A. Keller, S. Heinrich, and H. Niederreiter, editors, *Monte Carlo and Quasi-Monte Carlo Methods 2006*, pages 343–358. Springer, 2008b.

W. R. Gilks, S. Richardson, and D. J. Spiegelhalter, editors. *Markov Chain Monte Carlo in Practice*. Chapman & Hall/CRC, 1996.

P. J. Green. Reversible jump Markov Chain Monte Carlo computation and Bayesian model determination. *Biometrika*, 82(4):711–732, 1995.

S. L. Heston. A closed-form solution for options with stochastic volatility with applications to bond and currency options. *The Review of Financial Studies*, 6(2):327–343, 1993.

D. J. Higham. An algorithmic introduction to numerical simulation of stochastic differential equations. *SIAM Review*, 43(3):525–546, 2001.

J. Jacod and P. Protter. *Probability Essentials*. Springer, 2000.

I. Karatzas and S. E. Shreve. *Brownian Motion and Stochastic Calculus*, volume 113 of *Graduate Texts in Mathematics*. Springer, second edition, 1991.

W. J. Kennedy Jr and J. E. Gentle. *Statistical Computing*. Marcel Dekker, Inc., 1980.

J. F. C. Kingman. *Poisson Processes*, volume 3 of *Oxford Studies in Probability*. Clarendon Press, 1993.

P. E. Kloeden and E. Platen. *Numerical Solution of Stochastic Differential Equations*. Number 23 in Applications of Mathematics. Springer, 1999. Corrected Third Printing.

D. E. Knuth. *Seminumerical Algorithms*, volume 2 of *The Art of Computer Programming*. Addison-Wesley, second edition, 1981.

D. P. Kroese, T. Taimre, and Z. I. Botev. *Handbook of Monte Carlo Methods*. John Wiley & Sons, Ltd, 2011.

E. L. Lehmann and J. P. Romano. *Testing Statistical Hypotheses*. Springer, third edition, 2005.

X. Mao. *Stochastic Differential Equations and Applications*. Woodhead Publishing, second edition, 2007.

G. Marsaglia and W. W. Tsang. The ziggurat method for generating random variables. *Journal of Statistical Software*, 5(8):1–7, 2000.

M. Matsumoto and T. Nishimura. Mersenne twister: a 623-dimensionally equidistributed uniform pseudo-random number generator. *ACM Transactions on Modeling and Computer Simulation*, 8(1):3–30, 1998.

N. Metropolis. The beginning of the Monte Carlo method. *Los Alamos Science*, 125–130, 1987.

S. P. Meyn and R. L. Tweedie. *Markov Chains and Stochastic Stability*. Cambridge University Press, second edition, 2009.

P. Mörters and Y. Peres. *Brownian Motion*. Cambridge University Press, 2010.

J. R. Norris. *Markov Chains*. Cambridge University Press, 1997.

B. K. Øksendal. *Stochastic Differential Equations*. Springer, sixth edition, 2003.

R. K. Pathria. *Statistical Mechanics*. Elsevier, second edition, 1996.

M. Plischke and B. Bergersen. *Equilibrium Statistical Physics*. World Scientific, third edition, 2006.

V. Privman, editor. *Finite Size Scaling and Numerical Simulation of Statistical Systems*. World Scientific, 1990.

R Development Core Team. *R: A Language and Environment for Statistical Computing*. 2011.

B. D. Ripley. *Stochastic Simulation*. John Wiley & Sons, Ltd, 1987.

M. L. Rizzo. *Statistical Computing with R*. Chapman & Hall/CRC, 2008.

C. P. Robert and G. Casella. *Monte Carlo Statistical Methods*. Springer Texts in Statistics. Springer, second edition, 2004.

G. O. Roberts and J. S. Rosenthal. General state space Markov chains and MCMC algorithms. *Probability Surveys*, 1:20–71, 2004.

L. C. G. Rogers and D. Williams. *Diffusions, Markov Processes, and Martingales*, volume 2 of *Cambridge Mathematical Library*. Cambridge University Press, 2000. Ito calculus, Reprint of the second (1994) edition.

W. Rudin. *Real and Complex Analysis*. McGraw-Hill, third edition, 1987.

H. Scheffé. A useful convergence theorem for probability distributions. *Annals of Mathematical Statistics*, 18(3):434–438, 1947.

S. Tavaré, D. J. Balding, R. C. Griffiths, and P. Donnelly. Inferring coalescence times from DNA sequence data. *Genetics*, 145(2):505–518, 1997.

W. N. Venables and B. D. Ripley. *S Programming*. Springer, 2000.

W. N. Venables, D. M. Smith, and the R Development Core Team. An introduction to R. Available from http://cran.r-project.org/doc/manuals/R-intro.html, 2011 (accessed 14 May 2013).

J. von Neumann. Various techniques used in connection with random digits. In A. S. Householder, G. E. Forsythe and H. H. Germond, editors, *Monte Carlo Method*, volume 12 of *National Bureau of Standards Applied Mathematics Series*, pages 36–38. US Government Printing Office, 1951.

R. Waagepetersen and D. Sorensen. A tutorial on reversible jump MCMC with a view toward applications in QTL-mapping. *International Statistical Review*, 69:49–61, 2001.

D. Williams. *Weighing the Odds: A Course in Probability and Statistics*. Cambridge University Press, 2001.

G. Winkler. *Image Analysis, Random Fields and Dynamic Monte Carlo Methods*. Springer, 1995.

Index

ABC, *see* Approximate Bayesian
  Computation
acceptance probability
  independence sampler, 120
  Metropolis–Hastings method, 110,
   113
  random walk Metropolis, 117
  rejection sampling, 15
  RJMCMC, 163, 164
  thinning method, 65
antithetic paths, 247–248
antithetic variables, 88–93
  for SDEs, 247–248
Approximate Bayesian Computation,
  182–188
  with regression, 188–192
autocorrelation, 132–135

Bayes' rule, 139, 267
Bayesian inference, 72, 138–142,
  147–152, 172–179
bias, 76
  bootstrap estimates, 199–201
  Monte Carlo estimation, 97–99
  of discretised SDEs, 243–244
binomial distribution, 293
boolean values, 281
bootstrap confidence intervals, 203–208
  BC_a, 207
  simple, 205

bootstrap estimates, 192–208
  general, 196
  of confidence intervals, 203–208
  of the bias, 199–201
  of the standard error, 201–203
Box–Muller transform, 34
Brownian bridge, 220
Brownian motion, 214–221
  geometric, 221–223
  interpolation, 218–221
  Markov property, 216
  scaling property, 216
  simulation, 217–218
burn-in period, 130–132, 137, 151

Cauchy distribution, 37
CDF, *see* cumulative distribution
  function
change of variables, 33
χ^2-distribution, 6, 293
componentwise simulation, 48–50, 142
computational cost
  ABC, 184
  MCMC, 135
  Monte Carlo estimation, 75
  rejection sampling, 20, 22
  SDEs, 233, 235, 247
conditional density, 267
conditional distribution, 23–27, 183
conditional expectation, 194

confidence intervals, 83, 100–103, 245
    bootstrap, 203–208
continuous-time processes, 213–261
control variates, 93–96, 251
convergence diagnostics, 136–137
correlation, 90, 91–93, 95, 132
    auto-, 132–135
    of a sample, 97–99
critical region, 104
cumulative distribution function, 12,
    264

detailed balance condition, 114
diffusion coefficient, 225
discretisation error, 214
    for SDEs, 236–242
don't repeat yourself
drift, 225
DRY, see don't repeat yourself

effective sample size, 134
empirical distribution, 192
energy, 154
Euler–Maruyama scheme, 232–233
    for the Heston model, 256
events, 263
exponential distribution, 13, 265, 293

gamma distribution, 293
geometric Brownian motion, 221–223
geometric distribution, 11, 17, 20, 185
Gibbs measure, 154
Gibbs sampler, 142–159
    image processing, 157–159
    Ising model, 154–157
    parameter estimation, 147–152

half-normal distribution, 21
Harris recurrence, 127
Heston model, 255–259
hierarchical models, 45–50, 52, 147
hypothesis tests, 5, 103–106

i.i.d., 2
i.i.d. copies, 75

importance sampling, 84–88, 223
    for SDEs, 248–250
independence, 109, 266
independence sampler, 120–121
indicator function, 268
initial distribution
    for Markov chains, 50, 128
    for SDEs, 224
intensity function, 59, 161
inverse temperature, 154
inverse transform method, 12–15, 92
irreducible Markov chain, 127
Ising model, 154
Ito integral, 226–227
Ito's formula, 227–230

Jacobian matrix, 33, 34, 37, 164, 165

Kronecker delta, 168
kurtosis, 294

Lagrange multipliers, 246
law of large numbers, 269
    for Markov chains, 128–130
LCG, see linear congruential generator
Linear Congruential Generator, 2–4

marginal distribution, 48–50
Markov chain, 50–58, 109, 126–130
    continuous state space, 56–58
    discrete state space, 51–56
    Harris recurrent, 127
    initial distribution, 50, 128
    irreducible, 127
    law of large numbers, 128–130
    reversible, 112, 114
    time-homogeneous, 51
Markov Chain Monte Carlo, 109–180
    convergence, 126–138
    reversible jump, 159–179
Markov property
    of Brownian motion, 216
MCMC, see Markov Chain Monte
    Carlo

mean squared error, 76
  for antithetic variables, 90
  for control variates, 94
  for importance sampling, 85
  for MCMC, 130
  for Monte Carlo estimates, 77
  for SDEs, 243, 245–247
Metropolis–Hastings method, 110–126
  continuous state space, 110–113
  discrete state space, 113–116
  independence sampler, 120–121
  move types, 121–126, 163, 173–177
  random walk Metropolis, 116–119,
    129, 140–142
Milstein scheme, 234, 233–236
mixture distributions, 46–48, 147, 172
models
  continuous-time, 213–261
  hierarchical, 45–50, 52, 147
  statistical, 1, 41–68
Monte Carlo estimates, 69–108
  choice of sample size, 80–82
  error, 76–80, 82–84
  for integrals, 72
  for probabilities, 71, 81, 87, 91, 102,
    105, 244, 249
  for SDEs, 243–255
  multi-level, 250–255
  variance reduction, 84–96, 247–255
MSE, *see* mean squared error
multi-level Monte Carlo estimates,
  250–255

normal distribution, 265, 293
  generation, 23, 34
  half, 21
  multivariate, 41–45, 214
normalisation constant, 266

optimisation under constraints, 245
option pricing, 255–259

Pareto distribution, 162
partition function, 154

pixels, 153
plug-in estimator, 198
plug-in principle, 198
point estimators, 83, 97–100, 197–203
Poisson distribution, 58, 102, 162, 293
Poisson process, 58–67
  intensity function, 59
  thinning method, 65
posterior distribution, 73, 138–141,
    147, 158, 181
prior distribution, 138
PRNG, *see* pseudo random number
    generator
probability, 263–270
probability density, 264
probability distribution, 263
probability vector, 52
pseudo random number generator, 2

R programming, 271–297
random number generators, 2–8
random variables, 263, 292
  transformation of, 32–38
random walk, 50
random walk Metropolis, 116–119
ratio-of-uniforms method, 35–38
Rayleigh distribution, 14
rejection sampling, 15–32, 73, 120
  basic, 15–19, 182
  envelope, 19–23
  for conditional distributions, 23–27,
    150
resampling methods, 192–208
reversible jump Markov Chain Monte
    Carlo, 159–179
  dimension matching, 163–166, 175
  state space, 161, 172
  target distribution, 161, 173
  transitions, 162, 173
reversible Markov chain, 112, 114
RJMCMC, *see* reversible jump Markov
    Chain Monte Carlo
RMSE, *see* root-mean-square error
root-mean-square error, 79, 344

sample correlation, 97–99
scaling property
    of Brownian motion, 216
SDE, *see* stochastic differential
        equations
seed, 2, 8, 153, 293
semicircle distribution, 18
skewness, 105
slice sampler, 145
stability of discretisation schemes,
        241
standard error, 76, 201
    bootstrap estimates, 201–203
standard normal distribution, *see*
        normal distribution
state space
    Gibbs sampler, 142
    Markov chain, 50, 51, 56
    RJMCMC, 161
stationary density, 58
stationary distribution, 55
statistical computing, 1–262
statistical hypothesis tests, 5, 103–106
statistical inference, 96–106
    Bayesian, 138–142, 147–152,
        172–179
    bootstrap methods, 197–208
statistical models, 1, 41–68
stochastic analysis, 226–231
stochastic differential equations,
        224–242
    Euler–Maruyama scheme, 232–233
    initial distribution, 224
    Milstein scheme, 233–236
    Monte Carlo estimates, 243–255

strong error, 236–237
weak error, 237–240
stochastic integrals, 226–231
    time discretisation, 226
stochastic matrix, 52
Stratonovich integral, 230, 230–231
strong error, 236, 236–237
substitution rule, 33, 170
sufficient statistic, 183

test functions, 238
tests
    statistical, 5, 103–106
time discretisation, 214
    Brownian motion, 217
    for SDEs, 231–242
    stochastic integrals, 226
transformation
    of random variables, 32–38, 43
    of $\mathcal{U}[0, 1]$, 13
transition density, 57
transition kernel, 56
transition matrix, 51
truncation error, 276

uniform distribution, 4, 8, 27, 293
    discrete, 9, 293
    ratio-of-uniforms, 35

variance reduction methods, 84–96,
        247–255

weak error, 237–240
Wiener process, *see* Brownian motion
Wigner's semicircle distribution, 18